Edited by
Marek W. Urban

Handbook of Stimuli-Responsive Materials

Related Titles

Belfiore, L. A.

Physical Properties of Macromolecules

2010

ISBN: 978-0-470-22893-7

Kumar, Challa, S. S. R. (eds.)

Nanostructured Thin Films and Surfaces

Nanomaterials for the Life Sciences (vol 5)

2010

ISBN-13: 978-3-527-32155-1

Förch, R., Schönherr, H., Jenkins, A. T. A. (eds.)

Surface Design: Applications in Bioscience and Nanotechnology

2009

ISBN: 978-3-527-40789-7

Elias, H.-G.

Macromolecules

2788 pages in 4 volumes with approx. 1095 figures and approx. 580 tables

2009

ISBN: 978-3-527-31171-2

Zhao, Y., Ikeda, T. (eds.)

Smart Light-Responsive Materials

Azobenzene-Containing Polymers and Liquid Crystals

2009

ISBN: 978-0-470-17578-1

Vögtle, F., Richardt, G., Werner, N.

Dendrimer Chemistry

Concepts, Syntheses, Properties, Applications

2009

ISBN: 978-3-527-32066-0

Barner-Kowollik, C. (ed.)

Handbook of RAFT Polymerization

2008

ISBN: 978-3-527-31924-4

Schnabel, W.

Polymers and Light

Fundamentals and Technical Applications

2007

ISBN: 978-3-527-31866-7

Advincula, R. C., Brittain, W. J., Caster, K. C., Rühe, J. (eds.)

Polymer Brushes

Synthesis, Characterization, Applications

2004

ISBN: 978-3-527-31033-3

Edited by Marek W. Urban

Handbook of Stimuli-Responsive Materials

WILEY-VCH Verlag GmbH & Co. KGaA

The Editor

Prof. Dr. Marek W. Urban
University of Southern Mississippi
Professor of Polymer Science
118 College Drive
Hattiesburg, MS 39406
USA

All books published by **Wiley-VCH** are carefully produced. Nevertheless, authors, editors, and publisher do not warrant the information contained in these books, including this book, to be free of errors. Readers are advised to keep in mind that statements, data, illustrations, procedural details or other items may inadvertently be inaccurate.

Library of Congress Card No.: applied for

British Library Cataloguing-in-Publication Data
A catalogue record for this book is available from the British Library.

Bibliographic information published by the Deutsche Nationalbibliothek
The Deutsche Nationalbibliothek lists this publication in the Deutsche Nationalbibliografie; detailed bibliographic data are available on the Internet at <http://dnb.d-nb.de>.

© 2011 WILEY-VCH Verlag & Co. KGaA, Boschstr. 12, 69469 Weinheim, Germany

All rights reserved (including those of translation into other languages). No part of this book may be reproduced in any form – by photoprinting, microfilm, or any other means – nor transmitted or translated into a machine language without written permission from the publishers. Registered names, trademarks, etc. used in this book, even when not specifically marked as such, are not to be considered unprotected by law.

Typesetting Laserwords Private Ltd., Chennai, India
Printing and Binding Fabulous Printer Pte Ltd, Singapore
Cover Design Formgeber, Eppelheim

Printed in Singapore
Printed on acid-free paper

ISBN: 978-3-527-32700-3

Contents

Preface *XI*
List of Contributors *XV*

1 Synthetic and Physicochemical Aspects of Advanced Stimuli-Responsive Polymers *1*
Dirk Kuckling and Marek W. Urban
1.1 Introduction *1*
1.2 Controlled Free Radical Polymerization of Stimuli-Responsive Polymers *3*
1.3 Synthesis of Stimuli-Responsive Colloidal Dispersions *11*
1.4 Summary *21*
References *21*

2 Biological- and Field-Responsive Polymers: Expanding Potential in Smart Materials *27*
Debashish Roy, Jennifer N. Cambre, and Brent S. Sumerlin
2.1 Introduction *27*
2.2 Biologically Responsive Polymer Systems *28*
2.2.1 Glucose-Responsive Polymers *28*
2.2.1.1 Glucose-Responsive Systems Based on Glucose-GOx *28*
2.2.1.2 Glucose-Responsive Systems Based on ConA *29*
2.2.1.3 Glucose-Responsive Systems Based on Boronic Acid-Diol Complexation *30*
2.2.2 Enzyme-Responsive Polymers *32*
2.2.3 Antigen-Responsive Polymers *33*
2.2.4 Redox-/Thiol-Responsive Polymers *35*
2.3 Field-Responsive Polymers *39*
2.3.1 Electroresponsive Polymers *39*
2.3.2 Magnetoresponsive Polymers *40*
2.3.3 Ultrasound-Responsive Polymers *42*
2.3.4 Photoresponsive Polymers *43*
2.4 Conclusions *46*
References *47*

Handbook of Stimuli-Responsive Materials. Edited by Marek W. Urban
Copyright © 2011 WILEY-VCH Verlag GmbH & Co. KGaA, Weinheim
ISBN: 978-3-527-32700-3

3 Self-Oscillating Gels as Stimuli-Responsive Materials *59*
Anna C. Balazs, Pratyush Dayal, Olga Kuksenok, and Victor V. Yashin

- 3.1 Introduction *59*
- 3.2 Methodology *61*
- 3.2.1 Continuum Equations *61*
- 3.2.2 Formulation of the Gel Lattice Spring Model (gLSM) *63*
- 3.2.3 Model Parameters and Correspondence between Simulations and Experiments *67*
- 3.3 Results and Discussions *69*
- 3.3.1 Effect of Confinement on the Dynamics of the BZ Gels *69*
- 3.3.1.1 Linear Stability Analysis in Limiting Cases *69*
- 3.3.1.2 Oscillations Induced by the Release of Confinement *72*
- 3.3.1.3 Behavior of Partially Confined Samples *74*
- 3.3.2 Response of the BZ Gels to Nonuniform Illumination *77*
- 3.3.2.1 Modeling the Photosensitivity of the BZ Gels *77*
- 3.3.2.2 Autonomous Motion toward the Dark Region *78*
- 3.3.2.3 Light-Guided Motion along Complex Paths *81*
- 3.4 Conclusions *87*
- Acknowledgments *89*
- References *89*

4 Self-Repairing Polymeric Materials *93*
Biswajit Ghosh, Cathrin C. Corten, and Marek W. Urban

- 4.1 Introduction *93*
- 4.2 Damage and Repair Mechanisms in Polymers *97*
- 4.2.1 Dimensions of Damages and Repairs *99*
- 4.2.1.1 Angstrom-Level Repairs *99*
- 4.2.1.2 Nanometer-Level Repairs *107*
- 4.2.1.3 Micrometer-Level Repairs *110*
- 4.2.1.4 Millimeter-Level Repairs *112*
- 4.3 Summary *113*
- References *114*

5 Stimuli-Driven Assembly of Chromogenic Dye Molecules: a Versatile Approach for the Design of Responsive Polymers *117*
Brian Makowski, Jill Kunzelman, and Christoph Weder

- 5.1 Introduction *117*
- 5.2 Excimer-Forming Sensor Molecules *118*
- 5.3 Fluorescent Mechanochromic Sensors *121*
- 5.4 Thermochromic Sensors *129*
- 5.5 Chemical Sensing with Excimer-Forming Dyes *133*
- 5.6 Summary and Outlook *135*
- Acknowledgments *136*
- References *136*

6	**Switchable Surface Approaches** *139*	
	Aftin M. Ross, Himabindu Nandivada, and Joerg Lahann	
6.1	Introduction *139*	
6.2	Electroactive Materials *139*	
6.2.1	High-Density and Low-Density Self-Assembled Monolayers *139*	
6.2.2	Self-Assembled Monolayers with Hydroquinone Incorporation *141*	
6.3	Photoresponsive Materials *142*	
6.3.1	Molecules Containing Azobenzene Units *142*	
6.3.2	Molecules Containing Spiropyran Units *143*	
6.3.3	Photoresponsive Shape-Memory Polymers *144*	
6.4	pH-Responsive Materials *145*	
6.4.1	pH-Switchable Surfaces Based on Self-Assembled Monolayers (SAMs) *146*	
6.4.2	pH-Switchable Surfaces Based on Polymer Brushes *146*	
6.5	Thermoresponsive Materials *147*	
6.5.1	Temperature-Dependent Switching Based on Poly(N-iso-propylacrylamide) (PNIPAAm) *147*	
6.5.2	Temperature-Dependent Switching in Polymer Brushes *149*	
6.5.3	Thermoresponsive Shape-Memory Polymers *149*	
6.6	Switchable Surfaces Based on Supramolecular Shuttles *150*	
6.6.1	Rotaxane Shuttles *151*	
6.6.2	Switches Based on Catenanes *152*	
6.7	Switchable Surfaces Comprising DNA and Peptide Monolayers *153*	
6.7.1	DNA-Based Surface Switches *153*	
6.7.2	Switches Based on Aptamers *154*	
6.7.3	Switches Based on Helical Peptides *155*	
6.7.4	Switchable Surfaces of Elastinlike Polypeptides (ELP) *155*	
6.8	Summary *157*	
	References *157*	
7	**Layer-by-Layer Self-Assembled Multilayer Stimuli-Responsive Polymeric Films** *165*	
	Lei Zhai	
7.1	Introduction *165*	
7.2	Fabrication of Multilayer Polymer Coatings *167*	
7.3	Response of Multilayer Polymer Coatings to External Stimuli *169*	
7.3.1	Salt Concentration Change *169*	
7.3.2	pH Alternation *170*	
7.3.3	Temperature Variation *175*	
7.3.4	Light Radiation *178*	
7.3.5	Electrochemical Stimuli *181*	
7.3.6	Specific Molecular Interaction *183*	
7.4	Conclusion and Outlook *184*	
	References *185*	
	Further Reading *190*	

8	**Photorefractive Polymers** *191*	
	Kishore V. Chellapan, Rani Joseph, and Dhanya Ramachandran	
8.1	Introduction *191*	
8.2	The Photorefractive Effect in Polymers *193*	
8.2.1	Charge Generation and Transport in Polymers *193*	
8.2.2	Electro-Optic Response in Polymers *200*	
8.2.3	Formation of Photorefractive Gratings *204*	
8.3	The Two-Beam Coupling Effect *205*	
8.4	High-Performance Photorefractive Polymers *206*	
8.4.1	Guest–Host Polymer Systems *206*	
8.4.2	Functionalized Polymer Systems *211*	
8.5	Experimental Techniques *213*	
8.5.1	Photoconductivity and Electro-Optic Responses *213*	
8.5.2	Two-Beam Coupling *214*	
8.5.3	Diffraction Efficiency and Response Time *216*	
8.6	Conclusions *218*	
	References *218*	
9	**Photochromic Responses in Polymer Matrices** *223*	
	Dhanya Ramachandran and Marek W. Urban	
9.1	Introduction *223*	
9.2	Photochromic Polymeric Systems *224*	
9.2.1	Influence of Molecular Structures *224*	
9.2.2	Environmental Effects *226*	
9.2.3	Kinetics of Photoreactions *227*	
9.3	Photochromic Systems *229*	
9.3.1	Azobenzenes *229*	
9.3.2	Spiropyrans *231*	
9.3.3	Diarylethenes *234*	
9.3.4	Fulgides *237*	
9.4	Outlook of Photochromic Materials *238*	
	Acknowledgments *240*	
	References *240*	
10	**Covalent Bonding of Functional Coatings on Conductive Materials: the Electrochemical Approach** *247*	
	Michaël Cécius and Christine Jérôme	
10.1	Introduction *247*	
10.2	Electrodeposited Coatings *248*	
10.3	Electrografted Coatings *249*	
10.4	Compounds Requiring an Anodic Process *250*	
10.4.1	Aliphatic Primary Amines *250*	
10.4.2	Arylacetates *252*	
10.4.3	Primary Alcohols *253*	
10.4.4	Arylhydrazines *254*	

10.5	Compounds Requiring a Cathodic Process	*254*
10.5.1	Aryldiazonium Salts	*254*
10.5.2	(Meth)acrylic Monomers	*258*
10.5.3	Iodide Derivatives	*263*
10.6	Conclusions	*264*
	References	*265*

Index *269*

Preface

Nature is a source of inspiration for the design and development of new materials that are capable of responding to stimuli in a controllable and predictable fashion. These attributes are often manifested by nature's ability to reverse and to regenerate, commonly termed as stimuli-responsiveness. Although the concept of stimuli-responsiveness has been known for many years, the last decade, particularly, has witnessed a tremendous progress in this field. In 2002, the first International Symposium on Stimuli-Responsive Materials in Hattiesburg, USA, brought the international scientific community together and provided the first forum that has now matured into a major international conference that gathers scientists from around the world. Other conferences and meetings on similar topics followed, signifying strong scientific and technological interests in this continuously expanding field.

Inherent similarities as well as apparent differences between polymeric materials and entities produced by nature stimulated interest in stimuli-responsive polymeric materials. Although the similarities are obvious, with the common denominator being materials-functionality, what sets synthetic materials apart is their inability to respond to stimuli. Thus, significant interests and efforts are continuously directed toward synthesis of new materials and modification of the existing ones to achieve stimuli-responsive attributes. There are, however, significant challenges in mimicking of biological systems where structural and compositional gradients at various length scales are necessary for orchestrated and orderly responsive behaviors.

To tackle these challenges, numerous studies dealt with polymeric solutions, gels, surfaces and interfaces, but to lesser extent, with polymeric solids. These states of matter impose a different degree of restrictions on the mobility of polymeric segments or chains, thus making dimensional responsiveness more easily attainable for the systems with a higher solvent content and minimal energy inputs. Significantly greater challenges exist when designing chemically or physically cross-linked gels and solid polymeric networks that require maintaining their mechanical integrity. Restricted mobility within the network results from significant spatial limitations, thus imposing limits on obtaining stimuli-responsiveness. The challenge in designing these stimuli-responsive polymeric systems is to create networks capable of inducing minute molecular, yet orchestrated changes that lead to significant physicochemical responses upon external or internal stimuli.

Setting up the stage with an overview of synthetic and physicochemical aspects of advanced stimuli-responsive materials, the first few chapters of this volume provide a comprehensive coverage of biologically responsive systems, ranging from glucose, enzyme, and antigen responses to electro-, magneto-ultrasound-, and photoresponsive polymers, followed by modeling of dynamic processes in self-oscillating gels. Subsequent chapters focus on recent advances in self-healing materials in the context of their dimensions as well as assemblies of sensing and responsiveness of chromogenic dyes in polymer matrices. Switchable surfaces and their design using a variety of chemistries and morphologies, where pH, temperature, and electromagnetic radiation are the primary stimuli are discussed in the context of mechano-mutable attributers, followed by strategies in designing and fabrication of layer-by-layer self-assembly responsive films. The final chapters focus on photorefractive and photochromic polymers in the context of their chemical design and physicochemical attributes leading to photoconductivity, electro-optical, and photochromic responses, followed by electrochemical approaches giving raise to electrografted and electrodeposited coatings.

This volume presents selected recent developments in stimuli-responsive materials and is not meant to be inclusive. As dynamics of stimuli-responsiveness, this field is also dynamically evolving and future volumes will disseminate other aspects as they are discovered. This provides an opportunity to identify the challenges and needs for future research. Although there has been significant progress in the synthesis of precisely controllable polymerization methods leading to well-defined macromolecular blocks with stimuli-responsive characteristics, understanding the physical–chemical aspects of these systems remains a challenge. The area of particular interest is the synthetic generation of larger scale objects with diversified shapes and compositional gradients that are capable of responses. In this context, control of responsive ranges, the effect of solvent–solute interactions, as well as mechano-rheological behavior as a function of stimuli need further understanding.

These relationships are particularly significant in micro- and nanofluidics as well as in other aspects of polymer rheology. Although recent advances in the development of colloidal dispersions at sub-nano-diameter levels are promising, colloidal nanoparticles with versatile morphologies, shapes, and bioactive attributes are of particular interest.

Polymeric interfaces, although extensively studied from the perspective of structure–property relationships, represent an unprecedented opportunity for the development of new multicomponent composite systems with stimuli-responsive characteristics in spatially confined environments. The main challenges are the control and measurement of surface and interfacial density and control of the chain length of anchoring nano-objects with variable length scale responsiveness. Development of materials with new self-healing mechanisms and precise selectivity and self-repairing characteristics are of great interest. Even more challenging will be networks that exhibit photochromic responsiveness. Stimuli-responsive nanosurfaces will be particularly useful in the development of devices that resemble biologically active cilia with 3D actuation.

Enhanced mechanical integrity is essential to improve typically fragile polymeric gels and the balance between mechanical stability and rapid response times, reversibility, and processing conditions will be necessary for many new applications, in particular for biomedical systems. Further understanding of inclusive changes in polymer networks produced from natural building blocks, such as saccharides and aminoacids, nucleotides, and lipids will generate new avenues for regenerative medicine, where cell differentiation, membrane formation, neural network assemblies or other higher order hierarchical biocomponents may be produced.

Since dimensional changes in solid networks impose spatial and energetic limitations, design and formulation of heterogeneous structural features capable of charge transfer, ionization, or photoinduced conformational changes will be necessary. This can be achieved by combining in an orchestrated fashion low T_g and multistimuli-responsive monomers into one copolymer backbone with controllable architecture. Molecular components that exhibit displacements responding to sunlight will be possible if we can control separation of charges which quickly recombine thousands or millions times faster than the molecular motion. This may possibly be accomplished by designing molecular architectures capable of separating charges so that the "frozen" energy is used for going back and forth from one equilibrium to another, while retaining mechanical network integrity. Rotaxanes, spyropyrans, diarylethenes, fulgides, or azo-compounds represent selected examples of molecular entities that are capable of providing light-driven molecular motions. **These processes should be reversible and self reassembling, with an infinitely high "fatigue factor," that is the ability of infinitely long stimuli-responsiveness without physico-chemical changes.** There are other possibilities as well – the challenge will be to control kinetics to achieve stimuli-responsiveness, and, to overcome the barriers of biocompatibility, biodegradability, and nontoxicity. Reversibility and speed of stimuli-responsiveness are essential in each of the states, especially for solid networks, and the design of suitable chemical structures to control metastable equilibrium energy states will formulate conditions for the design of orchestrated heterogeneous networks.

Synthetic materials capable of responses to external or internal stimuli represent one of the most exciting and emerging areas of scientific interest and have many unexplored commercial applications. While there are many exciting challenges facing this continually evolving field and there are a number of opportunities in design, synthesis, and engineering of stimuli-responsive materials, nature will continue to serve as a supplier of endless inspiration. We hope that this volume will provide the readers with comprehensive overviews of selected areas ranging from synthetic aspects of to theoretical and physical insights into this rapidly growing field and, at the same time, open up a dialogue for new ideas and explorations.

Marek W. Urban

List of Contributors

Anna C. Balazs
University of Pittsburgh
Chemical Engineering
Department
1249 Benedum Hall
Pittsburgh, PA 15261
USA

Michaël Cécius
University of Liège
Center for Education and
Research on Macromolecules
(CERM)
Sart-Tilman Campus
B6, B-4000 Liège
Belgium

Jennifer N. Cambre
Southern Methodist University
Department of Chemistry
3215 Daniel Avenue
Dallas, TX 75275-0314
USA

Kishore V. Chellapan
Optical Microsystems Laboratory
Department of Electrical
Engineering
Koç University
Rumeli Feneri Yolu, Sariyer
34450, Istanbul - Turkey

Cathrin C. Corten
The University of
Southern Mississippi
School of Polymers and High
Performance Materials
118 College Drive
Hattiesburg, MS 39401
USA

Pratyush Dayal
University of Pittsburgh
Chemical Engineering
Department
1249 Benedum Hall
Pittsburgh, PA 15261
USA

Biswajit Ghosh
The University of
Southern Mississippi
School of Polymers and High
Performance Materials
118 College Drive
Hattiesburg, MS 9401
USA

Handbook of Stimuli-Responsive Materials. Edited by Marek W. Urban
Copyright © 2011 WILEY-VCH Verlag GmbH & Co. KGaA, Weinheim
ISBN: 978-3-527-32700-3

Christine Jérôme
University of Liège
Center for Education and
Research on Macromolecules
(CERM)
Sart-Tilman Campus
B6, B-4000 Liège
Belgium

Rani Joseph
Cochin University of Science
and Technology
Department of Polymer Science
and Rubber Technology
Cochin-22
Kochi 682 022, Kerala
India

Dirk Kuckling
University of Paderborn
Department of Chemistry
Warburger Str. 100
33098 Paderborn
Germany

Olga Kuksenok
University of Pittsburgh
Chemical Engineering
Department
1249 Benedum Hall
Pittsburgh, PA 15261
USA

Jill Kunzelman
PolyOne Corporation
33587 Walker Road
Avon Lake
OH 44012
USA

Joerg Lahann
University of Michigan
Department of Chemical
Engineering
3414 G.G. Brown
2300 Hayward Street
Ann Arbor, MI-48109
USA

Brian Makowski
University of Fribourg
Adolphe Merkle Institute and
Fribourg Center for
Nanomaterials
Route de l'Ancienne
Papeterie CP 209
CH-1723 Marly 1
Switzerland

and

Case Western Reserve University
Department of Macromolecular
Science and Engineering
2100 Adelbert Rd
Cleveland
OH
44106-7202
USA

Himabindu Nandivada
University of Michigan
Department of Chemical
Engineering
3414 G.G. Brown
2300 Hayward Street
Ann Arbor, MI-48109
USA

Dhanya Ramachandran
The University of
Southern Mississippi
School of Polymers and High
Performance Materials
Department of Polymer Science
118 College Drive
#10076, Hattiesburg, MS 39406
USA

Aftin M. Ross
University of Michigan
Department of Biomedical
Engineering
3414 G.G. Brown
2300 Hayward Street
Ann Arbor, MI-48109
USA

Debashish Roy
Southern Methodist University
Department of Chemistry
3215 Daniel Avenue
Dallas, TX 75275-0314
USA

Brent S. Sumerlin
Southern Methodist University
Department of Chemistry
3215 Daniel Avenue
Dallas, TX 75275-0314
USA

Marek W. Urban
The University of
Southern Mississippi
School of Polymers and High
Performance Materials
118 College Drive
Hattiesburg, MS 39406
USA

Christoph Weder
University of Fribourg
Adolphe Merkle Institute and
Fribourg Center for
Nanomaterials
Route de l'Ancienne
Papeterie CP 209
CH-1723, Marly 1
Switzerland

Victor V. Yashin
University of Pittsburgh
Chemical Engineering
Department
1249 Benedum Hall
Pittsburgh, PA 15261
USA

Lei Zhai
University of Central Florida
Department of Chemistry
400 Central Florida Boulevard
Chemistry Building (CH) 177
Orlando, FL 32816-2366
USA

1
Synthetic and Physicochemical Aspects of Advanced Stimuli-Responsive Polymers

Dirk Kuckling and Marek W. Urban

1.1
Introduction

Although the technological and scientific importance of functional polymers has been well established over the last few decades, currently much attention has been focused on stimuli-responsive polymers. This group of materials is of particular interest owing to their ability to respond to internal and/or external chemico-physical stimuli that is often manifested by the large macroscopic responses [1]. Stimuli-responsive polymers are also referred to as *smart*, *sensitive*, or *intelligent* polymers [2, 3], just to name a few. These terms are loosely used under the same stimuli-responsiveness umbrella attributed to selective polymer segments or the entire polymer backbones that exhibit stimuli-responsive characteristics. Notwithstanding the scientific challenges of designing stimuli-responsive polymers, the main technological interest is in the numerous applications ranging from reactive surfaces [4] to drug-delivery and separation systems [5], or from chemomechanical actuators [6] to other applications that have been extensively explored [7, 8].

In contrast to traditional polymers, in order to incorporate responsive components, it is necessary to copolymerize responsive blocks into a polymer or copolymer backbone [8]. For this reason, the preparation of well-defined block copolymers with different architectures is essential: for example, grafting amphiphilic blocks to a hydrophobic polymer backbone [9]. Using living anionic [10] and cationic polymerizations [11] as well as controlled radical polymerizations (CRPs) techniques [12], wide ranges of block copolymers were synthesized. However, the development of the CRP based on the concept of reversible chain termination minimizes the disadvantage of the free-radical polymerization, thus permitting the synthesis of well-defined block copolymer structures [13]. The growing demand for well-defined and functional soft materials in a nanoscale range has led to a significant increase of procedures that combine architectural control with the flexibility of incorporating functional groups. In view of these considerations, there has been a significant quest for elucidating a variety of controlled polymerization strategies, which resulted in nitroxide-mediated radical polymerization (NMRP) [14–16], atom transfer

Handbook of Stimuli-Responsive Materials. Edited by Marek W. Urban
Copyright © 2011 WILEY-VCH Verlag GmbH & Co. KGaA, Weinheim
ISBN: 978-3-527-32700-3

radical polymerization (ATRP) [17, 18], and reversible addition fragmentation chain transfer (RAFT) procedures [19, 20]. While details for each synthetic route are readily available in the literature, Figure 1.1 illustrates the basic principles governing these reactions, which are capable of producing well-defined homo and block copolymers of different architectures in solutions and on surfaces [21, 22]. While each synthetic route has its own attributes, in general, free-radical polymerization processes can be conducted using homogeneous or heterogeneous conditions. Ring-opening metathesis polymerization (ROMP) also provides a unique means of synthesizing well-defined copolymers [23–25]. For example, ROMP of norbornene derivatives leads to precisely controlled polydispersity (PDI), backbone configuration, and tacticity [26]. In particular, Ru-based ROMP appears to be a highly beneficial route for synthesizing a broad spectrum of copolymers with biological relevance. Precisely controlled peptide-pendant copolymers [27] and amino acid functionalized norbornenes containing ester carboxy groups [28, 29] are the prime examples. Taking advantage of the versatility of the ROMP process, bioactive and therapeutic polymers were also developed [30], including stimuli-responsive betaines [31] and acid–base sensitive phenanthroimidazole-based [32] and thiol-functional [33] polymers. In addition, notable synthesis of tunable, temperature-responsive polynobornenes with elastin peptide side chains was reported [34]. The first part of the chapter focuses primarily on homogeneous CRP, whereas the remaining sections outline heterogeneous colloidal synthesis and physicochemical aspects of stimuli responsiveness.

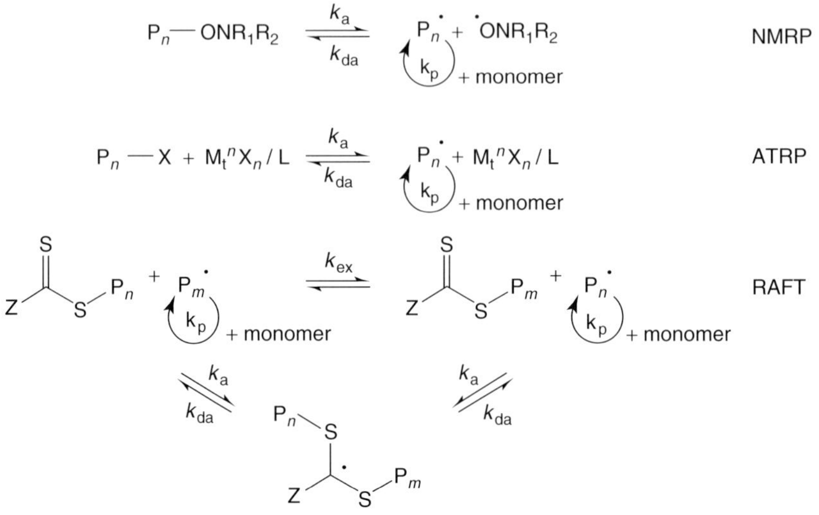

Figure 1.1 General mechanisms for controlled radical polymerization (CRP).

1.2
Controlled Free Radical Polymerization of Stimuli-Responsive Polymers

A CRP is a free-radical polymerization that displays a living character, that is, does not terminate or transfer, and is able to continue polymerization once the initial feed is exhausted by the addition of a monomer. However, termination reactions are inherent to a radical process, and modern CRP techniques seek to minimize such reactions, thus providing control over molecular weight and molecular weight distribution. More sophisticated CRP approaches incorporate many of the desirable features of traditional free-radical polymerization, such as compatibility with a wide range of monomers, tolerance of many functionalities, and facile reaction conditions. The control of molecular weight and molecular weight distribution has enabled access to complex architectures and site-specific functionality that were previously impossible to achieve via traditional free-radical polymerizations [35, 36].

The reversible deactivation of a growing radical chain can be achieved by stable (persistent) nitroxide radicals [37, 38]. Such radicals possess a structure similar to that of nitrogen monoxide. The single unpaired electron is delocalized over the nitrogen–oxygen bond. This delocalization as well as the captodative structure of the radical contributes to its stability. The deactivation occurs by the recombination of the radical chain end with such stable nitroxide. The formed C–O–N bond is thermolabile and can be cleaved at elevated temperatures (90–130 °C). Hence, the equilibrium between active and dormant species can be controlled by the reaction temperature. A recent major advance in nitroxide-mediated polymerization has been the development of a hydrido nitroxide, in which the presence of a hydrogen atom on the α-carbon leads to a significant increase in the range of vinyl monomers that undergo controlled polymerization [39]. Several nitroxides have been synthesized and they are illustrated in Scheme 1.1. The initiation of the reaction can be achieved by common initiators, such as azo-bisisobutyronitrile (AIBN) or benzoyl peroxide (BPO). An alternative approach is to use the so-called iniferter, which has the initiating and terminating moiety combined in one molecule. Using multifunctional iniferters, unique polymer structures (e.g., block, star, or graft copolymers) can be formed [40–42]. For example, telechelic poly(N-isopropylacrylamides) (PNIPAAms) could be synthesized via nitroxide-mediated controlled polymerization by introducing defined end-group moieties. Various functional groups, linked to the central nitroxide-initiator via a triazole moiety resulting from the so-called azide/alkyne-"click" reactions, were

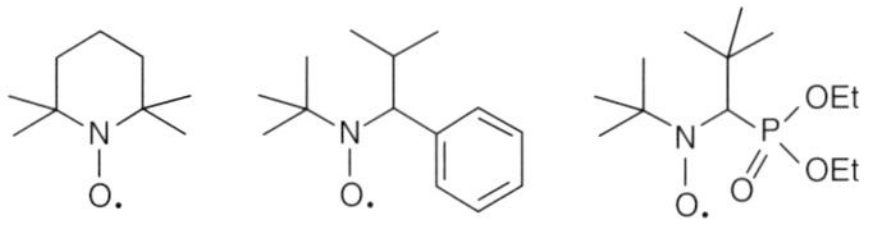

Scheme 1.1 Selected nitroxides suitable for the controlled polymerization of vinyl monomers.

probed with N-isopropylacrylamides (NIPAAms) and n-butyl acrylate (nBA) as monomers in terms of efficiency and livingness [43].

ATRP was developed by designing a proper catalyst (a transition metal compound and a ligand), using an initiator with an appropriate structure and adjusting the polymerization conditions. As a consequence, molecular weight during polymerization increased linearly with conversion and the polydispersities were typical for a living process [44]. The ATRP reaction mixture is, hence, a multicomponent system consisting of an initiator (mostly alkyl halogenides or chlorosulfonic acids), a transition metal catalyst, a ligand, a monomer, and if necessary a solvent and other compounds (an activator or a deactivator). Examples of initiators are illustrated in Scheme 1.2. The most significant part is the choice of the suitable catalyst/ligand system, which determines the equilibrium between dormant and active species. In the majority of studies, Cu is used as the catalyst, but the use of ruthenium, rhodium, palladium, nickel, and iron has also been reported [44]. Depending on the metal center, the ligands are nitrogen or phosphor compounds with a broad structural variety.

The choice of the initiator should be such that fast and quantitative initiation occurs. Under these conditions, all polymer chains grow at the same time. This is one prerequisite to obtain precise control of the molecular weight and a low PDI. Typical alkyl halogenides as initiators possess an acceptor substituent in the α-position to the C–X bond to weaken the C–X bond. In this case, a fast and selective transfer of the halogen atom from the initiator to the metal center can be achieved. In most cases, the halogen is chlorine or bromine.

ATRP can be performed in bulk as well as in solution. The use of a solvent is necessary if the polymer or catalyst complex is not soluble in the monomer. However, the solvent might have an influence on the ATRP process, changing the structure of the complex and enhancing its solubility, which is essential for establishing equilibrium between active and dormant species. The structure of the complex determines the rate and equilibrium of the transfer reaction. To control the polymerization, the transfer reaction between the solvent and the growing radical should not take place. The main advantage of ATRP is the tolerance of a variety of functional groups, enabling the polymerization of a large number of monomers under controlled conditions such as styrene, acrylate, methacrylate, acrylamide, and acrylnitrile derivatives. Currently, various efforts have been made to develop environmentally friendly ATRP processes [45].

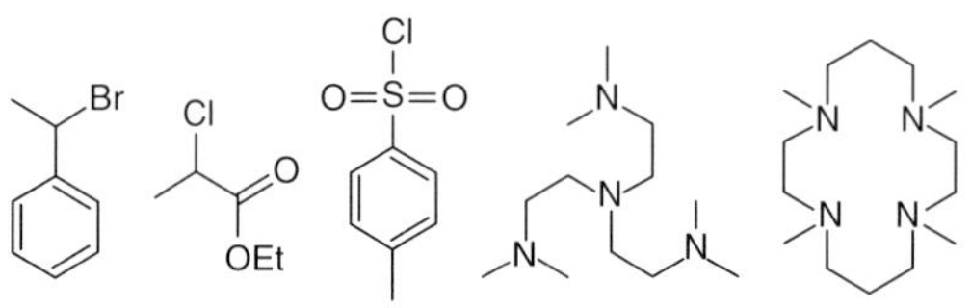

Scheme 1.2 Examples of initiators and ligands for ATRP.

For the synthesis of block copolymers, the reactivity of the macroinitiator has to be high enough to ensure fast reinitiation. However, the rate of reinitiation strongly depends on the halogen atom at the end of the macroinitiator. To maintain the controllability, a procedure for the halogen exchange has been developed [46]. The result of the formation of the second block also depends on the choice of the correct catalyst/ligand system. Hence, block formation has to be done in the correct order. For most systems, the catalyst complex as well as the solvent have to be tuned to fit the reactivity of the macroinitiator.

There are numerous examples using ATRP under different conditions (initiator, ligand, and catalyst) to form block copolymers based on substituted acrylates and methacrylates [47–53]. Amphiphilic random, gradient, and block copolymers of 2-(dimethylamino)ethyl methacrylate (DMAEMA) and n-butyl methacrylate (BMA) were synthesized by ATRP in water/2-propanol mixtures using a methoxy-poly(ethylene glycol) (MPEG) (M_n = 2000 g mol^{-1}) macroinitiator [54]. ATRP of dimethyl(1-ethoxycarbonyl)vinyl phosphate (DECVP) was performed in the presence of different catalyst systems and initiators, yielding polymers with controlled molecular weight and relatively low PDI (<1.5). PDECVP dissolves in water below 70 °C, but its critical solution temperature (T_c) depends on the polymer concentration [55]. Low-molecular-weight hydroxyethyl methacrylate (HEMA) oligomers prepared by ATRP (target degrees of polymerization, DP_n, less than 20) exhibited water solubility over a wide temperature range (no cloud point behavior). Furthermore, for actual DP_n's between 20 and 45, HEMA homopolymers exhibited inverse temperature solubility in dilute aqueous solution at pH 6.5, and their cloud points increased systematically as the DP_n was reduced. Statistical copolymerizations of HEMA with other comonomers such as glycerol monomethacrylate (GMA) and 2-hydroxypropyl methacrylate (HPMA) allowed the cloud point behavior to be manipulated. Finally, a range of novel HEMA-based block copolymers were synthesized, in which the HEMA block was either thermoresponsive or permanently hydrophilic, depending on its DP_n and the nature of the second block. Thus, diblock copolymer micelles with either hydroxylated cores or coronas could be prepared [56]. Poly(N-[(2,2-dimethyl-1,3-dioxolane)methyl]acrylamide) (PDMDOMAAm), a novel thermoresponsive polymer containing pendant dioxolane groups, was synthesized via ATRP. Water-soluble PDMDOMAAms with controlled molecular weight and narrow molecular weight distribution were obtained. The T_c of PDMDOMAAm was finely tuned over a wide temperature range by the partial hydrolysis of the acid labile dioxolane side group to form diol moieties (PDMDOMAAm diols). Unlike the traditional way of controlling T_c by copolymerization, the advantage of this method is that a series of thermoresponsive polymers with different T_cs can be prepared from a single batch of polymers with comparable molecular weight profiles [57].

Using ATRP catalyst system of tris-(2-dimethylaminoethyl)-amine (Me$_6$TREN) and Cu(I)chloride (CuCl), well-defined PNIPAAm could be synthesized at room temperature [58]. Narrow-dispersed PNIPAAms with well-controlled molecular weights and with end groups of varying hydrophobicity were synthesized in 2-propanol using the corresponding chloropropionate and chloropropionamide initiators. The choice of end groups affected the shape of the cloud point curves and the enthalpy of the phase transition [59]. A 2-chloropropionamide derivative featuring an azido group was used as the initiator to produce the end-functionalized PNIPAAm with an azido group. Subsequently, the "click" reaction between the azido end group and acetylene derivatives was demonstrated to produce PNIPAAm in which the end groups are modified by phenyl, 4-phenoxyphenyl, butyl, octyl, carboxylic acid, and hydroxymethyl groups. The resulting PNIPAAm derivatives show a T_c that ranges from 34.8 to 44.6 °C depending on the end group introduced [60]. Thermoresponsive polymers differing only in end functionalities induce phase transitions cooperatively only under dense-packed polymer brush conditions. This unique cooperative chain behavior in the hydrated micellar corona allows to regulate monodispersed micelle thermoresponse by blending well-defined diblock copolymers with thermoresponsive segments having hydrophobic and/or hydrophilic termini without any variations in critical micelle concentration (CMC) value or micelle size [61].

The syntheses of well-defined 7- and 21-arm PNIPAAm star polymers possessing β-cyclodextrin (β-CD) cores were achieved via the combination of ATRP and "click" reactions. A series of alkynyl terminally functionalized PNIPAAm (alkyne-PNIPAAm) linear precursors with varying DP_n were synthesized via ATRP of NIPAAm using propargyl 2-chloropropionate as the initiator. The subsequent "click" reactions of alkyne-PNIPAM with azido-β-CD led to the facile preparation of well-defined 7- and 21-arm star polymers [62].

Water-soluble poly(glycidol) (PGl) macroinitiators for ATRP have been prepared and their ability to form block copolymers with NIPAAm and 4-VP has been proven [63, 64]. On the basis of such polymers, a new method for the synthesis of smart nanohydrogels under additives-free conditions and at high solid content was investigated. The new core–shell nanohydrogels with cross-linked PNIPAAm core and hydrophilic PGl shell were obtained by photo cross-linking of PGl-*block*-PNIPAAm copolymers above their phase transition temperature. Figure 1.2 depicts reactions leading to these polymers [65]. Several graft copolymers are described, such as Chitosan-*graft*-PNIPAAm [66] and PNIPAAm-*graft*-Poly(2-vinyl pyridine) (P2VP) polymers, in previous reports [67]. Both polymers show a temperature and pH-sensitive phase behavior in aqueous solutions. Thermo- and pH-responsive micellization of poly(ethylene glycol)-β-P4VP-b-PNIPAAm in water was also studied. Micellization of the triblock copolymer, which was synthesized by sequential ATRP of 4VP and NIPAAm, occurred with combined stimulus of temperature and pH changes to form various morphological micelles [68]. Thermoresponsive materials with double-responsive AB-type diblock copolymers comprised of an NIPAAm segment and a poly(NIPAAm-co-(N-(hydroxymethyl) acrylamide) (HMAAm)) one were designed. Synthesized poly(NIPAAm-*co*-HMAAm)s showed sensitive thermoresponse, and the cloud point was completely tunable by the composition of HMAAm [69]. ATRP was also used to prepare thermosensitive cationic block copolymers of (3-acrylamidopropyl)-trimethylammonium chloride (AMPTMA) and NIPAAm with different block lengths [70]. Diblock copolymers poly(tetrahydrofuran-*block-tert*-butyl acrylate) (PTHF-*block*-PtBA) and

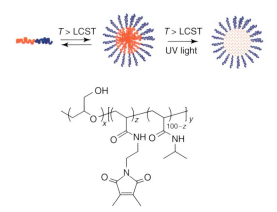

Figure 1.2 Synthetic approach used for preparation of PGl/PNIPAAm core-shell nano-hydrogels.

poly(tetrahydrofuran-b-1-ethoxyethyl acrylate) (PTHF-b-PEEA) were successfully synthesized by the dual initiator 4-hydroxybutyl-2-bromoisobutyrate (HBBIB). The isobutyrate and alcohol function of HBBIB were used for the ATRP of tBA (or EEA) and the living cationic ring-opening polymerization of THF, respectively. Hydrolysis or thermolysis of the aforementioned diblock copolymers results in amphiphilic pH-responsive copolymers poly(tetrahydrofuran-*block*-poly(acrylic acid) (PTHF-b-PAA) [71]. Cleavable block copolymers can be synthesized by a simple combination of the homopolymers synthesized by ATRP. Complementary reactive functionalities can be incorporated in these block copolymers that allow for the incorporation of additional functionalities in a postpolymerization step [72].

The RAFT process involves conventional radical polymerization in the presence of a suitable chain transfer agent (CTA). The degenerative transfer between the growing radicals and the CTAs provides controlled chain growth. A wide range of structurally diverse CTAs has been reported including dithioesters, trithiocarbonates, dithiocarbamates, and dithiocarbonates (xanthates), which are illustrated in Scheme 1.3 [73] The mechanism of the RAFT process is composed of the same three main steps as that of the conventional free-radical polymerization: initiation, propagation, and termination. Additionally, the propagation step in RAFT consists of two stages – the RAFT pre-equilibrium and the main RAFT equilibrium [35]. The first stage involves the activation of all added CTA along

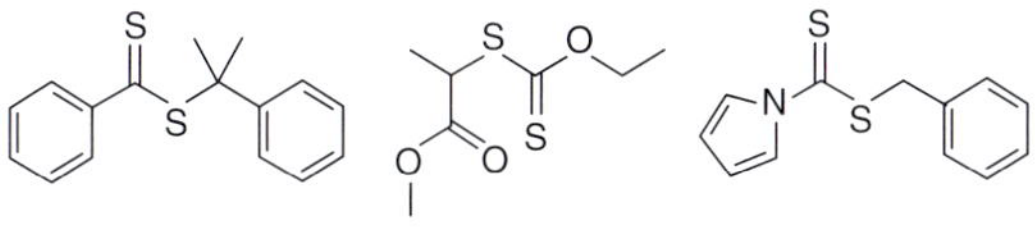

Scheme 1.3 Examples of thiocarbonylthio RAFT reagents.

with some degree of propagation, while the second stage consists of chain equilibrium and propagation. Because of the presence of the CTA and the subsequent degenerative transfer, the termination step is largely suppressed. The key to the structural control in the RAFT process is the careful selection of appropriate monomers, initiators, and CTAs. As polymers are synthesized by ATRP, the products from RAFT polymerization are colored due to the thiocarbonylthio end groups. However, these end groups can be readily removed by a posttreatment. A facile labeling technique was reported in which the telechelic thiocarbonylthio functionality of well-defined PNIPAAm prepared by room-temperature RAFT polymerization was first converted to the thiol and subsequently reacted with a maleimido-functional fluorescent dye [74]. Such an approach can be extended to the synthesis of hetero-telechelic a,ω biofunctionalized polymers [75, 76]. RAFT polymerization in the presence of a compound capable of both reversible chain transfer through a thiocarbonylthio moiety and propagation via a vinyl group led to highly branched copolymers by a method analogous to self-condensing vinyl copolymerization [77]. Highly branched PNIPAAm compounds were prepared by copolymerization of 3H-imidazole-4-carbodithioic acid 4-vinylbenzyl ester with NIPAAm [78]. NIPAAm star polymers were prepared using the four-armed RAFT agent pentaerythritoltetrakis(3-(S-benzyltrithiocarbonyl)-propionate) [79].

Acrylamides such as NIPAAm and N-ethylmethylacrylamide (EMA) or acrylamides containing proline and hydroxyproline moiety, N-acryloyl-L-proline (A-Pro-OH) and N-acryloyl-4-*trans*-L-proline (A-Hyp-OH), were readily polymerized by the RAFT process [80]. The latter case afforded well-defined amino-acid-based polymers [81]. A–B–A stereoblock polymers with atactic PNIPAAm as a hydrophilic block (either A or B) and a nonwater-soluble block consisting of isotactic PNIPAAm were also synthesized using RAFT polymerizations [82–84]. Using RAFT it was possible to obtain amphiphilic block copolymers of PNIPAAm (hydrophilic) and poly(styrene) (PS) or poly(*tert*-butylmethacrylate) (PtBMA) as the hydrophobic compounds [85]. The design of bisensitive narrowly distributed block copolymers consisting of NIPAAm and acrylic acid (AAc) was also feasible [86]. RAFT homopolymerization of 2-(diisopropylamino)ethyl methacrylate (DPA) and 2-(diethylamino)-ethyl methacrylate (DEA) and their random copolymerization were investigated. The random copolymers of DPA-ran-DEA were synthesized and used as macro-CTA to prepare poly(DPA-ran-DEA)-*block*-poly(N-(2-hydroxypropyl) methacrylamide) amphiphilic block copolymers [87]. Other amphiphilic block copolymers consist of PNIPAAm and of positively charged first- and second-generation dendronized polymethacrylates [88]. Novel double hydrophilic multiblock copolymers of N,N-dimethylacrylamide (DMAAm) and NIPAAm, m-PDMAAm$_p$–PNIPAAm$_q$, with varying DP$_n$s for PDMAAm and PNIPAAm sequences (p and q) were synthesized via consecutive RAFT polymerizations using polytrithiocarbonate as the CTA [89]. Thermosensitive association of a diblock copolymer consisting of poly(3-dimethyl(methacryloyloxyethyl) ammonium propane sulfonate) (PDMAEAPS), as an upper critical solution temperature (UCST) block, and poly(N,N-diethylacrylamide) (PDEAAm), as a lower critical solution temperature (LCST) block, has been investigated. Micelles form at temperatures both below the UCST and above the LCST of the blocks [90].

Monomers composed of a (meth)acrylate moiety connected to a short poly(ethylene) glycol (PEG) chain are versatile building blocks for the preparation of smart biorelevant materials. Many of these monomers are commercial and can be easily polymerized by CRP, allowing the synthesis of well-defined PEG-based macromolecular architectures such as amphiphilic block copolymers, dense polymer brushes, or biohybrids. Furthermore, the resulting polymers exhibit fascinating solution properties in an aqueous medium. Depending on the molecular structure of their monomer units, nonlinear PEG analogs can be either insoluble in water, readily soluble up to 100 °C, or thermoresponsive [91, 92]. The bromine chain ends of well-defined poly(oligo(ethylene glycol) acrylate) (POEGA) prepared using ATRP were successfully transformed into various functional end groups (w-hydroxy, w-amino, and w-Fmoc-amino acid) via a two-step pathway: (i) substitution of the bromine terminal atom by an azide function and (ii) 1,3-dipolar cycloaddition of the terminal azide and functional alkynes (propargyl alcohol, propargylamine, and N-R-(9-fluorenylmethyloxycarbonyl)-L-propargylglycine) [93, 94]. By this "click" chemistry, even cyclic polymers could be prepared [95, 96].

Monomers bearing an activated ester group can be polymerized under various controlled polymerization techniques, such as ATRP, NMRP, and RAFT polymerization. Combining the functionalization of polymers via polymeric-activated esters with these controlled polymerization techniques generates possibilities to realize highly functionalized polymer architectures [97]. Block copolymers containing stimuli-responsive segments provide important new opportunities for controlling the activity and aggregation properties of protein–polymer conjugates. A RAFT block copolymer PNIPAAm-*block*-PAAc was conjugated to streptavidin (SA) via the terminal biotin on the PNIPAAm block. The aggregation properties of the block copolymer–SA conjugate were very different from those of the free block copolymer. The outer-oriented hydrophilic block of PAA shields the intermolecular aggregation of the block copolymer–SA bioconjugate at pH values where the –COOH groups of PAA are significantly ionized [98]. PNIPAAm with imidazole end groups can be used to separate a histidine-tagged protein fragment directly from a crude cell lysate [99].

Amphiphilic diblock copolymers undergo a self-assembly micellar process in solvents that are selective for one of the blocks [100]. By choosing selective conditions for each block, conventional micelles and so-called inverse micelles can be formed. Examples of the so-called schizophrenic micelles were reported [101]. In this case hydrophilic AB diblock copolymers can form micelles in an aqueous solution, in which the A block forms the inner core and inverted micelles (with the B block forming the inner core) [102]. A diblock copolymer with two weak polybases, (poly-[2-(N-morpholino)ethyl methacrylate-*block*-2- and (diethyl amino)ethyl methacrylate) (PMEMA-*block*-DEAEMA), forms stable micelles with DEAEMA cores by adjusting the pH value of the solution. The formation of inverted micelles (MEMA core) was achieved by a "salting out" effect by adding electrolytes to the aqueous solution.

The synthesis of polyampholytes by using P2VP as a basic block was reported in several papers, for example, P2VP-*block*-poly(sodium-4-stryrenesulfonate) [103], P2VP-*block*-PAAc [104], and P2VP-*block*-PEO [100]. In this case, according to the corresponding pH value of the solution, it was possible to obtain precipitation, aggregation, or micellation. Recently, stimuli-responsive (pH-sensitive) block copolymers that self-assemble into vesicles without the addition of organic solvents have been reported [105]. Compared with pH-responsive materials, thermally responsive materials are advantageous for biological applications because of the stringent pH requirements in mammalian systems [106].

The behavior of double-responsive diblock copolymers of PNIPAAm-*block*-PAAc in aqueous solution is influenced by hydrogen-bonding interactions between the NIPAAm and AAc units [107]. This micellation behavior is often appealing to biomedical community for drug-delivery systems [108, 109]. Heterobifunctional block copolymers of PEG and PNIPAAm were synthesized by RAFT polymerization of NIPAAm using a macromolecular PEG-based CTA [110]. The synthesized block copolymers contained a carboxylic acid group from L-lysine at the focal point and a trithiocarbonate group at the terminus of the PNIPAAm block. The trithiocarbonate functionality was converted into a thiol group and used for conjugation of biotin to the end of the PNIPAAm block [111]. Alternatively, a biotinylated RAFT agent can be used [112]. Biotinylated copolymers that bind to the protein can be synthesized by ATRP as well [113, 114]. A series of well-defined PEO-*block*-PDMAEMA diblock copolymers were synthesized by ATRP techniques, followed by postpolymerization reactions to transform a portion of the tertiary amine groups of the PDMAEMA (poly(N,N-dimethylaminoethyl methacrylate) into phosphorozwitterions. Antiparasitic drugs used for the treatment of Leishmania were incorporated into the copolymer aggregates [115].

Current trends in the field of optical sensing include the development of dual sensors that respond simultaneously and independently to different stimuli [116]. In recent years, dual optical sensors have been reported, for example, for temperature and pH value, which would be beneficial, for example, to monitor chemical reactions and for biological diagnostics. The dual-sensitive polymeric material prepared by RAFT shows responsiveness in a temperature range from 10 to 20 °C and a pH range from 1 to 7 [117]. Actively controlled transport that is thermally switchable and size selective in a nanocapillary array membrane can be obtained by grafting PNIPAAm brushes onto the exterior surface of an Au-coated polycarbonate track-etched membrane. PNIPAAm brushes with 10–30 nm (dry film) thickness were grafted onto the Au surface through surface-initiated ATRP using a disulfide initiator [118]. Gold nanoparticles were prepared by the reduction of $HAuCl_4$ in the presence of thermosensitive PNIPAAm. Although thiol end-capped PNIPAM (poly(N-isopropylacrylamide) is known as a *macroligand effective* in stabilizing gold nanoparticles, this work showed that interactions between constitutive amides of PNIPAAm and gold are strong enough to protect gold nanoparticles against aggregation [119]. Highly stable hybrid unimolecular micelles with thermosensitive PNIPAAm shells incorporated with Ag nanoparticles were prepared *in situ* via a facile approach. Heating the hybrid unimolecular micellar solutions leads to the shrinkage of the PNIPAM shell and allows for tuning the relative spatial distances between neighboring Ag nanoparticles. These novel hybrid unimolecular micelles might be potential candidates for applications in sensors, catalysis, and optic/electronic devices [120].

1.3
Synthesis of Stimuli-Responsive Colloidal Dispersions

Colloidal dispersions represent one of the technologically important heterogeneous polymerization systems. Although colloidal particles have been synthesized for over five decades, complexities involved in synthetic aspects are related to their water-dispersive characteristics as well as the resulting particle morphologies. While the latter will determine film properties after and during colloidal particle coalescence, the former will affect solution characteristics. A schematic diagram illustrating reactions leading to the formation of colloidal particles is shown in Figure 1.3. While synthetic challenges arise from the complexity of synthetic conditions to accommodate a variety of monomers, there are opportunities for producing colloidal entities with a variety of shapes and morphologies.

The primary synthetic methods utilized for generating colloidal dispersions are suspension and emulsion polymerizations. Suspension polymerization occurs when monomers are suspended as a noncontinuous phase in a continuous aqueous medium and organic-soluble initiators facilitate polymerization by diffusion into the monomer droplets. Surfactants are typically utilized as stabilizing agents; however,

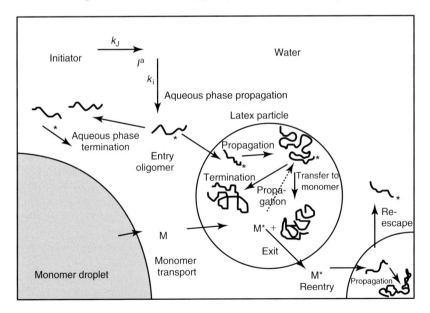

Figure 1.3 Schematic diagram illustrating principles of emulsion polymerization.

their concentration levels are too low to form micellar structures. Polymeric particles obtained by suspension polymerization range in size from 50 to 500 nm. Without proper agitation, the suspended particles remain in the continuous phase for a time period which may vary, depending upon experimental conditions.

Colloidal dispersions can also be synthesized using emulsion polymerization. Although similar to suspension polymerization, emulsion polymerization produces colloidal dispersions of typically smaller sizes while having increased stability within the continuous phase. As shown in Figure 1.3, emulsion polymerization proceeds through a free-radical polymerization process where monomers are polymerized in the presence of surfactant molecules suspended in an aqueous medium. The mechanism of the formation of polymer particles proceeds by two simultaneous processes. The first step is micellar nucleation that takes place when radicals (primary or oligomeric) enter from the aqueous phase into the micelles, followed by a homogeneous nucleation, which occurs when solution-polymerized oligomeric radicals become insoluble and are ultimately stabilized by free surfactant. When the concentration of surfactant molecules surpasses the CMC, highly ordered micellar structures are formed serving as polymerization loci for monomer molecules that have diffused through the continuous phase from the larger monomer droplets. At the same time, a water-soluble initiator is cleaved either thermally or photochemically, which initiates polymerization in the hydrophobic core of the micelle. As monomer continues to diffuse from the droplets and polymerize within micelles, surfactant molecules may be adsorbed from other micelles, solution, and monomer droplets [121–123]. Upon exhaustion of the monomer droplets, the reaction ceases, which is referred to as monomer starvation condition.

Emulsion polymerization reactions are typically conducted as a batch, semi-continuous, or continuous process. In a batch polymerization, water, initiator, surfactant, and monomer are incorporated into the reactor at the same time, and the cleavage of the initiator begins the polymerization process. This process provides limited control over particle nucleation and particle growth. For this reason, semicontinuous processes have been developed, providing better control over reaction conditions. During this process, one or more of monomers are polymerized over an extended period of time. To achieve control over the polymerization process, a small fraction of pre-emulsion, often referred to as a *seed*, can be utilized to induce polymerization, followed by the controlled addition of the remainder of the pre-emulsion. As discussed earlier, molecular weight can be controlled using CRP approaches discussed above, and colloidal size and shape control are major challenges. Dispersing agents are another significant component, which serve the dual purpose: stabilization of polymerization loci as well as stabilization of polymerized colloidal particles in an aqueous medium. One can envision that synthesis of larger sizes may be challenging if the glass transition temperature (T_g) of the copolymer is very low, particles may coagulate and precipitate out. Thus, from this prospective, an access of the free volume may not be beneficial, but as will be seen later, it will certainly help facilitating stimuli responsiveness [124, 125].

Because of the facile control of reaction kinetics, emulsion polymerization has found many applications and is often categorized according to the size of the resulting particles. While the most common are macroemulsions in the particle range of greater than 1 μm [126], miniemulsions are 50–200 nm [127, 128], and microemulsions are 10–100 nm [126]. Although there is a significant size overlap among these dispersions, there are major stability differences. While miniemulsions are generated by vigorous stirring and/or ultrasonication and are generally unstable, microemulsions are synthesized by spontaneous reactions of suitable monomers and are thermodynamically stable. For the latter, therefore, the monomers and their concentrations, dispersants, as well as other components must be carefully chosen to achieve desirable particle sizes. The use of reversible addition fragmentation chain transfer inverse microemulsion polymerization (RAFT-IMEP) resulted in the synthesis of well-controlled polymers retaining CTA functionality, which was accomplished by the design of pseudo-three-component phase diagrams consisting of hydrophilic monomer N,N-dimethylacrylamide, water, hexanes, nonionic surfactants, and a cosurfactant [129]. In contrast, using inverse miniemulsion photoinitiated polymerization of poly(N-isopropylacrylamide), high-quality composite quantum dot microspheres were also prepared [130]. Since the synthesis of hollow polymeric capsules offers the possibility of nano- or microreactors for a variety of applications, hollow polymeric capsules containing a hydrophilic liquid core were obtained in a simple one-pot miniemulsion process without the use of a sacrificial core and consist of polyurea, polythiourea, or polyurethane shells made by polycondensation at the interface of the droplets [131]. In addition, cross-linked nanoparticles of a random copolymer composed of methylmethacrylate (MMA), 4-vinylbenzyl chloride (VBC), and divinylbenzene (DVB) were also synthesized using RAFT miniemulsion polymerization in aqueous solutions [132]. Use of amphiphilic macro-RAFT agents in miniemulsion polymerization has been reported frequently. For example, RAFT-controlled/living *ab initio* emulsion polymerization by amphiphilic macro-RAFT self-assembly was reported [133] along with encapsulation of hydrophobic or hydrophilic pigments by emulsion polymerization using butyl acrylate/AAc random macro-RAFT copolymers [134]. In addition, poly(ethylene oxide) (PEO) macro-RAFT agent was used as both a stabilizer and a control agent in styrene polymerization in aqueous dispersed systems [135, 136]. Well-defined monodisperse NCs were also synthesized via this method [137]. In addition, PEO-RAFT interfacial inverse miniemulsion polymerization to synthesize the nanocapsules in a one-step process was reported [138].

Compared to homogeneous bulk or solution media, it might be a bit more complicated and challenging, but certainly beneficial, to extend ATRP to aqueous dispersions, thus opening avenues for commercial production of latex particles [139–141]. A new efficient initiation system, activators generated by electron transfer (AGET), significantly "face-lifted" the ATRP for heterogeneous systems [142–144], where instead of using activating Cu(I) catalyst that is sensitive to air, a higher oxidation state catalyst derived from a Cu(II) complex was utilized. Using this approach, thermally responsive random copolymers of di(ethylene glycol) methyl ether methacrylate (M(EO)2MA) and oligo(ethylene glycol)

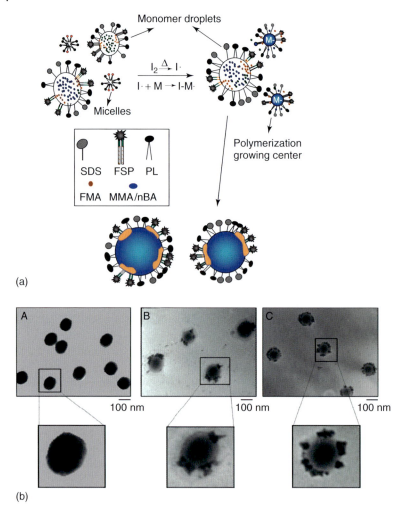

Figure 1.4 (a) Schematic diagram illustrating formation of mixed unimodal micelles and polymerization of non-spherical p-MMA/nBA/FMA colloidal particles; (b) TEM images of p-MMA/nBA (A); p-MMA/nBA/FMA (B and C).

methyl ether methacrylate (OEOMA) (Mn $\cong$ 300 or 475) with varies compositions were successfully synthesized via AGET ATRP in miniemulsion at 65 °C [145].

Heterogeneity of the colloidal reactions imposes several synthetic limitations. For example, the use of low-surface-tension monomers, such as fluoromonomers, has been a challenge. However, the use of biologically active phospholipids (PLs) as a dispersing agent provides synthetic conditions under which p-methyl methacrylate/n-butyl acrylate/heptadecafluorodecyl methacrylate (p-MMA/nBA/FMA) copolymer colloidal dispersions containing up to 15% w/w of heptadecafluorodecyl methacrylate (FMA) were copolymerized. These particles exhibit nonspherical particle morphologies and the choice of 1,2-dilauroyl-sn-glycero-3-phosphocholine (DLPC) was dictated by the fact that combining this PL with SDS/FSP surfactants results in the reduction of the overall surface tension of the aqueous phase from 72 to about 1–5 mN m^{-1} [146]. These conditions appear to be essential during polymerization of F-containing colloidal particles because lower surface tension not only facilitates efficient monomer transport to the polymerization loci but also provides stability of colloidal particles after synthesis. As shown earlier [147], DLPC in the presence of sodium dioctyl sulfosuccinate (SDOSS) forms unimodal micelles and consequently monodispersed particles are produced. In contrast, hydrogenated soybean phosphatidylcholine (HSPC) PL in the presence of SDOSS forms bimodal distribution of particles. Thus, the choice of the PL is crucial [148]. When monomers diffuse through an aqueous phase to the nucleation site, the reduced surface tension and monomer starvation conditions facilitate transport of higher quantities and polymerization of FMA into p-MMA/nBA particles. This is schematically illustrated in Figure 1.4. As MMA and nBA monomers initially migrate to the polymerization site, and upon initiation polymerize at the reactive site, monomer starvation conditions force FMA to migrate to the reactive site and diffuse to p-MMA/nBA copolymer core, which is facilitated by the presence of PL, which lowers the surface tension such that colloidal particles containing hydrophobic-lipophobic entities of p-FMA are stable and thus do not coagulate.

Combination of stimuli-responsive components with other monomers that provide higher free volume content is one of the prerequisites for preparing stimuli-responsive colloidal particles that form solid films. For example, in thermoplastic materials, it is desirable to copolymerize lower T_g components with stimuli-responsive species, as shown in Figure 1.5a. In thermosetting materials illustrated in Figure 1.5b, incorporation of an entity that responds to electromagnetic stimuli is a common approach and the challenge is to make stimuli-responsiveness reversible, and lower T_g facilitates spatial conditions for responsiveness because the presence of localized "voids" provides space for polymer chain rearrangements. Following this concept, poly(N-(DL)-(1-hydroxymethyl) propylmethacrylamide/n-butyl acrylate) (p(DL-HMPMA/nBA)) [12] and poly(2-(N,N'-dimethylamino)ethyl methacrylate/n-butyl acrylate) (p(DMAEMA/nBA)) [24] colloidal particles were synthesized, which upon coalescence retain their stimuli responsive properties. This is facilitated by the presence of low T_g nBA components. The temperature responsiveness is controlled by DL-HMPMA or DMAEMA components, while the lower T_g nBA component provides sufficient free volume for copolymer chain rearrangements. One interesting macroscopic phenomenon resulting from these combinations of monomers is detectable 3D changes observed in p(DL-HMPMA/nBA) and p(DMAEMA/nBA) films. While p(DL-HMPMA/nBA) shrink in the x–y plane and expand in the thickness (−) directions, p(DMAEMA/nBA) films shrink in all the directions at elevated temperatures. Such reversible dimensional differences in p(DL-HMPMA/nBA) and p(DMAEMA/nBA) copolymer films are attributed to preferential orientational

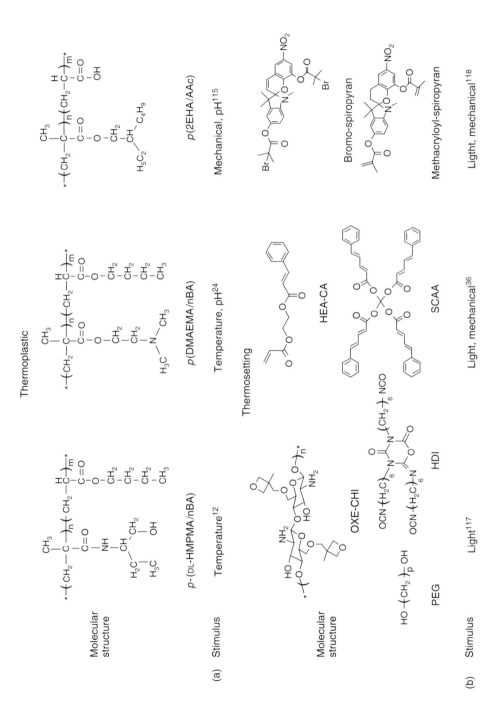

Figure 1.5 Examples of molecular structures of thermoplastic (A) and thermosetting (B) stimuli-responsive monomers.

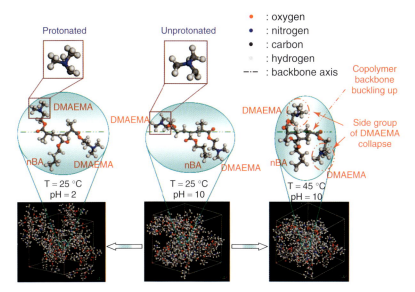

Figure 1.6 Computer simulation results of: p(DL-HMPMA/nBA) (A) and p(DMAEMA/nBA) (B).

changes of the side groups, as amide side groups in DL-HMPMA form preferential inter-/intramolecular interactions with itself or butyl ester pendant groups of nBA units, as compared to the ester side groups in DMAEMA. As a result, orientations of the side groups in p(DL-HMPMA/nBA) changes from preferentially parallel to perpendicular, which is responsible for the expansion in the z direction. Computer modeling results, shown in Figure 1.6, also confirmed the dimensional changes resulting from the buckling of copolymer backbone and a collapse of the DL-HMPMA or DMAEMA components leading to macroscopic volume changes of the entire network.

Designing polymer systems with different surface energy components afford another method to create responsive solid networks. Generally, in multicomponent networks, the component with a low surface energy is located at the top surface, while the component with the high surface energy is hidden beneath. Following this concept, poly(2-ethylhexyl acrylate/arylic acid) (p(2EHA/AAc)) [115] elastomers were developed, in which aliphatic portions with the low surface energy components are located at the top layer. Upon pressure load, chain rearrangements occur, which allow the high surface energy chains (acid portion) to migrate to the surface; this process is facilitated by hydrogen-bonding that leads to enhanced adhesion. Copolymers with long perfluoroalkyl and alkyl side chains are another example [119], which at low temperatures, phase separate perfluoroalkyl chains into liquid crystalline domains at the top layer, thus resulting in a low surface energy. At elevated temperatures, the liquid crystal domains become disordered and the perfluoroalkyl side chains mix with the aliphatic chains, which results in the increase in surface energy and the sharp decrease in tacticity.

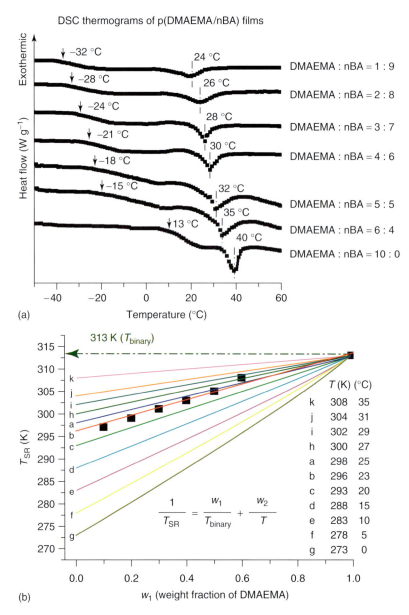

Figure 1.7 DSC thermograms of p(DMAEMA/nBA) copolymer films recorded for different DMAEMA/nBA weight ratios (A); Experimental T_{SR} values obtained from DSC measurements which allowed T_{SR} predictions for different T values plotted as a function of w_1: $1/T_{SR} = w_1/T_{binary} + w_2/T$.

1.3 Synthesis of Stimuli-Responsive Colloidal Dispersions

The first example of stimuli-responsive stratification of polymeric films, which would alter properties across the film thickness, was demonstrated on polyurethanes [149]. Other studies on colloidal dispersions also demonstrated that it is possible to control surface and interfaces responsiveness [150–154]. Heterogeneous stimuli-responsive polymer networks can also be generated by stratification across the film thickness [155]. Poly(methyl methacrylate/n-butyl acrylate/heptadeca-fluorodecylmethacrylate) (p(MMA/nBA/FMA) [154] colloidal dispersions is one example where the presence of bioactive dispersing agents PLs facilitates fluoromonomer copolymerization. These dispersions upon coalescence result in phase separation, where the FMA phase stratifies near the surface, thus resulting in ultralow static and kinetic coefficients of frictions, and MMA and nBA phases reside near the substrate. The degree of FMA stratification depends on the external temperature stimuli, which provides an opportunity for adjusting the film surface properties. Recent studies have shown that it is possible to entirely synthesize fluoro-surfactant free colloidal dispersions [156]. A similar process was utilized in the synthesis of colloidal particles that exhibit acorn shapes [157]. On the other hand, stimuli-responsive SS cross-linked redox responsive star polymer gels can be synthesized using ATRP in which SS cross-linked stars cleave under reducing conditions to form SH-functionalized soluble stars [158].

Stratification of the dispersing agents, such as SDOSS and PL during the film-formation processes, represents another stimuli-responsive heterogeneous network, where migration of individual species to the film–air (F–A) or film–substrate (F–S) interfaces is highly dependent on the external temperature, pH, and ionic strength. Furthermore, stratification of PLs leads to the formation of well-organized surface crystalline entities referred to as *surface-localized ionic clusters* (*SLICs*). Spectroscopic experiments combined with *ab initio* calculations [159] demonstrated that PL molecules recognize the MMA and nBA monomer boundaries along the copolymer backbone.

As mentioned above, the presence of free volume facilitates spatial conditions for polymer chain rearrangements. This is reflected in the T_g changes being controlled by the copolymer compositional changes. If stimuli-responsiveness results in conformational changes, these rearrangements should exhibit an endothermic character. This hypothesis has lead to the observation of compositional dependence of new endothermic stimuli-responsive transition (T_{SR}) [160] for stimuli-responsive polymeric solids [161]. As shown in Figure 1.7a, in addition to the lower T_g facilitating chain rearrangements, a series of DSC thermograms of p(DMAEMA/nBA) copolymer films recorded for different DMAEMA/nBA copolymer compositions show T_{SR}. As seen, similar to copolymer composition-dependent T_g transitions, the T_{SR} transitions also shift to higher temperatures as the amount of the stimuli-responsive DMAEMA component increases in the DMAEMA/nBA copolymer. On the basis of these experimental data, the following empirical relationship was established: $1/T_{SR} = w_1/T_{binary} + w_2/T_{form}$ or $1/T_{SR} = w_1(1/T_{binary} - 1/T) + 1/T$, where T_{SR} is the temperature of the stimuli-responsive transition, T_{binary} is the temperature of the stimuli-responsive homopolymer in a binary polymer–water equilibrium, w_1 and w_2 ($w_2 = 1 - w_1$) are weight fractions of each component in the copolymer,

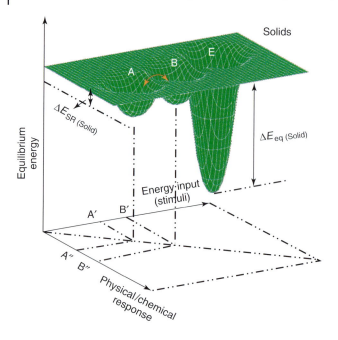

Figure 1.8 The relationship between equilibrium energy, stimuli energy input, and physical/chemical response in polymer solids.

and T is the film-formation temperature. As shown in Figure 1.7b, similar to the Fox equation [162] that allows predictions of the T_g for random copolymers, this relationship allows T_{SR} transition predictions in stimuli-responsive compositional solid films formed at different temperatures. While the T_g represents endothermic transitions due to segmental motion of the entire polymeric networks, for the T_{SR} to occur, free volume must be such that the local rearrangements of stimuli-responsive components are possible.

Because of spatial limitations resulting from a limited access of free volume, creating stimuli-responsive solid polymeric networks using colloidal processes represents a great challenge. For a given stimuli-responsive polymeric system, Figure 1.8 illustrates the relationship between the equilibrium energy (z axis), stimuli energy input (x axis), and chemical/physical response (y axis). Similar to polymeric gels, the minimum energy at equilibrium (E) provides stability of the network, whereas metastable energy states A and B facilitate stimuli responsiveness. In polymeric solids, the $\Delta E_{eq(solid)}$ at equilibrium is significantly greater compared to solutions, surfaces, and gels, thus resulting in a greater network integrity due to tightly entangled or cross-linked polymeric chains. As a consequence, the T_g is also relatively higher compared to other states. However, in solid polymeric networks, the entropic term (ΔS) does not contribute significantly to stimuli responsiveness because spatial mobility in these tighter networks is restricted. Polymer conformational changes as well as packing of responsive segments are significant contributors to the stimuli-responsive properties, and thermoplastic [163–165] and thermosetting [166–169] stimuli-responsive components of solids networks illustrated in Figure 1.5 represent only a fraction of what will become available in the future. There is an endless list of future applications and technologies that will take an advantage of these materials [170, 171].

1.4
Summary

Although the field of stimuli-responsive polymers is relatively new, and there are many recent advances being made, in particular in their synthesis, there are numerous opportunities available in creating new materials with stimuli-responsive attributes or in engineering unique nano-objects and devices in highly complex environments. Precise synthetic steps leading to an exact placement of a given monomer in a polymer backbone or as side groups have always been a challenge, and recent advances in ATRP, RAFT, and NMRP have created unprecedented opportunities for the creation of new stimuli-responsive polymers with well-defined molecular architectures. Furthermore, developments of colloidal synthesis resulting in different shape particles and copolymer morphologies provided an opportunity for new technological advances where stimuli-responsive attributes are maintained not only in solutions but also in a solid phase.

References

1. (a) Urban, M.W. (ed.) (2005) *Stimuli-Responsive Polymeric Films and Coatings*, ACS Symposium Series, Vol. 912, American Chemical Society and Oxford University Press; (b) Liu, F. and Urban, M.W. (2010) *Prog. Polym. Sci.*, **35**, 3.
2. Kuckling, D. (2009) *Colloid Polym. Sci.*, **287**, 881–891.
3. Hoffmann, A.S. and Stayton, P.S. (2004) *Macromol. Symp.*, **201**, 139–151.
4. Yakushiji, T., Sakai, K., Kikuchi, A., Aoyagi, T., Sakurai, Y., and Okano, T. (1998) *Langmuir*, **14**, 4657–4662.
5. Kopecek, J. (2009) *J. Polym. Sci., Part A: Polym. Chem.*, **47**, 5929–5946.
6. Kuckling, D., Richter, A., and Arndt, K.-F. (2003) *Macromol. Mater. Eng.*, **288**, 144–151.
7. Kuckling, D., Harmon, M.E., and Frank, C.W. (2003) *Langmuir*, **19**, 10660–10665.
8. Liu, R., Fraylich, M., and Saunders, B.R. (2009) *Colloid Polym. Sci.*, **287**, 627–643.
9. Bosman, A.W., Vestberg, R., Heumann, A., Frechet, J.M.J., and Hawker, C.J. (2003) *J. Am. Chem. Soc.*, **125**, 715–728.
10. Rempp, P. and Lutz, P. (1993) *Macromol. Chem. Macromol. Symp.*, **67**, 1–14.
11. Kennedy, J.P. and Ivan, B. (1991) *Designed Polymers by Carbocationic Macromolecular Engineering*, Carl Hanser Verlag, München.
12. Matyjaszewski, K. (1996) *Curr. Opin. Solid State Mater. Sci.*, **1**, 769–776.
13. Riess, G., Hurtrez, G., and Bahadur, P. (1985) Block copolymers, *Encyclopedia of Polymer Science and Engineering*, 2nd edn, John Wiley & Sons, Inc., New York, 201.

14. Matyjaszewski, K. (1998) in Controlled Radical Polymerization, ACS Symposium Series, Vol. 685 (ed. K. Matyjaszewski), American Chemical Society, Washington, DC.
15. Fischer, H. (1999) *J. Polym. Sci. Polym. Chem.*, **37**, 1885–1901.
16. Rodlert, M., Harth, E., Rees, I., and Hawker, C.J. (2000) *J. Polym. Sci., Part A: Polym. Chem.*, **38**, 4749–4763.
17. Matyjaszewski, K. (1995) *J. Phys. Org. Chem.*, **8**, 197–207.
18. Sawamoto, M. (1995) *Macromolecules*, **28**, 1721–1723.
19. Otsu, T., Matsumoto, A., and Tazaki, T. (1987) *Polym. Bull.*, **17**, 323–330.
20. Convertine, A.J., Ayres, N., Scales, C.W., Lowe, A.B., and McCornick, C.L. (2004) *Biomacromolecules*, **5**, 1177–1180.
21. Yu, K., Wang, H., Xue, L., and Han, Y. (2007) *Langmuir*, **23**, 1443–1452.
22. Zetterlund, P.B., Kagawa, Y., and Okubo, M. (2008) *Chem. Rev.*, **108**, 3747–3794.
23. Grubbs, R.H. and Tunas, W. (1989) *Science*, **243**, 907–915.
24. Trnka, T.M. and Grubbs, R.H. (2001) *Acc. Chem. Res.*, **43**, 18–24.
25. Schrock, R.R. and Hoveyda, A.H. (2003) *Angew. Chem., Int. Ed.*, **42**, 4592–4633.
26. Schrock, R.R. (1990) *Acc. Chem. Res.*, **23**, 158–165.
27. Maynard, H.D., Okada, S.Y., and Grubbs, R.H. (2000) *Macromolecules*, **33**, 6239–6248. *J. Am. Chem. Soc.*, (2001) **123**, 1275–1279.
28. Sutthasupa, S., Sanda, F., and Matsuda, T. (2008) *Macromolecules*, **41**, 305–311.
29. Sutthasupa, S., Shiotsuki, M., Matsuoka, H., Matsuda, T., and Sanda, F. (2010) *Macromolecules*, **43**, 1815–1822.
30. Smith, D., Pentzer, E.B., and Nguyen, S.T. (2007) *Polym. Rev.*, **47**, 419–459.
31. Rankin, D.A. and Lowe, A.B. (2008) *Macromolecules*, **41**, 614–622.
32. Noormofidi, N. and Slugovc, C. (2007) *Macromol. Chem. Phys.*, **208**, 1093–1100.
33. Hilf, S. and Kilbinger, A.F. (2009) *Macromolecules*, **42**, 4127–4133.
34. Conrad, R.M. and Grubbs, R.H. (2009) *Angew. Chem. Int. Ed.*, **48**, 8328–8330.
35. Boyer, C., Bulmus, V., Davis, T.P., Ladmiral, V., Liu, J., and Perrier, S. (2009) *Chem. Rev.*, **109**, 5402–5436.
36. Rosen, B.M. and Percec, V. (2009) *Chem. Rev.*, **109**, 5069–5119.
37. Malström, E.E. and Hawker, C.J. (1998) *Macromol. Chem. Phys.*, **199**, 923–935.
38. Sciannamea, V., Jerome, R., and Detrembleur, C. (2008) *Chem. Rev.*, **108**, 1104–1126.
39. Benoit, D., Hawker, C.J., Huang, E.E., Lin, Z., and Russell, T.P. (2000) *Macromolecules*, **33**, 1505–1507.
40. Hawker, C.J., Bosman, A.W., and Harth, E. (2001) *Chem. Rev.*, **101**, 3661–3688.
41. Fukuda, T., Terauchi, T., Goto, A., Tsujii, Y., and Miyamoto, T. (1996) *Macromolecules*, **29**, 3050–3052.
42. Benoit, D., Harth, E., Fox, P., Waymouth, R.M., and Hawker, C.J. (2000) *Macromolecules*, **33**, 363–370.
43. Binder, W.H., Gloger, D., Weinstabl, H., Allmaier, G., and Pittenauer, E. (2007) *Macromolecules*, **40**, 3097–3107.
44. Matyjaszewski, K. and Xia, J. (2001) *Chem. Rev.*, **101**, 2921–2990.
45. Tsarevsky, N.V. and Matyjaszewski, K. (2007) *Chem. Rev.*, **107**, 2270–2299.
46. Matyjaszewski, K., Shipp, D.A., Wang, J.-L., Grimaud, T., and Patten, T.E. (1998) *Macromolecules*, **31**, 6836–6840.
47. Moineau, G., Minet, M., Teyssie, P., and Jerome, R. (2000) *Macromol. Chem. Phys.*, **201**, 1108–1114.
48. Granel, C., Dubois, P., Jerome, R., and Teyssie, P. (1996) *Macromolecules*, **29**, 8576–8582.
49. Kotani, Y., Kato, M., Kamigaito, M., and Sawamoto, M. (1996) *Macromolecules*, **29**, 6979–6982.
50. Beers, K.L., Boo, S., Gaynor, S.G., and Matyjaszewski, K. (1999) *Macromolecules*, **32**, 5772–5776.
51. Wang, X., Luo, N., and Ying, S. (1999) *Polymer*, **40**, 4157–4161.
52. Xia, J., Zang, X., and Matyjaszewski, K. (1999) *Macromolecukes*, **32**, 3531–3533.
53. Zhang, Z.-B., Ying, S.-K., and Shi, Z.-Q. (1999) *Polymer*, **40**, 5439–5444.

54. Lee, S.B., Russell, A.J., and Matyjaszewski, K. (2003) *Biomacromolecules*, **4**, 1386–1393.
55. Huang, J. and Matyjaszewski, K. (2005) *Macromolecules*, **38**, 3577–3583.
56. Weaver, J.V.M., Bannister, I., Robinson, K.L., Bories-Azeau, X., Armes, S.P., Smallridge, M., and McKenna, P. (2004) *Macromolecules*, **37**, 2395–2403.
57. Zou, Y., Brooks, D.E., and Kizhakkedathu, J.N. (2008) *Macromolecules*, **41**, 5393–5405.
58. Masci, G., Giacomelli, L., and Crescenzi, V. (2004) *Macromol. Rapid Commun.*, **25**, 559–564.
59. Xia, Y., Burke, N.A.D., and Stöver, H.D.H. (2006) *Macromolecules*, **39**, 2275–2283.
60. Narumi, A., Fuchise, K., Kakuchi, R., Toda, A., Satoh, T., Kawaguchi, S., Sugiyama, K., Hirao, A., and Kakuchi, T. (2008) *Macromol. Rapid Commun.*, **29**, 1126–1133.
61. Nakayama, M. and Okano, T. (2008) *Macromolecules*, **41**, 504–507.
62. Xu, J. and Liu, S. (2009) *J. Polym. Sci., Part A: Polym. Chem.*, **47**, 404–419.
63. Mendrek, S., Mendrek, A., Adler, H.-J., Walach, W., Dworak, A., and Kuckling, D. (2008) *J. Polym. Sci., Part A: Polym. Chem.*, **46**, 2488–2499.
64. Mendrek, S., Mendrek, A., Adler, H.-J., Dworak, A., and Kuckling, D. (2009) *J. Polym. Sci. Part A: Polym. Chem.*, **47**, 1782–1794.
65. Mendrek, S., Mendrek, A., Adler, H.-J., Dworak, A., and Kuckling, D. (2009) *Macromolecules*, **42**, 9161–9169.
66. Lee, S.B., Ha, D.I., Cho, S.K., Kim, S.J., and Lee, Y.M. (2004) *J. Appl. Polym. Sci.*, **92**, 2612–2620.
67. Wohlrab, S. and Kuckling, D. (2001) *J. Polym. Sci., Part A: Polym. Chem.*, **39**, 3797–3804.
68. Zhang, W., Shi, L., Ma, R., An, Y., Xu, Y., and Wu, K. (2005) *Macromolecules*, **38**, 8850–8852.
69. Kotsuchibashi, Y., Kuboshima, Y., Yamamoto, K., and Aoyagi, T. (2008) *J. Polym. Sci., Part A: Polym. Chem.*, **46**, 6142–6150.
70. Patrizi, M.L., Diociaiuti, M., Capitani, D., and Masci, G. (2009) *Polymer*, **50**, 467–474.
71. Bernaerts, K.V., Willet, N., Van Camp, W., Jerome, R., and Du Prez, F.E. (2006) *Macromolecules*, **39**, 3760–3769.
72. Klaikherd, A., Ghosh, S., and Thayumanavan, S. (2007) *Macromolecules*, **40**, 8518–8520.
73. Lowe, A.B. and McCormick, C.L. (2007) *Prog. Polym. Sci.*, **32**, 283–351.
74. Scales, C.W., Convertine, A.J., and McCormick, C.L. (2006) *Biomacromolecules*, **7**, 1389–1392.
75. Roth, P.J., Jochum, F.D., Zentel, R., and Theato, P. (2010) *Biomacromolecules*, **11**, 238–244.
76. Segui, F., Qiu, X.-P., and Winnik, F.M. (2008) *J. Polym. Sci., Part A: Polym. Chem.*, **46**, 314–326.
77. Vogt, A.P. and Sumerlin, B.S. (2008) *Macromolecules*, **41**, 7368–7373.
78. Carter, S., Hunt, B., and Rimmer, S. (2005) *Macromolecules*, **38**, 4595–4603.
79. Plummer, R., Hill, D.J.T., and Whittaker, A.K. (2006) *Macromolecules*, **39**, 8379–8388.
80. Xu, J., Jiang, X., and Liu, S. (2008) *J. Polym. Sci., Part A: Polym. Chem.*, **46**, 60–69.
81. Mori, H., Kato, I., Matsuyama, M., and Endo, T. (2008) *Macromolecules*, **41**, 5604–5615.
82. Nuopponen, M., Kalliomäki, K., Laukkanen, A., Hietala, S., and Tenhu, H. (2008) *J. Polym. Sci., Part A: Polym. Chem.*, **46**, 38–46.
83. Hietala, S., Nuopponen, M., Kalliomäki, K., and Tenhu, H. (2008) *Macromolecules*, **41**, 2627–2631.
84. Kamigaito, M. and Satoh, K. (2008) *Macromolecules*, **41**, 269–276.
85. Nuopponen, M., Ojala, J., and Tenhu, H. (2004) *Polymer*, **45**, 3643–3650.
86. Schilli, C.M., Zhang, M., Rizzardo, E., Thang, S.H., Chong, Y.K., Edwards, F., Karlsson, G., and Müller, A.H.E. (2004) *Macromolecules*, **37**, 7861–7866.
87. Hu, Y.Q., Kim, M.S., Kim, B.S., and Lee, D.S. (2008) *J. Polym. Sci., Part A: Polym. Chem.*, **46**, 3740–3748.
88. Cheng, C., Schmidt, M., Zhang, A., and Schlüter, A.D. (2007) *Macromolecules*, **40**, 220–227.

89. Zhou, Y., Jiang, K., Song, Q., and Liu, S. (2007) *Langmuir*, **23**, 13076–13084.
90. Maeda, Y., Mochiduki, H., and Ikeda, I. (2004) *Macromol. Rapid Commun.*, **25**, 1330–1334.
91. Lutz, J.-F. (2008) *J. Polym. Sci., Part A: Polym. Chem.*, **46**, 3459–3470.
92. Skrabania, K., Kristen, J., Laschewsky, A., Akdemir, O., Hoth, A., and Lutz, J.-F. (2007) *Langmuir*, **23**, 84–93.
93. Lutz, J.-F., Börner, H.G., and Weichenhan, K. (2006) *Macromolecules*, **39**, 6376–6383.
94. Inglis, A.J., Sinnwell, S., Stenzel, M.H., and Barner-Kowollik, C. (2009) *Angew. Chem.*, **121**, 2447–2450.
95. Qiu, X.-P., Tanaka, F., and Winnik, F.M. (2007) *Macromolecules*, **40**, 7069–7071.
96. Xu, J., Ye, J., and Liu, S. (2007) *Macromolecules*, **40**, 9103–9110.
97. Theato, P. (2008) *J. Polym. Sci., Part A: Polym. Chem.*, **46**, 6677–6687.
98. Kulkarni, S., Schilli, C., Grin, B., Müller, A.H.E., Hoffman, A.S., and Stayton, P.S. (2006) *Biomacromolecules*, **7**, 2736–2741.
99. Carter, S., Rimmer, S., Rutkaite, R., Swanson, L., Fairclough, J.P.A., Sturdy, A., and Webb, M. (2006) *Biomacromolecules*, **7**, 1124–1130.
100. Prochazka, K., Martin, J.T., Webber, S.E., and Munk, P. (1996) *Macromolecules*, **29**, 6526–6530.
101. Liu, S., Billingham, N.C., and Armes, S.P. (2001) *Angew. Chem.*, **113**, 2390–2393.
102. Bütün, V., Billingham, N.C., and Armes, S.P. (1998) *J. Am. Chem. Soc.*, **120**, 11818–11819.
103. Varoqui, R., Tran, Q., and Pfefferkorn, E. (1979) *Macromolecules*, **12**, 831–835.
104. Kamachi, K., Kurihara, M., and Stille, J.K. (1972) *Macromolecules*, **5**, 161–167.
105. Du, J. and Armes, S.P. (2005) *J. Am. Chem. Soc.*, **127**, 12800–12801.
106. Li, Y., Lokitz, B.S., and McCormick, C.L. (2006) *Angew. Chem.*, **118**, 5924–5927.
107. Schilli, C.M., Zhang, M., Rizzardo, E., Thang, S.H., Chong, Y.K., Edwards, K., Karlsson, G., and Müller, A.H.E. (2004) *Macromolecules*, **37**, 7861–7866.
108. Satturwar, P., Eddine, M.N., Ravenelle, F., and Leroux, J.-C. (2007) *Eur. J. Pharm. Biopharm.*, **65**, 379–387.
109. Jones, M.-C. and Leroux, J.C. (1999) *Eur. J. Pharm. Biopharm.*, **48**, 101–111.
110. Zhao, J., Zhang, G., and Pispas, S. (2009) *J. Polym. Sci., Part A: Polym. Chem.*, **47**, 4099–4110.
111. You, Y.-Z. and Oupicky, D. (2007) *Biomacromolecules*, **8**, 98–105.
112. Hong, C.-Y. and Pan, C.-Y. (2006) *Macromolecules*, **39**, 3517–3524.
113. Vazquez-Dorbatt, V. and Maynard, H.D. (2006) *Biomacromolecules*, **7**, 2297–2302.
114. Bontempo, D. and Maynard, H.D. (2005) *J. Am. Chem. Soc.*, **127**, 6508–6509.
115. Karanikolopoulos, N., Pitsikalis, M., Hadjichristidis, N., Georgikopoulou, K., Calogeropoulou, T., and Dunlap, J.R. (2007) *Langmuir*, **23**, 4214–4224.
116. Nagl, S. and Wolfbeis, O.S. (2007) *Analyst*, **132**, 507–511.
117. Pietsch, C., Hoogenboom, R., and Schubert, U.S. (2009) *Angew. Chem.*, **121**, 5763–5766.
118. Lokuge, I., Wang, X., and Bohn, P.W. (2007) *Langmuir*, **23**, 305–311.
119. Aqil, A., Qiu, H., Greisch, J.-F., Jerome, R., De Pauw, E., and Jerome, C. (2008) *Polymer*, **49**, 1145–1153.
120. Xu, H., Xu, J., Zhu, Z., Liu, H., and Liu, S. (2006) *Macromolecules*, **39**, 8451–8455.
121. Flory, P. (1953) *Principles of Polymer Chemistry*, Cornell University Press, Ithaca.
122. Lovell, P.A. and El-Aasser, M.S. (eds) (1998) *Emulsion Polymerization and Emulsion Polymers*, John Wiley & Sons, Inc., New York.
123. Provder, T. and Urban, M.W. (eds) (1996) *Film Formation in Waterbourne Coatings*, ACS Symposium Series, Vol. 648, American Chemical Society, Washington, DC.
124. Liu, F. and Urban, M.W. (2010) *Prog. Pol. Sci.*, **35**, 3.
125. Urban, M.W. (2010) *McGraw-Hill Yearbook of Science & Technology*, McGraw-Hill, New York, pp. 362–365.

126. Shah, D.O. (ed.) (1985) ACS Symposium Series, Vol. 272, American Chemical Society, Washington, DC.
127. Chou, Y.J., El-Aasser, M.S., and Vanderhoff, J.W.J. (1980) *Dispersion Sci. Technol.*, **1**, 129–150.
128. Ugelstad, J., El-Aasser, M.S., and Vanderhoff, J.W.J. (1973) *Polym. Sci., Polym. Lett. Ed.*, **11**, 503–513.
129. Sogabe, A. and McCormick, C.L. (2009) *Macromolecules*, **42**, 5043–5052.
130. Janczewski, D., Tomczak, N., Han, M.-Y., and Vancso, G.J. (2009) *Macromolecules*, **42**, 1801–1804.
131. Crespy, D., Stark, M., Hoffmann-Richter, C., Ziener, U., and Landfester, K. (2007) *Macromolecules*, **40**, 3122–3135.
132. Feng, H., Zhao, Y., Pelletier, M., Dan, Y., and Zhao, Y. (2009) *Polymer*, **50**, 3470–3434.
133. Ferguson, C.J., Hughes, R.J., Nguyen, D., Pham, B.T.T., Gilbert, R.G., Serelis, A.K., Such, C.H., and Hawkett, B.S. (2005) *Macromolecules*, **38**, 2191–2204.
134. Nguyen, D., Zondanos, H.S., Farrugia, J.M., Serelis, A.K., Such, C.H., and Hawkett, B.S. (2008) *Langmuir*, **24**, 2140–2150.
135. dos Santos, A.M., Le Bris, T., Graillat, C., and Lansalot, M. (2009) *Macromolecules*, **42**, 946–956.
136. dos Santos, A.M., Pohn, J., Lansalot, M., and D'Agosto, F. (2007) *Macromol. Rapid Commun.*, **12**, 1325–1332.
137. Lu, F., Luo, Y., Li, B., and Zhao, Q. (2007) *Macromol. Rapid Commun.*, **28**, 868–874.
138. Lu, F., Luo, Y., Li, B., Zhao, Q., and Schork, F.J. (2010) *Macromolecules*, **43**, 568–571.
139. Qiu, J., Charleux, B., and Matyjaszewski, K. (2001) *Prog. Polym. Sci.*, **26**, 2083–2134.
140. Cunningham, M.F. (2008) *Prog. Polym. Sci.*, **33**, 365–398.
141. Min, K. and Matyjaszewski, K. (2009) *Cent. Eur. J. Chem.*, **7**, 657–674.
142. Jakubowski, W. and Matyjaszewski, K. (2005) *Macromolecules*, **38**, 4139–4146.
143. Gnanou, Y. and Hizal, G.J. (2004) *Polym. Sci., Part A: Polym. Chem.*, **42**, 351–359.
144. Min, K., Gao, H., and Matyjaszewski, K.J. (2005) *Am. Chem. Soc.*, **127**, 3825–3830.
145. Dong, H. and Matyjaszewski, K. (2010) *Macromolecules*, **43**, 4623–4628.
146. Pinazo, A., Wen, X., Liao, Y.C., Prosser, A.J., and Franses, E.I. (2002) *Langmuir*, **18**, 8888.
147. Lestage, D.J., Yu, M., and Urban, M.W. (2005) *Biomacromolecules*, **6**, 1561.
148. Lestage, D.J., Schleis, D.J., and Urban, M.W. (2004) *Langmuir*, **20**, 7027.
149. Urban, M.W. (2000) *Prog. Org. Coat.*, **40**, 195–202.
150. Zhao, Y. and Urban, M.W. (2000) *Macromolecules*, **33**, 7573–7581.
151. Lestage, D.J. and Urban, M.W. (2004) *Langmuir*, **20**, 6443–6449.
152. Dreher, W.R., Jarrett, W.L., and Urban, M.W. (2005) *Macromolecules*, **38**, 2205–2212.
153. Urban, M.W. (2006) *Polym. Rev.*, **46**, 329–339.
154. Misra, A., Jarrett, W.L., and Urban, M.W. (2007) *Macromolecules*, **40**, 6190–6198.
155. Urban, M.W. and Lestage, D.J. (2006) *Polym. Rev.*, **46**, 445–466.
156. Misra, A., Jarrett, W., and Urban, M.W. (2009) *Macromolecules*, **42**(20), 7828–7835.
157. Misra, A., Jarrett, W.L., and Urban, M.W. (2009) *Macromol. Rapid Commun.*, **31**, 119–127.
158. Kamada, J., Koynov, K., Corten, C., Juhari, A., Yoon, J.A., Urban, M.W., Balazs, A.C., and Matyjaszewski, K. (2010) *Macromolecules*, **43**, 4133–4139.
159. Yu, M., Urban, M.W., Sheng, Y., and Leszczynski, J. (2008) *Langmuir*, **24**, 10382–10389.
160. Liu, F. and Urban, M.W. (2009) *Macromolecules*, **42**, 2161–2167.
161. Liu, F., Jarret, W.L., and Urban, M.W. (2010) *Macromolecules*, **43**(12), 5330–5337.
162. Fox, T.G. (1956) *Bull. Am. Phys. Soc.*, **1**, 123.
163. Liu, F. and Urban, M.W. (2008) *Macromolecules*, **41**, 352–360.
164. Liu, F. and Urban, M.W. (2008) *Macromolecules*, **41**, 6531–6539.
165. Falsafi, A. and Tirrell, M. (2000) *Langmuir*, **16**, 1816–1824.

166. Lendlein, A., Jiang, H., Jünger, O., and Langer, R. (2005) *Nature*, **434**, 879–882.
167. Yu, Y., Nakano, M., and Ikeda, T. (2003) *Nature*, **425**, 145.
168. Ghosh, B. and Urban, M.W. (2009) *Science*, **323**, 1458–1460.
169. Crevoisier, G., Fabre, P., Corpart, J.M., and Leibler, L. (1999) *Science*, **285**, 1246–1249.
170. Cohen, M., Genzer, J., Luzinov, I., Mueller, M., Ober, C., Stamm, M., Szleifer, I., Zauscher, S., Urban, M.W., Winnik, F., and Minko, S. (2010) *Nat. Mater.*, **9**, 101–113.
171. Corten, C. and Urban, M.W. (2009) *Adv. Mater.*, **21**, 5011–5015.

2
Biological- and Field-Responsive Polymers: Expanding Potential in Smart Materials

Debashish Roy, Jennifer N. Cambre, and Brent S. Sumerlin

2.1
Introduction

Many of the most important substances in living systems are macromolecules that vary according to their surrounding environment. Synthetic (co)polymers can gain similar adaptive behavior such that their utility goes beyond providing structural support to allow active participation in a dynamic sense. Incorporating multiple copies of functional groups readily amenable to a change in character (e.g., charge, polarity, solvency) along a polymer backbone causes relatively minor changes in chemical structure to be synergistically amplified, which leads to transformations in macroscopic material properties. The "response" of a polymer can be defined in many ways. Responsive polymers in solution are typically classified as those that change their individual chain dimensions/size, secondary structure, solubility, or the degree of intermolecular association. In most cases, the physical or chemical event that causes these responses is limited to the formation or destruction of secondary forces (hydrogen bonding, hydrophobic effects, electrostatic interactions, etc.), simple reactions (e.g., acid–base reactions) of moieties pendant to the polymer backbone, and/or osmotic pressure differentials that result from such phenomena. In other systems, a response includes more significant alterations in the polymeric structure. This chapter includes both concepts with emphasis on those that hold promise in the areas of biomedical, sensing, and electronics applications.

Interest in stimuli-responsive polymers has persisted over many decades, and a great deal of work has been dedicated to devising examples of environmentally sensitive macromolecules that can be crafted into new smart materials. However, the overwhelming majority of reports in the literature describing stimuli-responsive polymers are dedicated to macromolecular systems sensitive to a few common stimuli, typically changes in pH, temperature, and electrolyte concentration. We aim at highlighting recent results and future trends of biological- and field-responsive polymers that have not yet been exploited to a similar extent. Many of the topics represent opportunities for making advances in biomedical fields, due to their specificity and the ability to respond to stimuli that are inherently present in biological systems. Synthetic polymers that adapt in response to specific interactions with biomacromolecules and small molecules associated with healthy or diseased states (e.g., glucose) may facilitate the application of smart polymers in drug delivery, diagnostics, sensing, separations, and so on. Additionally, it is often advantageous to capitalize on a stimulus specifically applied from an external source so that the location and rate of response can be easily adjusted. The ability to apply these types of stimuli in a noninvasive manner facilitates applications *in vivo*. The following discussion focuses on recent research that emphasizes these underutilized adaptive behaviors.

2.2
Biologically Responsive Polymer Systems

Smart polymers are becoming increasingly important for many biomedical applications. Whether for controlled drug delivery, biosensing/diagnostics, smart films/matrices for tissue engineering, or *in situ* construction of structural networks, it is often advantageous to employ polymers that respond to stimuli present in natural systems. In fact, many naturally occurring polymers alter their conformation and degree of self-assembly in response to the presence of specific chemical species in their surroundings. Recent attention has been devoted to endowing synthetic polymers with functionality [1] that allows responsive behavior when exposed to biological small molecules or biomacromolecules. In some cases, this arises from including common functional groups that interact with biologically relevant species, and in others adaptive behavior is the result of the synthetic polymer being conjugated to a biological component. Both of these concepts are discussed below.

2.2.1
Glucose-Responsive Polymers

While a variety of specific biological responses have been reported in the literature [2], polymers that respond to glucose have received considerable attention because of their potential application in both glucose-sensing and insulin delivery applications. Glucose-responsive polymeric systems are typically based on enzymatic oxidation of glucose by glucose oxidase (GOx), binding of glucose with concanavalin A (ConA), or reversible covalent bond formation between glucose and boronic acids.

2.2.1.1 Glucose-Responsive Systems Based on Glucose-GOx
The majority of reports detailing glucose-responsive polymers are based on the GOx-catalyzed reaction of glucose with oxygen. Typically, glucose sensitivity is not caused by direct interaction of glucose with the responsive polymer, but rather by the response of the polymer to by-products that result from the oxidation of glucose. The enzymatic action of GOx on glucose is highly specific and leads to by-products of gluconic acid and H_2O_2. The incorporation of a polymer that

responds to either of these small molecules can lead to a glucose-responsive system. Typically, a pH-responsive polymer is loaded or conjugated with GOx, and the gluconic acid by-product that results from the reaction with glucose induces a response in the pH-responsive macromolecule. Imanishi and coworkers reported the covalent modification of a cellulose film with GOx-conjugated poly(acrylic acid) (PAA) [3]. At neutral and high pH levels, the carboxylate units of the PAA chains were negatively charged and extended due to electrostatic repulsion, which resulted in the occlusion of the pores in the cellulose membrane. The gluconic acid that resulted from the addition of glucose led to a local pH reduction, protonation of the PAA carboxylate moieties, and concomitant collapse of the chains obscuring the membrane pores thus facilitating the release of entrapped insulin.

Peppas and coworkers exploited a similar concept to prepare glucose-responsive hydrogels [4]. Poly(methacrylic acid (PMAA)-*graft*-ethylene glycol) gels were synthesized in the presence of GOx. At neutral and high pH values, the gels were swollen by repulsion between negative charges on methacrylate units. The reduction in pH upon oxidation of glucose by GOx led to gel collapse. In addition to a reduction in electrostatic repulsion, the efficient response was attributed to enhanced hydrogen bonding between the carboxyl and ether groups of the ethylene glycol units. Similar hydrogels prepared by copolymerization of *N*-isopropylacrylamide (NIPAM) with methacrylic acid [5] or a sulfadimethoxine monomer with *N,N*-dimethylacrylamide (DMA) have also been reported [6]. As opposed to gluconic acid leading to chain collapse in carboxylate-containing polymers, the lowering of pH could also lead to chain expansion in the presence of a polybase [7]. Kost and coworkers examined the glucose-responsive nature of cross-linked poly(2-hydroxyethyl methacrylate-*co*-*N,N*-dimethylaminoethyl methacrylate) (poly(HEMA-*co*-DMAEMA) that contained entrapped GOx, catalase, and insulin [8]. A similar system has been reported by Narinesingh and coworkers [9]. The need for increased biocompatibility in GOx-based responsive materials has recently led to the incorporation of poly(ethylene glycol) (PEG) grafts [7, 10, 11] and other nontoxic, nonimmunogenic, biocompatible polymers, such as chitosan [11, 12].

2.2.1.2 Glucose-Responsive Systems Based on ConA

Another type of glucose-responsive system utilizes competitive binding of glucose with glycopolymer–lectin complexes [13]. Lectins are proteins that specifically bind carbohydrates. Because most lectins are multivalent, glycopolymers tend to cross-link and/or aggregate in their presence; however, this aggregation can be disrupted by introducing a competitively binding saccharide [14]. Numerous glucose-responsive materials that are based on competitive binding between lectins and glucose have been reported. The lectin most heavily employed to impart sensitivity to glucose is ConA.

Kim and coworkers reported the synthesis of monosubstituted conjugates of glucosyl-terminal poly(ethylene glycol) (G-PEG) and insulin [15]. The G-PEG-insulin conjugates were bound to ConA that was attached pendantly along a PEG-poly(vinylpyrrolidone-*co*-acrylic acid) backbone. When the concentration of glucose in the surrounding solution increased, competitive binding of glucose with

ConA led to displacement and release of the G-PEG-insulin conjugates. Sambanis et al. reported a ConA–glycogen gel that demonstrated a gel–sol transition in the presence of glucose as a result of preferred binding of ConA with free glucose over the glycogen-containing gel [16]. Hoffman and coworkers synthesized a glucose-responsive hydrogel via copolymerization of ConA vinyl macromonomers with a monomer modified with pendent glucose units [17]. The addition of free glucose caused the glycopolymer–ConA complex to dissociate and the gel to swell, with the degree of swelling depending on glucose concentration. Responsive systems in which ConA is entrapped in the hydrogel can lead to leakage of the protein and irreversible swelling. However, the copolymerized hydrogel described above prevented ConA leakage, and the hydrogels were demonstrated to be reversibly responsive. The response of these polymers was specific for glucose or mannose, while other sugars caused no response [18, 19].

2.2.1.3 Glucose-Responsive Systems Based on Boronic Acid-Diol Complexation

Another mechanism of glucose response relies on polymers composed of only synthetic components. The ability of boronic acids to reversibly complex with sugars has led to their being heavily employed as glucose sensors and as ligand moieties during chromatography [20]. Boronic acids are unique because their water solubility can be tuned by changes in pH or diol concentration (Scheme 2.1). In aqueous systems, boronic acids exist in equilibrium between an undissociated neutral form (**1**) and a dissociated anionic form (**2**) [2, 21–24]. In the presence of 1,2- or 1,3-diols, cyclic boronic esters between the neutral boronic acid and a diol are generally considered hydrolytically unstable [21]. On the other hand, the anionic form (**2**) reversibly binds with diols to form a boronate ester (**3**), shifting the equilibria to the anionic forms (**2** and **3**) [24]. Polymers containing neutral boronic acid groups are generally hydrophobic, whereas the anionic boronate groups impart water solubility [22]. As the concentration of glucose is increased, the ratio of the anionic forms (**2** and **3**) to the neutral form (**1**) increases, and the hydrophilicity of the system is enhanced. The solubility of boronic-acid-containing polymers is dependent not only on pH but also on the concentration of compatible diols in the surrounding medium [2].

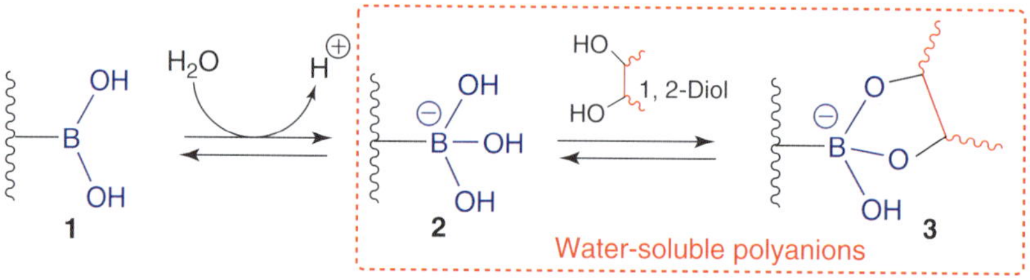

Scheme 2.1 Aqueous ionization equilibria of boronic acids. As the concentration of diol increases, the equilibria shift toward the anionic boronate forms of the boronic acid.

The majority of examples of boronic-acid-based responsive polymers have been hydrogels. Sakurai and coworkers reported the synthesis of a glucose-responsive hydrogel composed of terpolymers of 3-acrylamidophenylboronic acid (APBA), (N,N-dimethylamino)propylacrylamide (DMAPA), and DMA [25]. The boronic acid groups present in the terpolymer were allowed to complex with poly(vinyl alcohol) (PVA) at physiological pH. Addition of glucose led to competitive displacement of PVA, and the resulting decrease in cross-link density led to gel swelling. In similar systems, responsive hydrogels were made by complexing PVA with poly(N-vinyl-2-pyrrolidone-co-APBA) [26] or poly(DMA-co-3-methacrylamidophenylboronic acid-co-DMAPA-co-butyl methacrylate) [27]. Sakurai and coworkers reported another glucose-responsive polymeric hydrogel prepared via copolymerization of NIPAM with APBA and N,N-methylene-bis-acrylamide (MBA) as a cross-linker [23]. Glucose-responsive polymer gel particles were reported using a similar system [28], and poly(N-isopropylacrylamide) (PNIPAM)-based comb-type grafted hydrogels with rapid response to glucose concentration were recently reported by Xie and coworkers [29]. In addition to conventional hydrogels, several glucose-responsive microgels [30–33] and fluorescent nanospheres based on poly(NIPAM-co-APBA) [34] have been reported. Kataoka and coworkers recently reported the synthesis of a glucose-responsive copolymer of 4-(1,6-dioxo-2,5-diaza-7-oxamyl) phenylboronic acid (DDOPBA) ($pK_a \approx 7.8$) and NIPAM [35, 36].

Most glucose-responsive boronic-acid-based (co)polymers have been synthesized by conventional radical polymerization to yield random copolymers [37, 38], gels [31], or other cross-linked materials [39, 40]. To fully capitalize on the unique properties of boronic-acid-containing polymers, it is important to prepare well-defined (co)polymers with controlled molecular weights, narrow molecular weight distributions, and retained chain end functionalities [41, 42]. Controlled radical polymerization (CRP) methods including atom transfer radical polymerization (ATRP) [43–45] and reversible addition–fragmentation chain transfer (RAFT) polymerization [46–48] have been employed to prepare well-defined organoboron polymers [49–53]. We employed RAFT polymerization for the synthesis of well-defined boronic acid block copolymers by two different methods. The first approach involved the polymerization of the pinacol ester of 4-vinylphenylboronic acid, followed by a mild deprotection procedure [51]. We also reported the synthesis of well-defined block copolymers via direct RAFT polymerization of unprotected APBA [52, 53].

The block copolymers of hydrophilic DMA and responsive boronic acid monomers were hydrophilic and fully soluble above the pK_a of the boronic acid units. However, at pH < pK_a, the block copolymers self-assembled to form polymeric micelles. The boronic acid functionality within the hydrophobic core caused these micelles to dissociate with increasing pH or glucose concentration. Replacing the hydrophilic poly(N,N-dimethylacrylamide) (PDMA) block with temperature-responsive PNIPAM [54] led to triply responsive block copolymers that responded to changes in pH, glucose concentration, and temperature [53]. Depending on the combination of stimuli applied, these block copolymers were capable of forming micelles or reverse micelles.

2.2.2
Enzyme-Responsive Polymers

A relatively new area of research in stimuli-responsive polymeric systems is the design of materials that undergo macroscopic property changes when triggered by the selective catalytic actions of enzymes [55, 56]. Sensitivity of this type is unique because enzymes are highly selective in their reactivity, are operable under mild conditions present *in vivo*, and are vital components in many biological pathways. Enzyme-responsive materials are typically composed of an enzyme-sensitive substrate and another component that directs or controls interactions that lead to macroscopic transitions [56]. Catalytic action of the enzyme on the substrate can lead to changes in supramolecular architectures, swelling/collapse of gels, or the transformation of surface properties [56].

Hydrogels that are enzyme responsive can be used for the noninvasive formation of hydrogels *in situ*. Xu *et al.* reported the use of enzymatic dephosphorylation to induce a sol–gel transition. The small molecule fluorenylmethyloxycarbonyl(FMOC)-tyrosine phosphate was exposed to a phosphatase, and the resulting removal of phosphate groups led to a reduction in electrostatic repulsions, supramolecular assembly by π-stacking of the fluorenyl groups, and eventual gelation [57]. Another approach to convey enzyme sensitivity to a polymer is the incorporation of functional groups that react under enzymatic conditions. Exposure of the groups to a specific enzyme can lead to the creation of new covalent linkages that cause a change in macroscopic properties. Ulijin and coworkers used proteases to cause self-assembly of hydrogels via reversed hydrolysis (ligation) of peptides [58]. Transglutaminase has the ability to cross-link the side chains of lysine residues with glutamine residues within or across peptide chains [56]. This process was exploited for the synthesis of hydrogels of cross-linked functionalized PEG and lysine-containing polypeptides [59, 60]. Transglutaminase can similarly be used to cross-link naturally occurring polymers in the presence of cells [61]. Because some transglutaminase enzymes are only active in the presence of calcium ions, exposure to Ca^{2+} can also trigger enzymatic cross-linking [56]. Messersmith and coworkers used this phenomenon with a four-arm star-shaped PEG that contained a 20-residue fibrin peptide sequence at the end of each arm [62]. When the copolymer was mixed with Ca^{2+}-loaded liposomes designed to release their contents at body temperature, cross-linking of the peptide–PEG conjugates led to gels with potential applications as drug/gene delivery agents and tissue adhesives.

A wide variety of approaches have been employed to prepare hydrogels that respond to the presence of proteases. Typically, the hydrogel is exposed to a protease enzyme, and hydrolysis of protein- or peptide-based cross-linkers in the network leads to gel degradation and subsequent release of encapsulated contents. Hubbell and coworkers formed hydrogels in the presence of cells, using a Michael-type reaction between vinyl-sulfone-functionalized multiarmed telechelic PEG macromers and monocysteine adhesion peptides or bis-cysteine matrix metalloproteinases [63]. The resulting hydrogels locally degraded in response to cell-surface proteases, which led to gels with templated paths for cell migration. Moore and coworkers prepared chymotrypsin-responsive materials by incorporating a degradable CYKC tetrapeptide sequence as a cross-linker within polyacrylamide hydrogels [64]. The CYKC sequence contains a terminal cysteine conjugation site, a tyrosine residue that can be cleaved at the carboxyl side by chymotrypsin, and a lysine residue. When subjected to flowing or stationary solutions of α-chymotrypsin, the micron-sized gels rapidly dissolved due to the degradation of CYKC by α-chymotrypsin. Ulijin and coworkers reported protease-responsive hydrogels that potentially have applications for the removal of toxins or entrapment of drug molecules [55]. In this case, the response was caused by a change in osmotic pressure instead of cross-link degradation. Copolymer beads composed of acrylamide and PEG macromonomers were modified via an enzyme-cleavable tripeptide comprising combinations of glycine, phenylalanine, and positively charged arginine residues that imparted swelling due to electrostatic repulsions. Upon the addition of proteases, the tripeptide was cleaved, and the resulting loss of arginine groups led to a reduction in electrostatic repulsions and subsequent collapse of the hydrogel.

2.2.3
Antigen-Responsive Polymers

Antigen–antibody interactions are highly specific and are associated with complex immune responses that help recognize and neutralize foreign infection-causing objects in the body. Binding between antigens and antibodies relies on a variety of noncovalent interactions, such as hydrogen bonding, van der Waals forces, and electrostatic and hydrophobic interactions. Antibodies are employed in a number of immunological assays for the detection and measurement of biological and nonbiological substances [65], and the high affinity and specificity of their interactions with antigens can be harnessed to yield a variety of responsive synthetic polymeric systems. In most cases, antigen–antibody binding has been used to induce responses in hydrogels prepared by physically entrapping antibodies or antigens in networks, chemical conjugation of the antibody or antigen to the network, or using antigen–antibody pairs as reversible cross-linkers within networks [66]. Miyata *et al.* prepared antigen-sensitive hydrogels by coupling rabbit immunoglobulin G (Rabbit IgG) to *N*-succinimidylacrylate (NSA). The modified monomer was polymerized in the presence of goat anti-rabbit IgG as an antibody, acrylamide, and MBA, resulting in the formation of a hydrogel cross-linked both covalently and by antigen–antibody interactions. After the addition of rabbit IgG as a free antigen, competitive binding of the goat anti-rabbit IgG antibodies resulted in loss of the antigen–antibody cross-linkers and concomitant swelling of the hydrogel (Figure 2.1a) [67]. While this represented one of the first examples of an antigen-responsive polymeric system, the antibody was lost upon hydrogel

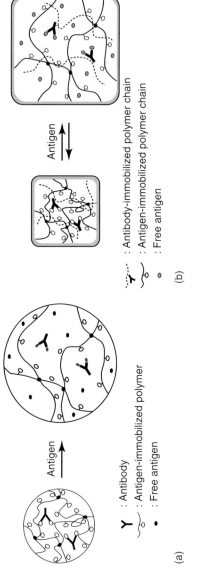

Figure 2.1 Nonreversible (a) and reversible (b) antigen-responsive hydrogels. The response in (a) is nonreversible as a result of loss of the antibody cross-links after addition of free antigen. Reversibility is possible in (b) because the antibody is covalently immobilized within the network. Reprinted by permission from Macmillan Publishers Ltd. [68], copyright (1999).

swelling, and antigen–antibody cross-links could not be reformed [65]. Miyata *et al.* later reported a similar hydrogel that was reversibly responsive [68]. The antigen (rabbit IgG) and antibody (goat anti-rabbit IgG) were each functionalized to contain vinyl groups. The modified antibody was polymerized with acrylamide, and the modified antigen was polymerized with acrylamide and MBA in the presence of the polymerized antibody to synthesize an antigen–antibody semi-interpenetrating network hydrogel (semi-interpenetrating polymer network (IPN)). Addition of free rabbit IgG antigen resulted in competitive binding with the antibodies in the hydrogel, which disrupted the antigen–antibody cross-links and led to swelling of the hydrogel (Figure 2.1b). Because both antigen and antibody were covalently bound within the semi-IPN, swelling was reversible. Stepwise changes in the antigen concentration induced pulsatile permeation of a protein through the network.

Antigen-responsive hydrogels based on polymerizable antibody Fab' fragments have also been reported [66]. The polymerizable Fab' fragment was copolymerized with NIPAM and MBA, and binding of antigens to the Fab' fragment caused reversible changes in volume. The response of the hydrogels was shown to be dependent on Fab' content, temperature, and pH. Hubble and coworkers reported an antigen-responsive hydrogel membrane based on a cross-linked dextran backbone grafted with both a fluorescein isothiocyanate (FITC) antigen and a sheep anti-FITC IgG antibody [69]. When free sodium fluorescein was added to the hydrogel, antibody–antigen cross-links were broken by competitive binding of the free antigen. The disruption of the cross-links resulted in reversible hydrogel swelling.

2.2.4
Redox-/Thiol-Responsive Polymers

Redox-/thiol-sensitive polymers are another class of responsive polymers with tremendous potential in the field of controlled drug delivery [70–76]. Interconversion of thiols and disulfides is an important reaction in many biological processes, playing an important role in the stability and rigidity of native proteins in living cells [77], and has been harnessed synthetically for various bioconjugation protocols [78, 79]. Because disulfide bonds are reversibly converted to thiols by exposure to various reducing agents and undergo disulfide exchange in the presence of other thiols, polymers containing disulfide linkages are both redox and thiol responsive (Scheme 2.2) [79, 80].

Glutathione (GSH) is the most abundant reducing agent in most cells [75], having a typical intracellular concentration of about 10 mM, whereas its concentration is only about 0.002 mM in the cellular exterior [81]. This significant variation in concentration has been used to design thiol- and redox-responsive delivery systems that release therapeutics upon entry into cells. Lee and coworkers synthesized polymeric micelles with shells cross-linked via thiol-reducible disulfide bonds to serve as biocompatible nanocarriers that preferentially release anticancer drugs in the reducing conditions characteristic of cancer tissues [82]. Redox-sensitive disulfide groups

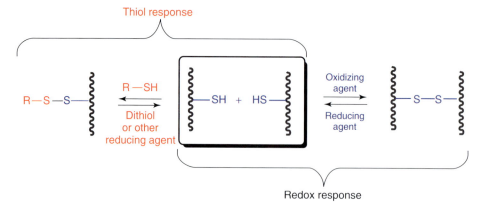

Scheme 2.2 Redox-/thiol-responsive behavior capable of being exploited in polymeric systems.

can also be directly introduced into side chains or backbones with an appropriate monomer, initiator, or chain transfer agent [75, 83–86]. Stayton and coworker synthesized a drug carrier by copolymerization of a pyridyl disulfide-containing acryloyl monomer with methacrylic acid and butyl acrylate [75]. The resulting polymers were thiol and pH sensitive and demonstrated membrane-disruptive properties useful for gene delivery. Tsarevsky and Matyjaszewski synthesized redox-/thiol-sensitive polymers using a disulfide-containing difunctional ATRP initiator [87], and later employed a disulfide-functional dimethacrylate monomer to synthesize redox sensitive nanogels via inverse miniemulsion ATRP [80, 88]. Incorporating cleavable disulfides within both micelle cores and shells is a feasible mechanism of stabilizing multimolecular aggregates. Stenzel *et al.* used RAFT to synthesize thiol-sensitive core cross-linked micelles consisting of poly(polyethylene glycol methyl ether methacrylate)-*block*-poly(5′-O-methacryloyluridine) and a dimethacrylate cross-linker [73]. The resulting nanoparticles demonstrated a drug release efficiency of up to 70% in 7 h in the presence of dithiothreitol (DTT). Liu and coworker also synthesized redox-responsive core cross-linked micelles of poly(ethylene oxide) (PEO)-*b*-poly(NIPAM-*co*-N-acryloxysuccinimide) via RAFT [89]. After reacting the micelles with cystamine, the disulfide bonds within the hydrophobic core were cleaved by treatment with DTT and re-formed again upon addition of cystamine as a thiol/disulfide exchange promoter. McCormick and coworkers synthesized disulfide-based cystamine-containing triblock copolymer micelles that were shell cross-linked [90]. Reversible cleavage of the micelles was accomplished with either DTT or tris(2-carboxyethyl)phosphine (TCEP), and the degraded micelles could be re-cross-linked using cystamine as a thiol exchanger. Shell cross-linked micelles of poly(L-cysteine)-*b*-poly(L-lactide) were also synthesized via ring-opening polymerization [91]. When treated with DTT, the cross-links within the micelle shells were lost, but when DTT was removed by dialysis, the shell cross-linked micelles were re-formed. Disulfide reduction can also be employed to induce morphological transitions of block copolymer aggregates in solution [92]. Polymeric micelles with interpolyelectrolyte-complexed cores composed of positively and negatively charged chains contained a PEG corona linked by disulfide bonds. Upon reduction with DTT, the PEG segments were detached from the micelles, leading to homopolymer complexes that adopted a vesicular structure due to the small curvature that resulted from the absence of shielding from PEG. The decreased free energy of the corona, due to the loss of the PEG chains from the polyion complex surface, is thought to be the driving force.

While the above-mentioned examples rely on block copolymer self-assembly into micelles or vesicles that can be subsequently (de)stabilized by thiol-disulfide chemistry, redox response can also be induced in aggregates that do not involve block copolymers. Thayumanavan and coworkers demonstrated that supramolecular polymer–surfactant complexes can form micelles susceptible to thiol-induced dissociation [93]. A polymer decorated with pendant carboxylates attached via disulfides was allowed to complex with a cationic surfactant. When micelles of the supramolecular complex were treated with GSH, cleavage of the polymer side chains resulted in aggregate dissociation and release of a model hydrophobic compound. Thiol-responsive triblocks composed of PNIPAM or poly(2-hydroxypropyl methacrylate) (PHPMA) and poly(2-(methacryloyloxy)ethylphosphorylcholine) (PMPC) have also been reported using the disulfide-based ATRP initiator. Because the integrity of the gels relied on disulfide linkages, controlled degradation could be induced under physiologically relevant conditions using GSH or DTT [71]. Rapid dissociation of PHPMA–PMPC–PHPMA triblock copolymer gels was observed under mild reducing conditions, indicating that materials prepared from such polymers may have potential applications as wound dressing [71]. Vogt and Sumerlin recently reported ABA triblocks copolymers capable of forming hydrogels that were both temperature and redox responsive (Scheme 2.3) [94]. A difunctional trithiocarbonate RAFT agent was used to synthesize symmetrical triblock copolymers of PNIPAM-b-PDMA-b-PNIPAM and poly(di(ethylene glycol)ethyl ether acrylate) (PDEGA)-b-PDMA-b-PDEGA. In each case, heating above the lower critical solution temperature (LCST) of the outer thermoresponsive block led to gelation. However, because the hydrophilic bridging PDMA chains contained labile trithiocarbonate linkages, aminolysis led to a free-flowing solution of thiol-decorated polymeric micelles. Under oxidizing conditions, the thiols on the termini of the micelle corona chains coupled to form disulfide-containing hydrogels. Reduction with either DTT or GSH led to gel degradation.

Monteiro and coworkers reported the synthesis and redox sensitivity of reversible cross-linked polystyrene networks. Aminolysis of RAFT-generated polymers led to telechelic thiol polymers that could reversibly assemble into responsive cyclic, bicyclic, multiblock, and network architectures [95, 96]. Star, block, and multiblock copolymers have been rendered thiol responsive by incorporating redox-labile linkages at block junction points. Liu *et al.* used RAFT to prepare thiol-sensitive biodegradable three-armed star polymers using both "core-first" and "arm-first" approaches [97]. Jeong and coworkers synthesized a disulfide-containing multiblock copolymer of PEO-b-poly(propylene oxide) (PPO-b-PEO) [74]. Drug release

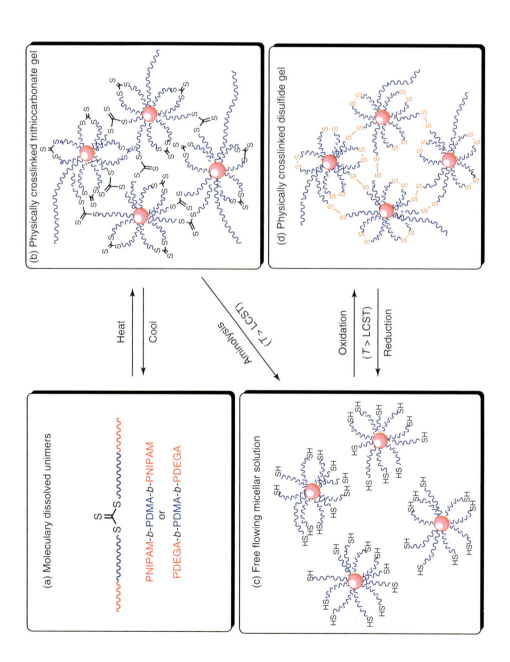

was significantly faster after treatment with GSH. Similarly, Oupicky and coworkers reported redox-responsive multiblock copolymers of PNIPAM and poly(N,N-dimethylaminoethyl methacrylate) (PDMAEMA) that were readily reduced to the constituent starting blocks due to the presence of a disulfide bridge between the blocks [70]. Hubbell and coworkers reported the design and characterization of block copolymers with hydrophilic PEG tethered to hydrophobic poly(propylene sulfide) (PPS) via a disulfide bridge [72, 98–100]. Because they should undergo sudden vesicular burst within the early endosome, polymeric vesicles assembled from such block copolymers may have promise for potential cytoplasmic delivery of biomolecular therapeutics, including peptides, proteins, oligonucleotides, and DNA. Tirelli and coworkers have also demonstrated a novel class of oxidation-responsive (co)polymers containing PPS [98–100]. PPS has a low T_g and is extremely hydrophobic until being oxidatively converted to the more polar poly(propylene sulfoxide) and ultimately poly(propylene sulfone). These novel materials may have potential applications as nanocontainers in drug delivery and biosensing.

2.3
Field-Responsive Polymers

Most methods of inducing a response in smart polymeric systems rely on kinetically restricted diffusion of the stimulus. Polymer gels that respond to changes in electrolyte concentration or pH require transport of externally introduced ions in the vicinity of the polymer backbone. As a result, the response of many traditional stimuli-responsive polymers is slow [101]. An alternative mechanism of stimulation that overcomes this issue is the application of electric, magnetic, sonic, or electromagnetic (photo/light) fields. In addition to having the ability to be applied or removed near instantaneously, some of these stimuli have the benefit of being directional, which can give rise to anisotropic deformation [102].

2.3.1
Electroresponsive Polymers

Electroresponsive polymers can lead to materials that swell, shrink, or bend in response to an electric field [103, 104]. Because electroresponsive polymers can

---

Scheme 2.3 Temperature and redox-responsive gelation of triblock copolymers prepared by RAFT. (a) Molecularly dissolved unimers of PNIPAM-b-PDMA-b-PNIPAM or PDEGA-b-PDMA-b-PDEGA. (b) Hydrogels formed upon heating above the LCST of the responsive PNIPAM or PDEGA blocks. (c) Free-flowing micellar solutions of PNIPAM-b-PDMA-SH or PDEGA-b-PDMA-SH resulting from trithiocarbonate aminolysis at $T >$ LCST. (d) Hydrogels formed from PNIPAM-b-PDMA-S-S-PDMA-b-PNIPAM or PDEGA-b-PDMA-S-S-PDMA-b-PDEGA upon oxidation of the thiol-terminated diblock aminolysis products [94]. Reproduced by permission of The Royal Society of Chemistry.

transform electrical energy into mechanical energy and have promising applications in biomechanics, artificial muscle actuation, sensing, energy transduction, sound dampening, chemical separations, and controlled drug delivery, these polymers are an increasingly important class of smart materials [104–106]. Gel deformation in an electric field is influenced by a number of factors, including variable osmotic pressure based on the voltage-induced motions of ions in the solution, pH, or salt concentration of the surrounding medium, position of the gel relative to the electrodes, thickness or shape of the gel, and the applied voltage [104, 107, 108]. For transforming the application of an electric field into a physical response, a polymer generally relies on collapse of a gel in an electric field, electrochemical reactions, electrically activated complex formation, ionic-polymer–metal interactions, electrorheological effects, or changes in electrophoretic mobility [103].

Most electroresponsive polymers have been investigated in the form of polyelectrolyte hydrogels [103, 109–114]. Polyelectrolyte gels deform under an electric field due to anisotropic swelling or deswelling as charged ions are directed toward the anode or cathode side of the gel [103]. Both synthetic and natural polymers have been employed. Naturally occurring polymers used to prepare electroresponsive materials include chitosan [109], chondroitin sulfate [115], hyaluronic acid [116], and alginate [113]. Major synthetic polymers have been prepared from vinyl alcohol [107, 112], allylamine [111], acrylonitrile [117], 2-acrylamido-2-methylpropane sulfonic acid [118], aniline [119], 2-hydroxyethyl methacrylate [120], methacrylic acid [112, 113], acrylic acid [121], and vinyl sulfonic acids [121]. Additionally, combinations of natural and synthetic components have been employed. Kim *et al.* prepared a semi-IPN hydrogel of poly(2-hydroxyethyl methacrylate) (PHEMA) and chitosan [109]. The response of the hydrogel to an electric field was investigated by measuring bending rate and angle, with these values increasing with ionic strength of the medium and applied voltage. In a recent study, Gao *et al.* reported an electroresponsive hydrogel based on PVA and poly(sodium maleate-*co*-sodium acrylate) [107]. In this case, hydrogel deformation increased when the concentration of NaCl or the electric voltage was increased.

Most polymers exhibiting electrosensitive behavior have been polyelectrolytes, but a few neutral polymers have also demonstrated electric field sensitivity in nonconducting media. Typically such systems require the presence of an additional charged or polarizable component with the ability to respond to an applied electric field. Zrinyi and coworkers synthesized a lightly cross-linked poly(dimethylsiloxane) (PDMS) gel containing electrosensitive colloidal TiO_2 particles (Figure 2.2) [103]. Significant and rapid deformation of the gel in silicon oil was observed. Because the TiO_2 particles were unable to exit the matrix, the force acting on the particles was transferred to the surrounding polymer, which resulted in gel deformation.

2.3.2
Magnetoresponsive Polymers

Polymers that respond to the presence or absence of magnetic fields can exist as free chains in solution, can be immobilized to surfaces, or can be cross-linked

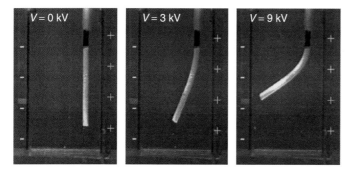

Figure 2.2 Electroresponse of a PDMS gel loaded with 10% TiO_2 as a function of uniform field strength. Reprinted from [103]. Copyright (2000) with permission from Elsevier.

within networks. The majority of reports in the literature involve the latter and describe the rapid response of magnetoresponsive gels swollen with complex fluids [101, 102, 122–124]. Generally, inorganic magnetic (nano)particles are physically entrapped within or covalently immobilized to a 3D cross-linked network [125], leading to materials with distortion in shape and size that occurs reversibly and instantaneously in the presence of a nonuniform magnetic field [123, 125, 126]. In this case, the magnetophoretic force [102] acting on the polymeric material as a result of the magnetic susceptibility of the particles has led to their increased use as soft biomimetic actuators, sensors, cancer therapy agents, artificial muscles, switches, separation media, membranes, and drug-delivery systems [102, 125, 127–129]. In uniform magnetic fields, a different phenomenon occurs. In this case, there is a lack of magnetic-field–particle interactions, but particle–particle interactions result from the creation of induced magnetic dipoles. Particle assembly within the surrounding polymer matrix can lead to significant transformations in material properties. Incorporation of magnetic nanoparticles (Fe_3O_4) within PNIPAM-based microgels is well documented in the literature [102, 130, 131]. Zrinyi and coworkers investigated magnetoresponsive polymer gel beads prepared by incorporating the magnetic nanoparticles into PNIPAM and PVA hydrogels [102, 132]. In uniform magnetic fields, the gel beads assembled into straight chainlike structures. However, when a nonuniform field was applied, aggregation was observed. Incorporation of the magnetic particles into cylindrical-shaped polymer gels yielded materials that mimicked muscular contraction by undergoing rapid and controllable changes in shape when exposed to a magnetic field.

Most examples of magnetoresponsive polymer systems involve noncovalent interactions between polymer chains and magnetic particles [125, 127, 128]. However, recent synthetic advances have facilitated covalent immobilization of polymer chains directly to the surface of magnetic particles. Pyun *et al.* have employed nitroxide-mediated radical polymerization (NMP) to prepare well-defined polymeric surfactants that stabilize magnetic nanoparticles. By casting nanoparticle dispersions from organic media onto surfaces, a diverse range of mesoscale morphologies have been observed, ranging from randomly entangled chains, field aligned 1D mesostructures, and nematiclike liquid crystal colloidal assemblies [124, 133, 134]. After ligand exchange to obtain a dispersion of polyacrylonitrile-stabilized ferromagnetic Co nanoparticles, a film was cast under an applied magnetic field [133]. Stabilization and pyrolysis of the resulting nanoparticle chains led to 1D carbon nanoparticle chains with cobalt inclusions. Ihara and coworkers recently reported a novel route to prepare magnetoresponsive gels via surface initiated ATRP [135]. Iron nanoparticles served as cross-linkers, eliminating the need for a conventional cross-linking agent.

2.3.3
Ultrasound-Responsive Polymers

Many of the suggested applications for stimuli-responsive polymers rely on the controlled release of therapeutic compounds at a specified rate or location within the body. In some cases, the change in environmental conditions necessary to impart responsive drug release occurs by passive or active migration of the polymeric carrier to an area of the body having conditions that encourage release (e.g., pH-responsive polymers can release drugs at specific points in the digestive tract or within endosomes due to inherent differences in pH). However, the application of systems that rely on other stimuli may be complicated by the inability to locally apply the stimulus at the targeted site. A change in temperature could lead to release in thermoresponsive polymer carriers *in vitro*, but localized heating and cooling *in vivo* is not always trivial at sites deep within the body. However, ultrasound is an effective stimulus that can be applied on demand externally and has proved effective at inducing drug release within the body [136]. Employing ultrasound-responsive polymers for controlled drug delivery is attractive because the method is noninvasive and has been successfully used in other areas of medical treatment and diagnostics [137].

It is difficult to pinpoint the specific macromolecular characteristics that allow response to ultrasound. Many polymers of varying composition, architecture, polarity, T_g, and so on, have been demonstrated to alter their behavior when exposed to ultrasonic stimulation. As opposed to magnetoresponsive polymers, foreign additives are not required [136]. However, polymers existing in a few specific physical states are mostly considered. Polymeric systems that respond to ultrasound have generally been gels or other nonswollen macroscopic solids, polymeric micelles, or layer-by-layer (LbL) coated microbubbles.

The ultrasound-induced release rate of incorporated materials from biodegradable polyglycolides, polylactides, and poly[bis(*p*-carboxyphenoxy)alkane anhydrides with sebacic acid and nonbiodegradable ethylene–vinyl acetate copolymers has been extensively studied by Kost and coworkers [138–140]. Each of these systems exhibited release kinetics that were significantly enhanced with increasing ultrasound intensity. Ultrasound significantly accelerates the rate of degradation in biodegradable polymers and enhances permeation through nonerodible polymers

[141, 142]. Poly(lactide-co-glycolide) microspheres and poly(HEMA-co-DMAEMA) hydrogels have been investigated as ultrasound-responsive drug-delivery systems [139, 142]. Miyazaki et al. reported an ethylene–vinyl alcohol copolymer system that was capable of delivering insulin on demand at increased rates in diabetic rats by external ultrasound irradiation [143]. Stoodley and coworkers reported ultrasound-responsive antibiotic release from PHEMA hydrogels [137]. The hydrogels were coated to minimize passive release with a layer of C_{12}-methylene chains by reaction with dodecyl isocyanate. In the absence of ultrasound, the antibiotic was retained within the polymer matrix; however, the application of low-intensity ultrasound led to disruption of the alkane coating and subsequent drug release. Kooiman et al. recently reported a similar system composed of ultrasound-responsive polymeric microcapsules with a shell of fluorinated end-capped poly(L-lactic acid) [144].

Acoustic destabilization of supramolecular polymer assemblies has also been considered [145–150]. Generally, polymeric micelles are induced to dissociate or adopt loosely associated morphologies when exposed to ultrasound [148]. Low-frequency ultrasound also enhances local cellular uptake of drugs, indicating that this approach could prove useful for delivery, provided precautions are taken to prevent cavitational damage to vital structures in the body. Pitt and coworkers [145, 149, 150] have demonstrated that hydrophobic drugs can be incorporated into PEO-b-PPO-b-PEO micelles and that release could be induced with low-frequency ultrasound.

Another method by which polymers can contribute to ultrasound responsiveness is through the coating and stabilization of microbubbles that result from sonication. Cavitation and microbubble implosion generate local shock waves and microjets that temporarily perforate membranes to facilitate cell entry [151, 152]. This process of "sonoporation" has proved effective *in vitro* and *in vivo*. De Smedt and coworkers demonstrated that microbubbles resulting from ultrasonic irradiation can be directly coated with a polymer shell and that the resulting structures can potentially be used for gene delivery applications [151]. Perfluorcarbon-gas-filled microbubbles were coated via an LbL approach with positively charged poly(allylamine hydrochloride) and negatively charged DNA, and the ultrasound responsiveness, physical properties, DNA binding/protection of the particles were investigated. Binding DNA on the microbubble surface ensured it was proximal to the site of cell membrane poration, which enhanced the probability of DNA entering the cell by the generated microjets.

2.3.4
Photoresponsive Polymers

Photoresponsive polymers change their properties when irradiated with light of appropriate wavelength [153, 154]. Typically these changes are the result of light-induced structural transformations of specific functional groups along the

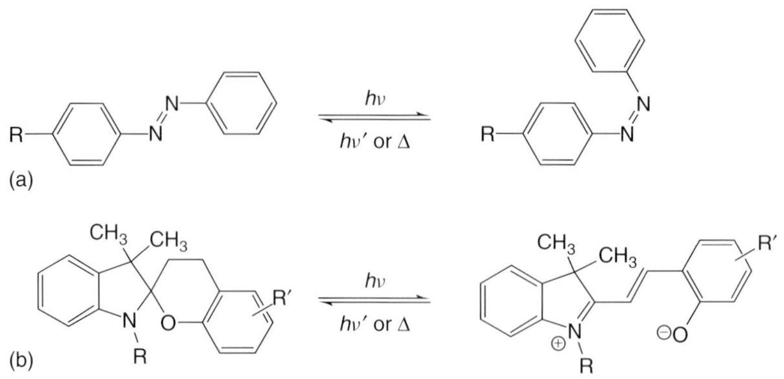

Scheme 2.4 Reversible photoinduced transformations of (a) azobenzene and (b) spiropyran derivatives.

polymer backbone or side chains [155–157]. Possible applications of photoresponsive polymers include reversible optical storage, polymer viscosity control, photomechanical transduction/actuation, protein bioactivity modulation, tissue engineering, and drug delivery [155, 158–163]. An important aspect of photosensitive polymer systems is that irradiation is a straightforward, noninvasive mechanism to induce responsive behavior. Photoresponsive polymers have been investigated for many years, but there has been a recent expansion in efforts to involve increasingly complex macromolecular architectures.

The most well studied examples of photoresponsive polymers are those that contain azobenzene groups [153, 164, 165]. Azobenzene is a well-known chromophore with a light-induced cis-to-trans isomerization that is accompanied by a fast and complete change in electronic structure, geometric shape, and polarity (Scheme 2.4) [155, 166]. By incorporating azobenzene derivatives, materials with variable shape, polarity, and self-assembly behavior can be obtained [167]. Wang and coworkers described photoinduced deformation of epoxy-based azobenzene-containing polymer colloids [168, 169]. Depending on the wavelength of irradiation, these photoresponsive colloids transformed morphology from spheres to spindles and finally to rods [169]. Tirrell and coworkers prepared photochromic derivatives of elastinlike polypeptides [poly(VPGVG)] where VPGVG is valine–proline–glycine–valine–glycine [170]. The inherent thermoresponsive nature of the polymers could be tuned by incorporating one azobenzene moiety for every 30 amino acid residues. Alonso *et al.* also reported the photoresponsive properties of azobenzene and spiropyran derivatives of elastinlike polypeptides [171, 172]. The change in polarity that accompanies isomerization has led to azobenzene-containing block copolymers being used to prepare photoresponsive micelles and vesicles [173–175]. Azo chromophores have been incorporated into many other polymeric systems, including poly(N-hydroxypropyl methacrylamide) (PHPMAm) [176], PAA [177], PDMAEMA [178, 179], and PNIPAM [180]. Photoresponsive dendrimers based on azobenzene derivatives have also been reported [181, 182].

In addition to enabling photoswitching of bulk material properties [155], surface properties [183, 184], and polymeric aggregates in solution [185], azobenzene groups can be employed to control supramolecular assembly of polymers. Ghadiri and coworkers reported a new photoresponsive peptide system composed of two ring-shaped cyclic peptides tethered by an azobenzene moiety [186, 187]. When the azobenzene group was in the trans state, the cyclic peptide units demonstrated intermolecular hydrogen bonding to yield extended linear chains. Ultraviolet (UV)-induced isomerization to the cis state led to intramolecular hydrogen bonding and depolymerization of the supramolecular complex. Similar photoresponsive supramolecular polymers based on diarylethene chromophores have also been reported [188].

The noninvasive nature in which light can be applied to a polymer solution offers significant potential for the application of photoresponsive polymers in biological systems. Hoffman and coworkers employed azobenzene-containing polymers as switches to reversibly activate enzymes in response to distinct wavelengths of light [159]. The authors used the same concept to reversibly control biotin binding by site-specific conjugation to streptavidin [158]. A polymer–streptavidin conjugate successfully bound biotin under UV irradiation, but upon exposure to visible light, the polymer collapsed to block biotin association.

The responsive nature of azobenzenes has been used to generate photomechanical effects and even macroscopic motion [185, 189]. Photocontraction of azobenzene-containing polymer films leads to changes in film thickness [190] and bending/unbending motions [191, 192] that can bring about macroscopic 3D motion [193]. Ikeda and coworkers developed a photoresponsive laminated film consisting of an azobenzene-containing cross-linked liquid–crystalline polymer and flexible polyethylene [193]. The laminated film was fully extended and flat under UV light, but recovered its original bent shape when irradiated with visible light. This type of photoresponsive polymeric film may lead to devices capable of converting light energy into mechanical work to induce motion in the absence of a wired connection to a power source.

Chromophores other than azobenzene have also been used to impart photoresponsive behavior. For instance, spiropyran derivatives can be incorporated terminally [194] or pendantly [161, 179, 195–197] to bring about light sensitivity. Spiropyran groups are relatively nonpolar, but irradiation with the appropriate wavelength of light leads to a zwitterionic isomer with a significantly increased dipole moment (Scheme 2.4b). The isomerization can be reversed by irradiating with visible light. This concept has been employed to prepare a variety of spiropyran-containing photoresponsive polymers, including PAA [161], PHPMAm [196], and PNIPAM [195, 198]. Matyjaszewski and coworkers used the phototunable change in polarity to prepare polymeric micelles with responsive spiropyran-containing blocks [197]. A block copolymer of PEO and a spiropyran-containing methacrylate monomer was prepared by ATRP, and the polymeric micelles formed in aqueous solution of the resulting polymer were completely disrupted by UV irradiation and regenerated by irradiation with visible light. Laschewsky modified PHPMAm by copolymerization with a monomer containing the photoreactive cinnamate moiety

[199]. Photoisomerization of pendant trans-cinnamate groups to cis-cinnamate groups yielded polymers with increased polarity and higher cloud points. Unlike other frequently used photoreactive groups, the cinnamate group is easily incorporated, chemically inert, and exhibits thermal stability in both the trans- and cis-forms.

While most examples of photoresponsive behavior rely on light-induced isomerization, another viable approach capitalizes on polarity changes that result from cleavage of photolabile bonds. Such photo(solvo)lysis reactions have been used to impart responsive behavior to methacrylate polymers. Zhao and coworkers investigated light-responsive amphiphilic diblock copolymers of PEO and a methacrylate monomer with photolabile pyrenylmethyl chromophores pendantly attached [200]. When exposed to UV irradiation, cleavage of the pyrenylmethyl esters caused the pyrene-containing hydrophobic methacrylate units to be converted to hydrophilic methacrylic acid units, which resulted in micelle dissociation. However, while photosolvolysis of pyrenylmethyl esters requires nucleophilic solvents, photolysis of polymers containing 2-nitrobenzyl side chain moieties can take place both in solution and in the solid state [201].

Most of the above-mentioned examples require isomerization induced by UV irradiation. However, UV and visible light are readily absorbed by the skin, so these responsive systems have potential limitations for many biomedical applications. However, infrared radiation penetrates skin with less risk of damage and might be more applicable for photoactivation of drug carriers within a living system. Therefore, Frechet and coworkers investigated block copolymers of PEG hydrophilic and 2-diazo-1,2-naphthoquinone. Infrared irradiation led to micelle dissociation as a result of the photoinduced rearrangement of the 1,2-napto-quinone units to yield anionic/hydrophilic 3-indenecarboxylate groups [202]. Responsive polymeric micelles based on diblock copolymers of PEG and poly(2-nitrobenzyl methacrylate) have also been reported in which the hydrophobic methacrylate block was converted to hydrophilic PMAA upon irradiation with near-infrared light [201].

2.4 Conclusions

Polymeric materials that respond to a wide variety of new stimuli are continually being developed. Inspiration is often provided by insight into the mechanisms of feedback-controlled biological processes. Indeed, synthetic polymers that adapt in response to specific interactions with naturally occurring molecules obviously have potential in controlled release and imaging applications. Similarly, macromolecules with the ability to respond to noninvasive stimuli (e.g., field-responsive polymers) may facilitate development of new therapeutic strategies. In many cases, hybrid materials that respond to multiple stimuli are beneficial. Despite numerous advances, many challenges and opportunities exist within the field of stimuli-responsive polymers. It is likely that other underutilized stimuli will take on greater roles in the next generation of smart materials.

References

1. Theato, P. (2008) Synthesis of well-defined polymeric activated esters. *J. Polym. Sci. Part A: Polym. Chem.*, **46**, 6677–6687.
2. Miyata, T., Uragami, T., and Nakamae, K. (2002) Biomolecule-sensitive hydrogels. *Adv. Drug Delivery Rev.*, **54**, 79–98.
3. Ito, Y., Casolaro, M., Kono, K., and Imanishi, Y. (1989) An insulin-releasing system that is responsive to glucose. *J. Control. Release*, **10**, 195–203.
4. Hassan, C.M., Doyle, F.J., and Peppas, N.A. (1997) Dynamic behavior of glucose-responsive poly(methacrylic acid-g-ethylene glycol) hydrogels. *Macromolecules*, **30**, 6166–6173.
5. Huang, H.Y., Shaw, J., Yip, C., and Wu, X.Y. (2008) Microdomain pH gradient and kinetics inside composite polymeric membranes of pH and glucose sensitivity. *Pharm. Res.*, **25**, 1150–1157.
6. Kang, S.I. and Bae, Y.H. (2003) A sulfonamide based glucose-responsive hydrogel with covalently immobilized glucose oxidase and catalase. *J. Control. Release*, **86**, 115–121.
7. Podual, K., Doyle, F.J. III, and Peppas, N.A. (2000) Dynamic behavior of glucose oxidase-containing microparticles of poly(ethylene glycol)-grafted cationic hydrogels in an environment of changing pH. *Biomaterials*, **21**, 1439–1450.
8. Traitel, T., Cohen, Y., and Kost, J. (2000) Characterization of glucose-sensitive insulin release systems in simulated in vivo conditions. *Biomaterials*, **21**, 1679–1687.
9. Guiseppi-Elie, A., Brahim, S.I., and Narinesingh, D. (2002) A chemically synthesized artificial pancreas: release of insulin from glucose-responsive hydrogels. *Adv. Mater.*, **14**, 743–746.
10. Schwartz, L.M. and Peppas, N.A. (1998) Novel poly(ethylene glycol)-grafted, cationic hydrogels: preparation, characterization, and diffusive properties. *Polymer*, **39**, 6057–6066.
11. Ravaine, V., Ancia, C., and Catargi, B. (2008) Chemically controlled closed-loop insulin delivery. *J. Control. Release*, **132**, 2–11.
12. Kashyap, N., Viswanad, B., Sharma, G., Bhardwaj, V., Ramarao, P., and Kumar, M.N.V.R. (2007) Design and evaluation of biodegradable, biosensitive in situ gelling system for pulsatile delivery of insulin. *Biomaterials*, **28**, 2051–2060.
13. Brownlee, M. and Cerami, A. (1979) A glucose-controlled insulin-delivery system: semisynthetic insulin bound to lectin. *Science*, **206**, 1190–1191.
14. Gil, E.S. and Hudson, S.M. (2004) Stimuli-responsive polymers and their bioconjugates. *Prog. Polym. Sci.*, **29**, 1173–1222.
15. Liu, F., Song, S.C., Mix, D., Baudy, M., and Kim, S.W. (1997) Glucose-induced release of glycosylpoly(ethylene glycol) insulin bound to a soluble conjugate of concanavalin A. *Bioconjug. Chem.*, **8**, 664–672.
16. Cheng, S.Y., Gross, J., and Sambanis, A. (2004) Hybrid pancreatic tissue substitute consisting of recombinant insulin-figreting cells and glucose-responsive material. *Biotechnol. Bioeng.*, **87**, 863–873.
17. Miyata, T., Jikihara, A., Nakamae, K., and Hoffman, A.S. (2004) Preparation of reversibly glucose-responsive hydrogels by covalent immobilization of lectin in polymer networks having pendent glucose. *J. Biomater. Sci., Polym. Ed.*, **15**, 1085–1098.
18. Miyata, T., Jikihara, A., and Hoffman, A.S. (1996) Preparation of poly(2-glucosyloxyethyl methacrylate)-concanavalin A complex hydrogel and its glucose-sensitivity. *Macromol. Chem. Phys.*, **197**, 1135–1146.
19. Nakamae, K., Miyata, T., Jikihara, A., and Hoffman, A.S. (1994) Formation of poly(glucosyloxyethyl methacrylate)-Concanavalin A complex and its glucose-sensitivity. *J. Biomater. Sci., Polym. Ed.*, **6**, 79–90.
20. James, T.D. and Shinkai, S. (2002) Artificial receptors as chemosensors for carbohydrates. *Top. Curr. Chem.*, **218**, 159–200.

21. Lorand, J.P. and Edwards, J.O. (1959) Polyol complexes and structure of the benzeneboronate ion. *J. Org. Chem.*, **24**, 769–774.
22. Kataoka, K., Miyazaki, H., Okano, T., and Sakurai, Y. (1994) Sensitive glucose-induced change of the lower critical solution temperature of poly[N,N-(dimethylacrylamide)-*co*-3-(acrylamido)-phenylboronic acid] in physiological saline. *Macromolecules*, **27**, 1061–1062.
23. Kataoka, K., Miyazaki, H., Bunya, M., Okano, T., and Sakurai, Y. (1998) Totally synthetic polymer gels responding to external glucose concentration: their preparation and application to on-off regulation of insulin release. *J. Am. Chem. Soc.*, **120**, 12694–12695.
24. Springsteen, G. and Wang, B. (2002) A detailed examination of boronic acid-diol complexation. *Tetrahedron*, **58**, 5291–5300.
25. Hisamitsu, I., Kataoka, K., Okano, T., and Sakurai, Y. (1997) Glucose-responsive gel from phenylborate polymer and poly(vinyl alcohol): prompt response at physiological pH through the interaction of borate with amino group in the gel. *Pharm. Res.*, **14**, 289–293.
26. Kitano, S., Koyama, Y., Kataoka, K., Okano, T., and Sakurai, Y. (1992) A novel drug delivery system utilizing a glucose responsive polymer complex between poly(vinyl alcohol) and poly(N-vinyl-2-pyrrolidone) with a phenylboronic acid moiety. *J. Control. Release*, **19**, 162–170.
27. Kikuchi, A., Suzuki, K., Okabayashi, O., Hoshino, H., Kataoka, K., Sakurai, Y. *et al.* (1996) Glucose-sensing electrode coated with polymer complex gel containing phenylboronic acid. *Anal. Chem.*, **68**, 823–828.
28. Matsumoto, A., Kurata, T., Shiino, D., and Kataoka, K. (2004) Swelling and shrinking kinetics of totally synthetic, glucose-responsive polymer gel bearing phenylborate derivative as a glucose-sensing moiety. *Macromolecules*, **37**, 1502–1510.
29. Zhang, S.B., Chu, L.Y., Xu, D., Zhang, J., Ju, X.J., and Xie, R. (2008) Poly(N-isopropylacrylamide)-based comb-type grafted hydrogel with rapid response to blood glucose concentration change at physiological temperature. *Polym. Adv. Technol.*, **19**, 937–943.
30. Zhang, Y., Guan, Y., and Zhou, S. (2006) Synthesis and volume phase transitions of glucose-sensitive microgels. *Biomacromolecules*, **7**, 3196–3201.
31. Ge, H., Ding, Y., Ma, C., and Zhang, G. (2006) Temperature-controlled release of diols from N-isopropylacrylamide-co-acrylamidophenylboronic acid microgels. *J. Phys. Chem. B*, **110**, 20635–20639.
32. Lapeyre, V., Gosse, I., Chevreux, S., and Ravaine, V. (2006) Monodispersed glucose-responsive microgels operating at physiological salinity. *Biomacromolecules*, **7**, 3356–3363.
33. Hoare, T. and Pelton, R. (2008) Charge-switching, amphoteric glucose-responsive microgels with physiological swelling activity. *Biomacromolecules*, **9**, 733–740.
34. Zenkl, G., Mayr, T., and Klimant, I. (2008) Sugar-responsive fluorescent nanospheres. *Macromol. Biosci.*, **8**, 146–152.
35. Matsumoto, A., Ikeda, S., Harada, A., and Kataoka, K. (2003) Glucose-responsive polymer bearing a novel phenylborate derivative as a glucose-sensing moiety operating at physiological pH conditions. *Biomacromolecules*, **4**, 1410–1416.
36. Matsumoto, A., Yoshida, R., and Kataoka, K. (2004) Glucose-responsive polymer gel bearing phenylborate derivative as a glucose-sensing moiety operating at physiological pH. *Biomacromolecules*, **5**, 1038–1045.
37. Pellon, J., Schwind, L.H., Guinard, M.J., and Thomas, W.M. (1961) Polymerization of vinyl monomers containing boron. II. p-vinylbenzeneboronic acid. *J. Polym. Sci.*, **55**, 161–167.
38. Shiomori, K., Ivanov, A.E., Galaev, I.Y., Kawano, Y., and Mattiasson, B. (2004) Thermoresponsive properties of sugar sensitive copolymer of N-isopropylacrylamide and

3-(acrylamido)phenylboronic acid. *Macromol. Chem. Phys.*, **205**, 27–34.

39. Lennarz, W.J. and Snyder, H.R. (1960) Arylboronic acids. III. Preparation and polymerization of p-vinylbenzeneboronic acid. *J. Am. Chem. Soc.*, **82**, 2169–2171.

40. Letsinger, R.L. and Hamilton, S.B. (1959) Organoboron compounds. X. Popcorn polymers and highly cross-linked vinyl polymers containing boron. *J. Am. Chem. Soc.*, **81**, 3009–3012.

41. Matyjaszewski, K. and Davis, T. (2002) *Handbook of Radical Polymerization*, Wiley Interscience, Hoboken, NJ.

42. Braunecker, W.A. and Matyjaszewski, K. (2007) Controlled/living radical polymerization: features, developments, and perspectives. *Prog. Polym. Sci.*, **32**, 93–146.

43. Wang, J.S. and Matyjaszewski, K. (1995) Controlled/''living'' radical polymerization. Atom transfer radical polymerization in the presence of transition-metal complexes. *J. Am. Chem. Soc.*, **117**, 5614–5615.

44. Kato, M., Kamigaito, M., Sawamoto, M., and Higashimura, T. (1995) Polymerization of methyl methacrylate with the carbon tetrachloride/dichlorotris(triphenylphosphine)ruthenium(II)/methylaluminum bis(2,6-di-tert-butylphenoxide) initiation system: possibility of living radical polymerization. *Macromolecules*, **28**, 1721–1723.

45. Vogt, A.P. and Sumerlin, B.S. (2006) An efficient route to macromonomers via ATRP and click chemistry. *Macromolecules*, **39**, 5286–5292.

46. Chiefari, J., Chong, Y.K., Ercole, F., Krstina, J., Jeffery, J., Le, T.P.T. et al. (1998) Living free-radical polymerization by reversible addition-fragmentation chain transfer: the RAFT process. *Macromolecules*, **31**, 5559–5562.

47. Perrier, S. and Pittaya, T. (2005) Macromolecular design via reversible addition-fragmentation chain transfer (RAFT)/xanthates (MADIX) polymerization. *J. Polym. Sci., Part A: Polym. Chem.*, **43**, 5347–5393.

48. De, P., Gondi, S.R., and Sumerlin, B.S. (2008) Folate-conjugated thermoresponsive block copolymers: highly efficient conjugation and solution self-assembly. *Biomacromolecules*, **9**, 1064–1070.

49. Qin, Y., Cheng, G., Achara, O., Parab, K., and Jäkle, F. (2004) A new route to organoboron polymers via highly selective polymer modification reactions. *Macromolecules*, **37**, 7123–7131.

50. Yang, Q., Guanglou, C., Anand, S., and Jäkle, F. (2002) Well-defined boron-containing polymeric lewis acids. *J. Am. Chem. Soc.*, **124**, 12672–12673.

51. Cambre, J.N., Roy, D., Gondi, S.R., and Sumerlin, B.S. (2007) Facile strategy to well-defined water-soluble boronic acid (co)polymers. *J. Am. Chem. Soc.*, **129**, 10348–10349.

52. Roy, D., Cambre, J.N., and Sumerlin, B.S. (2008) Sugar-responsive block copolymers by direct RAFT polymerization of unprotected boronic acid monomers. *Chem. Commun.*, 2477–2479.

53. Roy, D., Cambre, J.N., and Sumerlin, B.S. (2009) Triply-responsive boronic acid block copolymers: solution self-assembly induced by changes in temperature, pH, or sugar concentration. *Chem. Commun.*, **16**, 2106–2108.

54. Schild, H.G. (1992) Poly(N-isopropylacrylamide): experiment, theory, and application. *Prog. Polym. Sci.*, **17**, 163–249.

55. Thornton, P.D., McConnell, G., and Ulijin, R.V. (2005) Enzyme responsive polymer hydrogel beads. *Chem. Commun.*, 5913–5915.

56. Ulijin, R.V. (2006) Enzyme-responsive materials: a new class of smart biomaterials. *J. Mater. Chem.*, **16**, 2217–2225.

57. Yang, Z., Gu, H., Fu, D., Gao, P., Lam, J.K., and Xu, B. (2004) Enzymatic formation of supramolecular hydrogels. *Adv. Mater.*, **16**, 1440–1444.

58. Toledano, S., Williams, R.J., Jayawarna, V., and Ulijin, R.V. (2006) Enzyme-triggered self-assembly of peptide hydrogels via reversed hydrolysis. *J. Am. Chem. Soc.*, **128**, 1070–1071.

59. Griffith, L.G. and Sperinde, J.J. (1997) Synthesis and characterization of enzymatically cross-linked poly(ethylene glycol) hydrogels. *Macromolecules*, **30**, 5255–5264.
60. Sperinde, J.J. and Griffith, L.G. (2000) Control and prediction of gelation kinetics in enzymatically cross-linked poly(ethylene glycol) hydrogels. *Macromolecules*, **33**, 5476–5480.
61. Chen, T., Small, D.A., McDermott, M.K., Bentley, W.E., and Payne, G.F. (2003) Enzymatic methods for in situ cell entrapment and cell release. *Biomacromolecules*, **4**, 1558–1563.
62. Sanborn, T.J., Messersmith, P.B., and Barron, A.E. (2002) In situ crosslinking of a biomimetic peptide-PEG hydrogel via thermally triggered activation of factor XIII. *Biomaterials*, **23**, 2703–2710.
63. Lutolf, M.P., Raeber, G.P., Zisch, A.H., Tirelli, N., and Hubbell, J.A. (2003) Cell-responsive synthetic hydrogels. *Adv. Mater.*, **15**, 888–892.
64. Plunkett, K.N., Berkowski, K.L., and Moore, J.S. (2005) Chymotrypsin responsive hydrogels: application of a disulfide exchange protocol for the preparation of methacrylamide containing peptides. *Biomacromolecules*, **6**, 632–637.
65. Miyata, T. and Uragami, T. (2001) in *Polymeric Biomaterials*, 2nd edn (ed. S. Dumitriu), CRC Press, New York, pp. 959–974.
66. Lu, Z., Kopeckova, P., and Kopecek, J. (2003) Antigen responsive hydrogels basedon polymerizable antibody Fab fragment. *Macromol. Biosci.*, **3**, 296–300.
67. Miyata, T., Asami, N., and Uragami, T. (1999) Preparation of an antigen-sensitive hydrogel using antigen-antibody bindings. *Macromolecules*, **32**, 2082–2084.
68. Miyata, T., Asami, N., and Uragami, T. (1999) A reversibly antigen-responsive hydrogel. *Nature*, **399**, 766–769.
69. Zhang, R., Bowyer, A., Eisenthal, R., and Hubble, J. (2007) A smart membrane based on an antigen-responsive hydrogel. *Biotechnol. Bioeng.*, **97**, 976–984.
70. You, Y.-Z., Zhou, Q.-H., Manickam, D.S., Wan, L., Mao, G.-Z., and Oupicky, D. (2007) Dually responsive multiblock copolymers via reversible addition-fragmentation chain transfer polymerization: synthesis of temperature- and redox-responsive copolymers of poly(N-isopropylacrylamide) and poly(2-(dimethylamino)ethyl methacrylate). *Macromolecules*, **40**, 8617–8624.
71. Madsen, J., Armes, S.P., Bertal, K., Lomas, H., MacNeil, S., and Lewis, A.L. (2008) Biocompatible wound dressings based on chemically degradable triblock copolymer hydrogels. *Biomacromolecules*, **9**, 2265–2275.
72. Cerritelli, S., Velluto, D., and Hubbell, J.A. (2007) PEG-SS-PPS: reduction-sensitive disulfide block copolymer vesicles for intracellular drug delivery. *Biomacromolecules*, **8**, 1966–1972.
73. Zhang, L., Liu, W., Lin, L., Chen, D., and Stenzel, M.H. (2008) Degradable disulfide core-cross-linked micelles as a drug delivery system prepared from vinyl functionalized nucleosides via the RAFT process. *Biomacromolecules*, **9**, 3321–3331.
74. Sun, K.H., Sohn, Y.S., and Jeong, B. (2006) Thermogelling poly(ethylene oxide-b-propylene oxide-b-ethylene oxide) disulfide multiblock copolymer as a thiol-sensitive degradable polymer. *Biomacromolecules*, **7**, 2871–2877.
75. Bulmus, V., Woodward, M., Lin, L., Murthy, N., Stayton, P., and Hoffman, A. (2003) A new pH-responsive and glutathione-reactive, endosomal membrane-disruptive polymeric carrier for intracellular delivery of biomolecular drugs. *J. Control. Release*, **93**, 105–120.
76. Meng, F., Hennink, W.E., and Zhong, Z. (2009) Reduction-sensitive polymers and bioconjugates for biomedical applications. *Biomaterials*, **30**, 2180–2198.
77. Castellani, O.F., Martinez, E.N., and Anon, M.C. (1999) Role of disulfide bonds upon the structural stability of an amaranth globulin. *J. Agric. Food Chem.*, **47**, 3001–3008.

78. Saito, G., Swanson, J.A., and Lee, K.-D. (2003) Drug delivery strategy utilizing conjugation via reversible disulfide linkages: role and site of cellular reducing activities. *Adv. Drug Deliv. Rev.*, **55**, 199–215.
79. Jocelyn, P.C. (1987) Chemical reduction of disulfides. *Methods Enzymol.*, **143**, 246–256.
80. Oh, J.K., Siegwart, D.J., Lee, H.-I., Sherwood, G., Peteanu, L., Hollinger, J.O. et al. (2007) Biodegradable nanogels prepared by atom transfer radical polymerization as potential drug delivery carriers: synthesis, biodegradation, in vitro release, and bioconjugation. *J. Am. Chem. Soc.*, **129**, 5939–5945.
81. Jones, D.P., Carlson, J.L., Samiec, P.S., Sternberg, P. Jr., Mody, V.C. Jr., Reed, R.L. et al. (1998) Glutathione measurement in human plasma evaluation of sample collection, storage and derivatization conditions for analysis of dansyl derivatives by HPLC. *Clin. Chim. Acta*, **275**, 175–184.
82. Koo, A.N., Lee, H.J., Kim, S.E., Chang, J.H., Park, C., Kim, C. et al. (2008) Disulfide-cross-linked PEG-poly(amino acid)s copolymer micelles for glutathione-mediated intracellular drug delivery. *Chem. Commun.*, 6570–6572.
83. Tsarevsky, N.V. and Matyjaszewski, K. (2002) Reversible redox cleavage/coupling of polystyrene with disulfide or thiol groups prepared by atom transfer radical polymerization. *Macromolecules*, **35**, 9009–9014.
84. Wong, L., Boyer, C., Jia, Z., Zareie, H.M., Davis, T.P., and Bulmus, V. (2008) Synthesis of versatile thiol-reactive polymer scaffolds via RAFT polymerization. *Biomacromolecules*, **9**, 1934–1944.
85. Liu, J., Bulmus, V., Barner-Kowollik, C., Stenzel, M.H., and Davis, T.P. (2007) Direct synthesis of pyridyl disulfide-terminated polymers by RAFT polymerization. *Macromol. Rapid Commun.*, **28**, 305–314.
86. Klaikherd, A., Ghosh, S., and Thayumanavan, S. (2007) A facile method for the synthesis of cleavable block copolymers from ATRP-based homopolymers. *Macromolecules*, **40**, 8518–8520.
87. Tsarevsky, N.V. and Matyjaszewski, K. (2005) Combining atom transfer radical polymerization and disulfide/thiol redox chemistry: a route to well-defined (Bio)degradable polymeric materials. *Macromolecules*, **38**, 3087–3092.
88. Oh, J.K., Tang, C., Gao, H., Tsarevsky, N.V., and Matyjaszewski, K. (2006) Inverse miniemulsion ATRP: a new method for synthesis and functionalization of well-defined water-soluble/cross-linked polymeric particles. *J. Am. Chem. Soc.*, **128**, 5578–5584.
89. Zhang, J., Jiang, X., Zhang, Y., Li, Y., and Liu, S. (2007) Facile fabrication of reversible core cross-linked micelles possessing thermosensitive swellability. *Macromolecules*, **40**, 9125–9132.
90. Li, Y., Lokitz, B.S., Armes, S.P., and McCormick, C.L. (2006) Synthesis of reversible shell cross-linked micelles for controlled release of bioactive agents. *Macromolecules*, **39**, 2726–2728.
91. Sun, J., Chen, X., Lu, T., Liu, S., Tian, H., Guo, Z. et al. (2008) Formation of reversible shell cross-linked micelles from the biodegradable amphiphilic diblock copolymer poly(L-cysteine)-block-poly(L-lactide). *Langmuir*, **24**, 10099–10106.
92. Dong, W.-F., Kishimura, A., Anraku, Y., Chuanoi, S., and Kataoka, K. (2009) Monodispersed polymeric nanocapsules: spontaneous evolution and morphology transition from reducible hetero-PEG PICmicelles by controlled degradation. *J. Am. Chem. Soc.*, **131**, 3804–3805.
93. Ghosh, S., Yesilyurt, V., Savariar, E.N., Irvin, K., and Thayumanavan, S. (2009) Redox, ionic strength, and pH sensitive supramolecular polymer assemblies. *J. Polym. Sci., Part A: Polym. Chem.*, **47**, 1052–1060.
94. Vogt, A.P. and Sumerlin, B.S. (2009) Temperature and redox responsive hydrogels from ABA triblock copolymers prepared by RAFT polymerization. *Soft Matter*, **5**, 2347–2351.

95. Gemici, H., Legge, T.M., Whittaker, M., Monteiro, M.J., and Perrier, S. (2007) Original approach to multi-block copolymers via reversible addition-fragmentation chain transfer polymerization. *J. Polym. Sci., Part A: Polym. Chem.*, **45**, 2334–2340.

96. Whittaker, M.R., Goh, Y.-K., Gemici, H., Legge, T.M., Perrier, S., and Monteiro, M.J. (2006) Synthesis of monocyclic and linear polystyrene using the reversible coupling/cleavage of thiol/disulfide groups. *Macromolecules*, **39**, 9028–9034.

97. Liu, J., Liu, H., Jia, Z., Bulmus, V., and Davis, T. (2008) An approach to biodegradable star polymeric architectures using disulfide coupling. *Chem. Commun.*, 6582–6584.

98. Napoli, A., Valentini, M., Tirelli, N., Müller, M., and Hubbell, J.A. (2004) Oxidation-responsive polymeric vesicles. *Nat. Mater.*, **3**, 183–189.

99. Napoli, A., Tirelli, N., Wehrli, E., and Hubbell, J.A. (2002) Lyotropic behavior in water of amphiphilic ABA triblock copolymers based on poly(propylene sulfide) and poly(ethylene glycol). *Langmuir*, **18**, 8324–8329.

100. Napoli, A., Tirelli, N., Kilcher, G., and Hubbell, A. (2001) New synthetic methodologies for amphiphilic multi-block copolymers of ethylene glycol and propylene sulfide. *Macromolecules*, **34**, 8913–8917.

101. Zrinyi, M. (1997) Magnetic-field-sensitive polymer gels. *Trends Polym. Sci.*, **5**, 280–285.

102. Zrinyi, M. (2000) Intelligent polymer gels controlled by magnetic fields. *Colloid. Polym. Sci.*, **278**, 98–103.

103. Filipcsei, G., Feher, J., and Zrinyi, M. (2000) Electric field sensitive neutral polymer gels. *J. Mol. Struct.*, **554**, 109–117.

104. Shiga, T. (1997) Deformation and viscoelastic behavior of polymer gels in electric fields. *Adv. Polym. Sci.*, **134**, 131–163.

105. Kulkarni, R.V. and Biswanath, S. (2007) Electrically responsive smart hydrogels in drug delivery: a review. *J. Appl. Biomater. Biomech.*, **5**, 125–139.

106. Ramanathan, S. and Block, L.H. (2001) The use of chitosan gels as matrixes for electrically-modulated drug delivery. *J. Control. Release*, **70**, 109–123.

107. Gao, Y., Xu, S., Wu, R., Wang, J., and Wei, J. (2008) Preparation and characteristic of electric stimuli responsive hydrogel composed of polyvinyl alcohol/poly (sodium maleate-co-sodium acrylate). *J. Appl. Polym. Sci.*, **107**, 391–395.

108. Shiga, T. and Kurauchi, T. (1990) Deformation of polyelectrolyte gels under the influence of electric field. *J. Appl. Polym. Sci.*, **39**, 2305–2320.

109. Kim, S.J., Kim, H.I., Shin, S.R., and Kim, S.I. (2004) Electrical behavior of chitosan and poly(hydroxyethyl methacrylate) hydrogel in the contact system. *J. Appl. Polym. Sci.*, **92**, 915–919.

110. Kim, S.J., Park, S.J., Lee, S.M., Lee, Y.M., Kim, H.C., and Kim, S.I. (2003) Electroactive characteristics of interpenetrating polymer network hydrogels composed of poly(vinyl alcohol) and poly(N-isopropylacrylamide). *J. Appl. Polym. Sci.*, **89**, 890–894.

111. Kim, S.J., Park, S.J., Shin, M.-S., and Kim, S.I. (2002) Characteristics of electrical responsive chitosan/polyallylamine interpenetrating polymer network hydrogel. *J. Appl. Polym. Sci.*, **86**, 2290–2295.

112. Kim, S.J., Yoon, S.G., Lee, S.M., Lee, S.H., and Kim, S.I. (2004) Electrical sensitivity behavior of a hydrogel composed of poly(methacrylic acid)/poly(vinyl alcohol). *J. Appl. Polym. Sci.*, **91**, 3613–3617.

113. Kim, S.J., Yoon, S.G., Lee, Y.H., and Kim, S.I. (2004) Bending behavior of hydrogels composed of poly(methacrylic acid) and alginate by electrical stimulus. *Polym. Int.*, **53**, 1456–1460.

114. Tanaka, T., Nishio, I., Sun, S.T., and Ueno-Nishio, S. (1982) Collapse of gels in an electric field. *Science*, **218**, 467–469.

115. Jensen, M., Birch Hansen, P., Murdan, S., Frokjaer, S., and Florence, A.T. (2002) Loading into

and electro-stimulated release of peptides and proteins from chondroitin 4-sulphate hydrogels. *Eur. J. Pharm. Sci.*, **15**, 139–148.
116. Sutani, K., Kaetsu, I., and Uchida, K. (2001) The synthesis and the electric-responsiveness of hydrogels entrapping natural polyelectrolyte. *Radiat. Phys. Chem.*, **61**, 49–54.
117. Kim, S.J., Shin, S.R., Lee, J.H., Lee, S.H., and Kim, S.I. (2003) Electrical response characterization of chitosan/polyacrylonitrile hydrogel in NaCl solutions. *J. Appl. Polym. Sci.*, **90**, 91–96.
118. Lin, S.-B., Yuan, C.-H., Ke, A.-R., and Quan, Z.-L. (2008) Electrical response characterization of PVA-P(AA/AMPS) IPN hydrogels in aqueous Na_2SO_4 solution. *Sens. Actuators, B.*, **B134**, 281–286.
119. Kim, H.I., Gu, B.K., Shin, M.K., Park, S.-J., Yoon, S.-G., Kim, I.-Y. et al. (2005) Electrical response characterization of interpenetrating polymer network hydrogels as an actuator. *Proc. SPIE Int. Soc. Opt. Eng.*, **5759**, 447–453.
120. Kim, S.J., Shin, S.R., Lee, S.M., Kim, I.Y., and Kim, S.I. (2004) Electromechanical properties of hydrogels based on chitosan and poly(hydroxyethyl methacrylate) in NaCl solution. *Smart Mater. Struct.*, **13**, 1036–1039.
121. El-Hag Ali, A., Abd El-Rehim, H.A., Hegazy, E.-S.A., and Ghobashy, M.M. (2006) Synthesis and electrical response of acrylic acid/vinyl sulfonic acid hydrogels prepared by g-irradiation. *Radiat. Phys. Chem.*, **75**, 1041–1046.
122. Filipcsei, G., Csetneki, I., Szilagyi, A., and Zrinyi, M. (2007) Magnetic field-responsive smart polymer composites. *Adv. Polym. Sci.*, **206**, 137–189.
123. Barsi, L., Buki, A., Szabo, D., and Zrinyi, M. (1996) Gels with magnetic properties. *Prog. Colloid. Polym. Sci.*, **102**, 57–63.
124. Pyun, J. (2007) Nanocomposite materials from functional polymers and magnetic colloids. *Polym. Rev.*, **47**, 231–263.
125. Szabo, D., Szeghy, G., and Zrinyi, M. (1998) Shape transition of magnetic field sensitive polymer gels. *Macromolecules*, **31**, 6541–6548.
126. Zrinyi, M., Barsi, L., Szabo, D., and Kilian, H.G. (1997) Direct observation of abrupt shape transition in ferrogels induced by nonuniform magnetic field. *J. Chem. Phys.*, **106**, 5685–5692.
127. Babincova, M., Leszczynska, D., Sourivong, P., Cicmanec, P., and Babinec, P. (2001) Superparamagnetic gel as a novel material for electromagnetically induced hyperthermia. *J. Magn. Magn. Mater.*, **225**, 109–112.
128. Starodoubtsev, S.G., Saenko, E.V., Khokhlov, A.R., Volkov, V.V., Dembo, K.A., Klechkovskaya, V.V. et al. (2003) Poly(acrylamide) gels with embedded magnetite nanoparticles. *Microelectron. Eng.*, **69**, 324–329.
129. Sewell, M.K., Fugit, K.D., Ankareddi, I., Zhang, C., Hampel, M.L., Kim, D.-H. et al. (2008) Magnetothermally-triggered drug delivery using hydrogels with imbedded cobalt ferrite, iron platinum or manganese ferrite nanoparticles. *PMSE Prepr.*, **98**, 694–695.
130. Zhang, J., Xu, S., and Kumacheva, E. (2004) Polymer microgels: reactors for semiconductor, metal, and magnetic nanoparticles. *J. Am. Chem. Soc.*, **126**, 7908–7914.
131. Brugger, B. and Richtering, W. (2007) Magnetic, thermosensitive microgels as stimuli-responsive emulsifiers allowing for remote control of separability and stability of oil in water-emulsions. *Adv. Mater.*, **19**, 2973–2978.
132. Xulu, P.M., Filipcsei, G., and Zrinyi, M. (2000) Preparation and responsive properties of magnetically soft poly(N-isopropylacrylamide) gels. *Macromolecules*, **33**, 1716–1719.
133. Bowles, S.E., Wu, W., Kowalewski, T., Schalnat, M.C., Davis, R.J., Pemberton, J.E. et al. (2007) Magnetic assembly and pyrolysis of functional ferromagnetic colloids into one-dimensional carbon nanostructures. *J. Am. Chem. Soc.*, **129**, 8694–8695.
134. Korth, B.D., Keng, P., Shim, I., Bowles, S.E., Tang, C., Kowalewski, T. et al. (2006) Polymer-coated ferromagnetic

135. Czaun, M., Hevesi, L., Takafuji, M., and Ihara, H. (2008) A novel approach to magneto-responsive polymeric gels assisted by iron nanoparticles as nano cross-linkers. *Chem. Commun.*, 2124–2126.
136. Kost, J., and Langer, R. (1992) Responsive polymer systems for controlled delivery of therapeutics. *Trends Biotechnol.*, **10**, 127–131.
137. Norris, P., Noble, M., Francolini, I., Vinogradov, A.M., Stewart, P.S., Ratner, B.D. et al. (2005) Ultrasonically controlled release of ciprofloxacin from self-assembled coatings on poly(2-hydroxyethyl methacrylate) hydrogels for Pseudomonas aeruginosa biofilm prevention. *Antimicrob. Agents Chemother.*, **49**, 4272–4279.
138. Kost, J., Leong, K., and Langer, R. (1989) Ultrasound-enhanced polymer degradation and release of incorporated substances. *Proc. Natl. Acad. Sci. U.S.A.*, **86**, 7663–7666.
139. Lavon, I. and Kost, J. (1998) Mass transport enhancement by ultrasound in non-degradable polymeric controlled release systems. *J. Control. Release*, **54**, 1–7.
140. Liu, L.S., Kost, J., D'Emanuele, A., and Langer, R. (1992) Experimental approach to elucidate the mechanism of ultrasound-enhanced polymer erosion and release of incorporated substances. *Macromolecules*, **25**, 123–128.
141. Kost, J., Leong, K., and Langer, R. (1988) Ultrasonically controlled polymeric drug delivery. *Makromol. Chem., Macromol. Symp.*, **19**, 275–285.
142. Kost, J., Liu, L.S., Gabelnick, H., and Langer, R. (1994) Ultrasound as a potential trigger to terminate the activity of contraceptive delivery implants. *J. Control. Release*, **30**, 77–81.
143. Miyazaki, S., Yokouchi, C., and Takada, M. (1988) External control of drug release: controlled release of insulin from a hydrophilic polymer implant by ultrasound irradiation in diabetic rats. *J. Pharm. Pharmacol*, **40**, 716–717.
144. Kooiman, K., Boehmer, M.R., Emmer, M., Vos, H.J., Chlon, C., Shi, W.T. et al. (2009) Oil-filled polymer microcapsules for ultrasound-mediated delivery of lipophilic drugs. *J. Control. Release*, **133**, 109–118.
145. Husseini, G.A., Myrup, G.D., Pitt, W.G., Christensen, D.A., and Rapoport, N.Y. (2000) Factors affecting acoustically triggered release of drugs from polymeric micelles. *J. Control. Release*, **69**, 43–52.
146. Marin, A., Muniruzzaman, M., and Rapoport, N. (2001) Mechanism of the ultrasonic activation of micellar drug delivery. *J. Control. Release*, **75**, 69–81.
147. Marin, A., Muniruzzaman, M., and Rapoport, N. (2001) Acoustic activation of drug delivery from polymeric micelles: effect of pulsed ultrasound. *J. Control. Release*, **71**, 239–249.
148. Rapoport, N.Y., Christensen, D.A., Fain, H.D., Barrows, L., and Gao, Z. (2004) Ultrasound-triggered drug targeting of tumors in vitro and in vivo. *Ultrasonics*, **42**, 943–950.
149. Husseini, G.A., Rapoport, N.Y., Christensen, D.A., Pruitt, J.D., and Pitt, W.G. (2002) Kinetics of ultrasonic release of doxorubicin from pluronic P105 micelles. *Colloids Surf., B*, **24**, 253–264.
150. Munshi, N., Rapoport, N., and Pitt, W.G. (1997) Ultrasonic activated drug delivery from Pluronic P-105 micelles. *Cancer Lett.*, **118**, 13–19.
151. Lentacker, I., De Geest, B.G., Vandenbroucke, R.E., Peeters, L., Demeester, J., De Smedt, S.C. et al. (2006) Ultrasound-responsive polymer-coated microbubbles that bind and protect DNA. *Langmuir*, **22**, 7273–7278.
152. Marmottant, P. and Hilgenfeldt, S. (2003) Controlled vesicle deformation and lysis by single oscillating bubbles. *Nature*, **423**, 153.
153. Kumar, G.S. and Neckers, D.C. (1989) Photochemistry of azobenzene-containing polymers. *Chem. Rev.*, **89**, 1915–1925.
154. Dai, S., Ravi, P., and Tam, K.C. (2009) Thermo-and photo-responsive

polymeric systems. *Soft Matter*, **5**, 2513–2533.
155. Natansohn, A. and Rochon, P. (2002) Photoinduced motions in azo-containing polymers. *Chem. Rev.*, **102**, 4139–4175.
156. Irie, M. and Ikeda, T. (1997) in *Functional Monomers and Polymers*, 2nd edn (eds K. Takemoto, R. Ottenbrite, and M. Kamachi), Marcel Dekker, New York, pp. 65–116.
157. He, J., Tong, X., and Zhao, Y. (2009) Photoresponsive nanogels based on photocontrollable cross-links. *Macromolecules*, **42**, 4845–4852.
158. Shimoboji, T., Ding, Z.L., Stayton, P.S., and Hoffman, A.S. (2002) Photoswitching of ligand association with a photoresponsive polymer-protein conjugate. *Bioconjugate Chem.*, **13**, 915–919.
159. Shimoboji, T., Larenas, E., Fowler, T., Kulkarni, S., Hoffman, A.S., and Stayton, P.S. (2002) Photoresponsive polymer-enzyme switches. *Proc. Natl. Acad. Sci. U.S.A.*, **99**, 16592–16596.
160. Ruhmann, R. (1997) Polymers for optical storage – photoresponsive polymers by structure variation in side-group polymethacrylates. *Polym. Int.*, **43**, 103–108.
161. Moniruzzaman, M., Sabey, C.J., and Fernando, G.F. (2007) Photoresponsive polymers: an investigation of their photoinduced temperature changes during photoviscosity measurements. *Polymer*, **48**, 255–263.
162. Nagasaki, T. (2008) Photoresponsive polymeric materials for drug delivery systems: double targeting with photoresponsive polymers. *Drug Deliv. Syst.*, **23**, 637–643.
163. Leclerc, E., Furukawa, K.S., Miyata, F., Sakai, Y., Ushida, T., and Fujii, T. (2004) Fabrication of microstructures in photosensitive biodegradable polymers for tissue engineering applications. *Biomaterials*, **25**, 4683–4690.
164. Zhao, Y. and He, J. (2009) Azobenzene-containing block copolymers: the interplay of light and morphology enables new functions. *Soft Matter*, **5**, 2686–2693.
165. Jochum, F.D. and Theato, P. (2009) Temperature and light sensitive copolymers containing azobenzene moieties prepared via a polymer analogous reaction. *Polymer*, **50**, 3079–3085.
166. Matejka, L., Ilavsky, M., Dusek, K., and Wichterle, O. (1981) Photomechanical effects in crosslinked photochromic polymers. *Polymer*, **22**, 1511–1515.
167. Viswanathan, N.K., Kim, D.Y., Bian, S., Williams, J., Liu, W., Li, L. et al. (1999) Surface relief structures on azo polymer films. *J. Mater. Chem.*, **9**, 1941–1955.
168. Deng, Y., Li, N., He, Y., and Wang, X. (2007) Hybrid colloids composed of two amphiphilic Azo polymers: fabrication, characterization, and photoresponsive properties. *Macromolecules*, **40**, 6669–6678.
169. Li, Y., He, Y., Tong, X., and Wang, X. (2005) Photoinduced deformation of amphiphilic azo polymer colloidal spheres. *J. Am. Chem. Soc.*, **127**, 2402–2403.
170. Strzegowski, L.A., Martinez, M.B., Gowda, D.C., Urry, D.W., and Tirrell, D.A. (1994) Photomodulation of the inverse temperature transition of a modified elastin poly(pentapeptide). *J. Am. Chem. Soc.*, **116**, 813–814.
171. Alonso, M., Reboto, V., Guiscardo, L., San Martin, A., and Rodriguez-Cabello, J.C. (2000) Spiropyran derivative of an elastin-like bioelastic polymer: photoresponsive molecular machine to convert sunlight into mechanical work. *Macromolecules*, **33**, 9480–9482.
172. Alonso, M., Reboto, V., Guiscardo, L., Mate, V., and Rodriguez-Cabello, J.C. (2001) Novel photoresponsive p-phenylazobenzene derivative of an elastin-like polymer with enhanced control of azobenzene content and without pH sensitiveness. *Macromolecules*, **34**, 8072–8077.
173. Wang, G., Tong, X., and Zhao, Y. (2004) Preparation of azobenzene-containing amphiphilic diblock copolymers for light-responsive micellar aggregates. *Macromolecules*, **37**, 8911–8917.
174. Tong, X., Wang, G., Soldera, A., and Zhao, Y. (2005) How can azobenzene

block copolymer vesicles be dissociated and reformed by light? *J. Phys. Chem. B*, **109**, 20281–20287.
175. Liu, X. and Jiang, M. (2006) Optical switching of self-assembly: micellization and micelle-hollow-sphere transition of hydrogen-bonded polymers. *Angew. Chem. Int. Ed.*, **45**, 3846–3850.
176. Sugiyama, K. and Sono, K. (2001) Characterization of photo- and thermoresponsible amphiphilic copolymers having azobenzene moieties as side groups. *J. Appl. Polym. Sci.*, **81**, 3056–3063.
177. Khoukh, S., Oda, R., Labrot, T., Perrin, P., and Tribet, C. (2007) Light-responsive hydrophobic association of azobenzene-modified poly(acrylic acid) with neutral surfactants. *Langmuir*, **23**, 94–104.
178. Ravi, P., Sin, S.L., Gan, L.H., Gan, Y.Y., Tam, K.C., Xia, X.L. et al. (2005) New water soluble azobenzene-containing diblock copolymers: synthesis and aggregation behavior. *Polymer*, **46**, 137–146.
179. Lee, H.-I., Pietrasik, J., and Matyjaszewski, K. (2006) Phototunable temperature-responsive molecular brushes prepared by ATRP. *Macromolecules*, **39**, 3914–3920.
180. Desponds, A. and Freitag, R. (2003) Synthesis and characterization of photoresponsive N-isopropylacrylamide cotelomers. *Langmuir*, **19**, 6261–6270.
181. Junge, D.M. and McGrath, D.V. (1997) Photoresponsive dendrimers. *Chem. Commun.*, 857–858.
182. Liao, L.-X., Stellacci, F., and McGrath, D.V. (2004) Photoswitchable flexible and shape-persistent dendrimers: comparison of the interplay between a photochromic azobenzene core and dendrimer structure. *J. Am. Chem. Soc.*, **126**, 2181–2185.
183. Delaire, J.A. and Nakatani, K. (2000) Linear and nonlinear optical properties of photochromic molecules and materials. *Chem. Rev.*, **100**, 1817–1845.
184. Yager, K.G. and Barrett, C.J. (2001) All-optical patterning of azo polymer films. *Curr. Opin. Solid State Mater. Sci.*, **5**, 487–494.
185. Yager, K.G. and Barrett, C.J. (2006) Novel photo-switching using azobenzene functional materials. *J. Photochem. Photobiol. A: Chem.*, **182**, 250–261.
186. Vollmer, M.S., Clark, T.D., Steinem, C., and Ghadiri, M.R. (1999) Photoswitchable hydrogen-bonding in self-organized cylindrical peptide systems. *Angew. Chem. Int. Ed.*, **38**, 1598–1601.
187. Steinem, C., Janshoff, A., Vollmer, M.S., and Ghadiri, M.R. (1999) Reversible photoisomerization of self-organized cylindrical peptide assemblies at air-water and solid interfaces. *Langmuir*, **15**, 3956–3964.
188. Takeshita, M., Hayashi, M., Kadota, S., Mohammed, K.H., and Yamato, T. (2005) Photoreversible supramolecular polymer formation. *Chem. Commun.*, 761–763.
189. Hugel, T., Holland Nolan, B.A., Moroder, L., Seitz, M., and Gaub Hermann, E. (2002) Single-molecule optomechanical cycle. *Science*, **296**, 1103–1106.
190. Tanchak, O.M. and Barrett, C.J. (2005) Light-induced reversible volume changes in thin films of azo polymers: the photomechanical effect. *Macromolecules*, **38**, 10566–10570.
191. Yu, Y., Nakano, M., and Ikeda, T. (2003) Photomechanics: directed bending of a polymer film by light. *Nature*, **425**, 145.
192. Ikeda, T., Nakano, M., Yu, Y., Tsutsumi, O., and Kanazawa, A. (2003) Anisotropic bending and unbending behavior of azobenzene liquid-crystalline gels by light exposure. *Adv. Mater.*, **15**, 201–205.
193. Yamada, M., Kondo, M., Miyasato, R., Naka, Y., Mamiya, J.-I., Kinoshita, M. et al. (2009) Photomobile polymer materials-various three-dimensional movements. *J. Mater. Chem.*, **19**, 60–62.
194. Such, G.K., Evans, R.A., and Davis, T.P. (2004) Control of photochromism through local environment effects using living radical polymerization (ATRP). *Macromolecules*, **37**, 9664–9666.

195. Ivanov, A.E., Eremeev, N.L., Wahlund, P.O., Galaev, I.Y., and Mattiasson, B. (2002) Photosensitive copolymer of N-isopropylacrylamide and methacryloyl derivative of spirobenzopyran. *Polymer*, **43**, 3819–3823.

196. Konak, C., Rathi, R.C., Kopeckova, P., and Kopecek, J. (1997) Photoregulated association of water-soluble copolymers with spirobenzopyran-containing side chains. *Macromolecules*, **30**, 5553–5556.

197. Lee, H.-I., Wu, W., Oh, J.K., Mueller, L., Sherwood, G., Peteanu, L. *et al.* (2007) Light-induced reversible formation of polymeric micelles. *Angew. Chem. Int. Ed.*, **46**, 2453–2457.

198. Shiraishi, Y., Miyamoto, R., and Hirai, T. (2009) Spiropyran-conjugated thermoresponsive copolymer as a colorimetric thermometer with linear and reversible color change. *Org. Lett.*, **11**, 1571–1574.

199. Laschewsky, A. and Rekai, E.D. (2000) Photochemical modification of the lower critical solution temperature of cinnamoylated poly(N-2-hydroxypropylmethacrylamide) in water. *Macromol. Rapid Commun.*, **21**, 937–940.

200. Jiang, J., Tong, X., and Zhao, Y. (2005) A new design for light-breakable polymer micelles. *J. Am. Chem. Soc.*, **127**, 8290–8291.

201. Jiang, J., Tong, X., Morris, D., and Zhao, Y. (2006) Toward photocontrolled release using light-dissociable block copolymer micelles. *Macromolecules*, **39**, 4633–4640.

202. Goodwin, A.P., Mynar, J.L., Ma, Y., Fleming, G.R., and Frechet, J.M.J. (2005) Synthetic micelle sensitive to IR light via a two-photon process. *J. Am. Chem. Soc.*, **127**, 9952–9953.

3
Self-Oscillating Gels as Stimuli-Responsive Materials

Anna C. Balazs, Pratyush Dayal, Olga Kuksenok, and Victor V. Yashin

3.1
Introduction

One of the challenges in designing synthetic biomimetic systems is that of creating macroscopic objects that can sense adverse environmental conditions and then respond to these conditions in specified ways. Recently, we have developed theoretical and computational models for chemoresponsive polymer gels and, through these models, we have designed such "smart," adaptive systems. Our efforts are focused on a particular class of responsive gels, namely, those undergoing the Belousov–Zhabotinsky (BZ) reaction. Such BZ gels exhibit self-sustained pulsations, which can be harnessed to perform mechanical work. To impart the desired biomimetic functionality to these BZ gels, we must establish effective routes for controlling their spatiotemporal behavior. Herein, we focus on two examples of BZ gels where external stimuli are used to control the behavior of the system. In one of the examples, we describe how light can be harnessed to regulate the self-sustained motion of millimeter-sized BZ gel "worms." By tailoring the arrangement of illuminated and nonilluminated regions, we direct the movement of these worms along complex paths, guiding them to bend, reorient, and turn. Notably, the path and the direction of the gel's motion can be dynamically and remotely reconfigured. Hence, our findings can be utilized to design intelligent, autonomously moving "soft robots" that can be reprogrammed to move "on demand" to a specific target location and to remain at this location for a chosen period of time. In another example, we consider confining walls to be the external agent that stimulates changes in the properties of the system. These studies are important because in many potential applications the BZ gels would be localized on a surface or confined between walls. We found that the removal of just a portion of a confining surface can drive a previously stationary domain into an oscillatory state. Hence, dynamic changes in the nature of the confinement can provide an effective means of tuning the gel's mechanical action and, thus, provides a route for tailoring the functionality of the material.

Polymer gels constitute ideal responsive materials since a variety of external stimuli can be used to induce large-scale, periodic changes in the volume and shape

Handbook of Stimuli-Responsive Materials. Edited by Marek W. Urban
Copyright © 2011 WILEY-VCH Verlag GmbH & Co. KGaA, Weinheim
ISBN: 978-3-527-32700-3

of the material [1–5]. Because of these rhythmic structural changes, oscillating gels can perform sustained mechanical work and, thus, have been used to create microactuators [6] and pulsatile drug delivery devices [3, 7]. There is, however, one class of oscillating gels where the driving force for these structural changes does not come from an external stimulus, but rather from an internal chemical reaction; these materials are referred to as *BZ gels* because the polymer network undergoes the oscillating chemical reaction known as the *Belousov–Zhabotinsky* reaction [8, 9]. Discovered in the 1950s, the original BZ reaction occurred in a fluid and the temporal oscillations and spatial concentration patterns produced via this reaction provided researchers with an exceptional "laboratory" for systematically exploring systems far from equilibrium [10, 11]. In an important step toward harnessing this oscillatory behavior for technological applications, Yoshida *et al.* [12–17] incorporated the BZ chemistry into responsive polymer gels and, thus for the first time, used this reaction to power a mechanical action.

A vital aspect of these BZ gels is that the ruthenium (Ru) catalysts are covalently bonded to the polymer chains; the BZ reaction causes Ru to undergo a periodic oxidation and reduction that dynamically alters the hydrophilicity of the polymers and, thereby, drives the rhythmic swelling and deswelling of the gel [12–16]. Because of this mechanism, millimeter-sized gels can oscillate autonomously for hours [15] and can be "refueled" by simply adding more reactants to the surrounding solution. From a fundamental viewpoint, the BZ gels constitute an intriguing example of a far from equilibrium system where nonlinear chemical kinetics is coupled to the elastodynamics of a polymer network.

Such autonomously functioning, active materials could prove highly useful in a range of technological applications. A significant challenge, however, is that of establishing routes for controlling the spatiotemporal behavior of these BZ gels so that the chemomechanical waves propagate in controllable, user-specified patterns. It is in this context that external stimuli could play a valuable role. Herein, we describe our recent work on modeling BZ gels [18–22] and specifically focus on two examples where external stimuli are used to control the behavior of the system.

In one of the examples, we describe how light can be harnessed to regulate the self-sustained motion of millimeter-sized BZ gel "worms" [21]. By tailoring the arrangement of illuminated and nonilluminated regions, we direct the movement of these worms along complex paths, guiding them to bend, reorient, and turn [22]. Notably, the path and the direction of the gel's motion can be dynamically and remotely reconfigured (as opposed to being fixed, for example, by a pattern on an underlying surface).

In our second example, we consider confining walls to be the external agent that stimulates changes in the properties of the system. In many potential applications, the BZ gels would be localized on a surface or confined between walls. Thus, it becomes necessary to establish how the confining, hard walls could be utilized to manipulate the pattern formation and yield well-defined behavior. By undertaking these studies, we found that the removal of just a portion of a wall can drive a previously stationary domain into an oscillatory state and, in this manner, we can create materials with controlled arrangements of pulsatile and nonpulsatile behavior [20]. Thus, our findings provide new guidelines for tailoring the functionality of self-oscillating gels. On a fundamental level, the findings reveal how the behavior of a nonequilibrium system with coupled chemical and mechanical degrees of freedom is affected by confinement.

3.2 Methodology

3.2.1 Continuum Equations

The dynamics of this complex BZ gel system can be described [19, 23] using a modified version of the original Oregonator model for the BZ reaction in solution [24] and equations for the elastodynamics of the polymer network. The Oregonator model describes the kinetics of the BZ reaction in terms of the dimensionless concentrations of the oxidized catalyst, v, and the key reaction intermediate (the activator), u. In the modified version, the equations explicitly depend on the volume fraction of polymer, ϕ, which acts as a neutral diluent. The resulting equations for the gel dynamics are as follows [18, 19]:

$$\frac{d_p \phi}{dt} = -\phi \nabla \cdot \mathbf{v}^{(p)} \tag{3.1}$$

$$\frac{d_p v}{dt} = -v \nabla \cdot \mathbf{v}^{(p)} + \varepsilon G(u, v, \phi) \tag{3.2}$$

$$\frac{d_p u}{dt} = -u \nabla \cdot \mathbf{v}^{(p)} + \nabla \cdot \left[\mathbf{v}^{(p)} \frac{u}{1-\phi} \right] + \nabla \cdot \left[(1-\phi) \nabla \frac{u}{1-\phi} \right] + F(u, v, \phi) \tag{3.3}$$

Here, $d_p/dt \equiv \partial/\partial t + \mathbf{v}^{(p)} \cdot \nabla$ denotes the material time derivative, where $\mathbf{v}^{(p)}$ is the velocity of the polymer network. The functions $G(u, v, \phi)$ and $F(u, v, \phi)$ describe the BZ reaction occurring within the gel [25]:

$$G(u, v, \phi) = (1-\phi)^2 u - (1-\phi)v \tag{3.4}$$

$$F(u, v, \phi) = (1-\phi)^2 u - u^2 - (1-\phi)fv\frac{u - q(1-\phi)^2}{u + q(1-\phi)^2} \tag{3.5}$$

The parameters f, q, and ε in the above equations have the same meaning as in the original Oregonator model. The stoichiometric parameter f effectively controls the concentration of oxidized catalyst, v, in the steady state and affects the amplitude of the oscillations in the oscillatory state [26].

Equation 3.1 is the continuity equation for the polymer; Equation 3.2 describes the evolution of v due to the BZ reaction and the transport of the catalyst with the movement of the polymer network. It may be recalled that the catalyst is tethered to the chains and, thus, does not diffuse through the solution. Equation 3.3 characterizes the changes in u due to the BZ reaction, the transport of this activator with the solvent, and the diffusion of the activator within the solvent. For simplicity, we assumed that it is solely the polymer–solvent interdiffusion that contributes to the gel dynamics; hence, in Equation 3.3, we took into account that $\phi \mathbf{v}^{(p)} + (1 - \phi)\mathbf{v}^{(s)} = 0$, where $\mathbf{v}^{(s)}$ is the solvent velocity [19].

The dynamics of the polymer network is assumed to be purely relaxational, so that the forces acting on the deformed gel are balanced by the frictional drag due to the motion of the solvent [27]. Thus, we can write [19]

$$\mathbf{v}^{(p)} = \Lambda_0 (1 - \phi)(\phi/\phi_0)^{-3/2} \nabla \cdot \hat{\boldsymbol{\sigma}} \tag{3.6}$$

Here, Λ_0 is the dimensionless kinetic coefficient, and $\hat{\boldsymbol{\sigma}}$ is the dimensionless stress tensor measured in units of $v_0^{-1} T$, where v_0 is the volume of a monomeric unit, T is the temperature measured in energy units, and ϕ_0 is the volume fraction of polymer in the undeformed state. The factor $(\phi/\phi_0)^{-3/2}$ in Equation 3.6 takes into account the ϕ-dependence of the polymer–solvent friction in swollen polymer gels [27].

The stress tensor can be derived [19] from the free energy density of the deformed gel, U, which consists of the elastic energy density associated with the deformations, U_{el}, and the polymer–solvent interaction energy density, U_{FH}. $U = U_{el}(I_1, I_3) + U_{FH}(I_3)$ where $I_1 = \text{tr}\hat{\mathbf{B}}$ and $I_3 = \det \hat{\mathbf{B}}$ are the invariants of the left Cauchy–Green (Finger) strain tensor $\hat{\mathbf{B}}$ [28]. The invariant I_3 characterizes the volumetric changes in the deformed gel [28]. The local volume fractions of polymer in the deformed and undeformed states are related in the following way: $\phi = \phi_0 I_3^{-1/2}$.

The elastic energy contribution U_{el} describes the rubber elasticity of the crosslinked polymer chains and is proportional to the cross-link density, c_0, the number density of elastic strands in the undeformed polymer network. We use the Flory model [29] to specify U_{el}:

$$U_{el} = \frac{c_0 v_0}{2}(I_1 - 3 - \ln I_3^{1/2}) \tag{3.7}$$

The energy of the polymer–solvent interaction is taken to be of the following Flory–Huggins form [19]:

$$U_{FH} = \sqrt{I_3}[(1 - \phi)\ln(1 - \phi) + \chi_{FH}(\phi)\phi(1 - \phi) - \chi^* v(1 - \phi)] \tag{3.8}$$

where $\chi^* > 0$ describes the hydrating effect of the metal-ion catalyst and captures the coupling between the gel dynamics and the BZ reaction, and $\chi_{FH}(\phi)$ is the polymer–solvent interaction parameter. In Equation 3.8, the coefficient $I_3^{1/2}$ appears in front of the conventional Flory–Huggins energy because the energy density is defined per unit volume of gel in the undeformed state. Finally, using Equations 3.7 and 3.8, one can derive the following constitutive equation for the chemoresponsive polymer gels [19]:

$$\hat{\boldsymbol{\sigma}} = -P(\phi, v)\hat{\mathbf{I}} + c_0 v_0 \frac{\phi}{\phi_0}\hat{\mathbf{B}} \tag{3.9}$$

where $\hat{\mathbf{I}}$ is the unit tensor, and the isotropic pressure $P(\phi, v)$ is defined as

$$P(\phi, v) = -[\phi + \ln(1 - \phi) + \chi(\phi)\phi^2] + c_0 v_0 \phi (2\phi_0)^{-1} + \chi^* v \phi \tag{3.10}$$

where the interaction parameter $\chi(\phi) = \chi_0 + \chi_1 \phi$ is derived from the Flory–Huggins parameter, $\chi_{FH}(\phi)$ [30]. Specifically, the relationship between $\chi(\phi)$ and $\chi_{FH}(\phi)$ is given as $\chi(\phi) = \chi_{FH}(\phi) - (1 - \phi) \partial \chi_{FH}(\phi)/\partial \phi$. We note that the value of $\chi(\phi)$ coincides with the Flory–Huggins interaction parameter only when there is no dependence on the polymer volume fraction, that is, when $\chi_{FH} = \chi_0$.

The gel attains a steady state if the elastic stresses are balanced by the osmotic pressure and, simultaneously, the reaction exhibits a stationary regime. For small, uniformly swollen gel samples, such stationary solutions $(\phi_{st}, u_{st}, v_{st})$ are found by solving the following equations:

$$c_0 v_0 \left[\left(\frac{\phi_{st}}{\phi_0} \right)^{1/3} - \frac{\phi_{st}}{2\phi_0} \right] = \pi_{osm}(\phi_{st}, v_{st}); \quad F(u_{st}, v_{st}, \phi_{st}) = 0;$$

$$G(u_{st}, v_{st}, \phi_{st}) = 0 \tag{3.11}$$

The left-hand side of the first equation in Equation 3.11 represents an elastic stress [19].

3.2.2
Formulation of the Gel Lattice Spring Model (gLSM)

To study the dynamic behavior of the BZ gels, we numerically integrate Equations 3.1–3.3 in three dimensions using our recently developed "gel lattice spring model" (gLSM) [18, 19]. This method combines a finite element approach for the spatial discretization of the elastodynamic equations and a finite difference approximation for the reaction and diffusion terms. We briefly describe the 3D formulation of the gLSM below; for more details of the model, we refer the reader to Ref. [18].

We represent a 3D reactive, deformable gel by a set of general lineal hexahedral elements [31, 32] (Figure 3.1a). Initially, the sample is undeformed and consists of $(L_x - 1) \times (L_y - 1) \times (L_z - 1)$ identical cubic elements (see inset in Figure 3.1a); here L_i is the number of nodes in the i-direction, where $i = x, y, z$. The linear size of the elements in the undistorted state is given by Δ. (In the simulations presented below, we set $\Delta = 1$.) For a homogeneous BZ gel in the undeformed state, the polymer and crosslinks are uniformly distributed over the gel sample, that is, each undeformed element is characterized by the same volume fraction ϕ_0 and crosslink density c_0. Upon deformation, the elements move together with the polymer network so that the amount of polymer and number of crosslinks within each hexahedral element remain equal to their initial values.

Each element is labeled by the vector $\mathbf{m} = (i, j, k)$, and the element nodes are numbered by the index $n = 1 - 8$ and characterized by the coordinates $\mathbf{r}_n(\mathbf{m})$ [18]. Within each element $\mathbf{m}$, the concentrations of the dissolved reagent, $u(\mathbf{m})$, the oxidized metal-ion catalyst, $v(\mathbf{m})$, and the volume fraction of polymer, $\phi(\mathbf{m})$, are taken to be spatially uniform. The value of $\phi(\mathbf{m})$ is related to the volume of the element, $V(\mathbf{m})$, as $\phi(\mathbf{m}) = \Delta^3 \phi_0 / V(\mathbf{m})$. Within each element, we define a local

64 | *3 Self-Oscillating Gels as Stimuli-Responsive Materials*

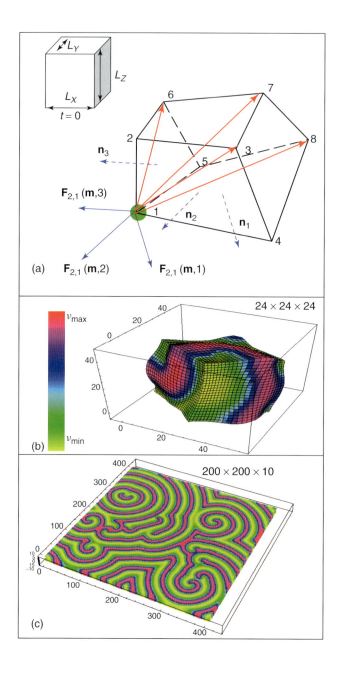

coordinate system and calculate the coordinates within the element **m** in this local coordinate system through the values of the nodal coordinates, $\mathbf{r}_n(\mathbf{m})$, and a set of "shape functions," as detailed in Refs. [31, 32]. We perform all the necessary volume and surface integrations within each linear hexahedral element in this local coordinate system [18].

We assume that the dynamics of the polymer network is purely relaxational; hence, we find the velocity of node n of the element **m** is proportional to the force acting on this node, $\mathbf{F}_n(\mathbf{m})$, that is, [18, 19]

$$\frac{d\mathbf{r}_n(\mathbf{m})}{dt} = M_n(\mathbf{m})\mathbf{F}_n(\mathbf{m}) \tag{3.12}$$

where $M_n(\mathbf{m})$ is the nodal mobility that depends on the volume fractions of polymer in the adjacent elements [18].

The total force acting on each node contains contributions from the elastic and osmotic properties of the system. We have shown [18] that the total force acting on node n of the element **m** consists of two contributions, $\mathbf{F}_n(\mathbf{m}) = \mathbf{F}_{1,n}(\mathbf{m}) + \mathbf{F}_{2,n}(\mathbf{m})$. The first term, $\mathbf{F}_{1,n}(\mathbf{m})$, describes the neo-Hookean elasticity contribution to the energy of the system and can be expressed as a combination of the linear springlike forces [18]:

$$\mathbf{F}_{1,n}(\mathbf{m}) = \frac{c_0 v_0 \Delta}{12} \left(\sum_{NN(\mathbf{m}')} w(n',n)[\mathbf{r}_{n'}(\mathbf{m}') - \mathbf{r}_n(\mathbf{m})] + \sum_{NNN(\mathbf{m}')} [\mathbf{r}_{n'}(\mathbf{m}') - \mathbf{r}_n(\mathbf{m})] \right) \tag{3.13}$$

Here, $\sum_{NN(\mathbf{m}')}$ and $\sum_{NNN(\mathbf{m}')}$ represent the respective summations over all the next-nearest neighbor nodal pairs and next-next-nearest neighbor nodal pairs belonging to all the neighboring elements **m'** adjacent to node n of the element **m**. Above, $w(n',n) = 2$ if n and n' belong to an internal face and $w(n',n) = 1$ if n and n' belong to a boundary face [18]. Unlike the situation for purely 2D deformations [19], there is no contribution from the interaction between nearest neighbors in Equation 3.13.

Figure 3.1 (a) Schematic of the 3D element. For each node, we provide its numbering within the element (1 : 8). The entire sample consists of $L_x \times L_y \times L_z$ nodes. In the underfomed state, the set of indexes $i = 1...L_x, j = 1...L_y$, and $k = 1...L_z$ defines the position of the nodes in x-, y-, and z-directions, respectively (see an inset in the top left corner). Forces acting on the node 1 (marked by the green circle) of the element $\mathbf{m} = (i,j,k)$ are marked by red and blue arrows. The red arrows mark the springlike elastic forces acting between the node 1 and the next-nearest and next-next-nearest neighbors within the same element **m**. The blue arrows mark contributions to nodal forces from the isotropic pressure within this element. (b and c) Snapshots of oscillations in BZ gels observed using gLSM model. The stoichiometric factor in BZ reaction is $f = 0.68$. The size of the sample is $24 \times 24 \times 24$ nodes in (b) and $200 \times 200 \times 10$ in (c). The minimum and maximum values for the color bar in (b) are $v_{min} = 8 \times 10^{-4}$ and $v_{max} = 0.4166$, respectively. We note that we use the same color bar in the following images, whereas the values of v_{min} and v_{max} are given separately for each figure.

The second contribution to the force acting on node n of the element $\mathbf{m}$, $\mathbf{F}_{2,n}(\mathbf{m})$, can be written as [18]

$$\mathbf{F}_{2,n}(\mathbf{m}) = \frac{1}{4}\sum_{\mathbf{m}'} P[\phi(\mathbf{m}'), v(\mathbf{m}')][\mathbf{n}_1(\mathbf{m}')S_1(\mathbf{m}') + \mathbf{n}_2(\mathbf{m}')S_2(\mathbf{m}') + \mathbf{n}_3(\mathbf{m}')S_3(\mathbf{m}')] \quad (3.14)$$

In the above equation, the summation is performed over all the neighboring elements $\mathbf{m}'$ that include the node n of the element $\mathbf{m}$. The pressure within each element, $P[\phi(\mathbf{m}'), v(\mathbf{m}')]$, is calculated according to Equation 3.10. In Equation 3.14, the vector $\mathbf{n}_l(\mathbf{m}')$ is the outward normal to the face l of element $\mathbf{m}'$, and S_l is the area of this face [18]. The vectors $\mathbf{n}_l(\mathbf{m})$ are shown in Figure 3.1a for element $\mathbf{m}$ that includes node $n = 1$.

Both contributions to the force acting on node $n = 1$ of the element $\mathbf{m}$ from within this element are shown schematically in Figure 3.1a. The springlike forces between node $n = 1$ and neighboring nodes are marked by red arrows, while the forces $\mathbf{F}_{2,n}(\mathbf{m})$ are depicted by the blue arrows. We emphasize that the total force acting on node n of the element $\mathbf{m}$ includes similar contributions from each of the neighboring elements containing this node. If the forces acting on the node n of the element $\mathbf{m}$ are known, we can calculate its velocity in the overdamped regime as [19]

$$\frac{d\mathbf{r}_n(\mathbf{m})}{dt} = M_n(\mathbf{m})\left(F_{1,n}(\mathbf{m}) + F_{2,n}(\mathbf{m})\right) \quad (3.15)$$

where $M_n(\mathbf{m})$ is the mobility of the node, given by

$$M_n(\mathbf{m}) = 8\frac{\Lambda_0\sqrt{\phi_0}}{\Delta^3}\frac{(1- <\phi(\mathbf{m})>_n)}{\sqrt{<\phi(\mathbf{m})>_n}} \quad (3.16)$$

Here $<\phi(\mathbf{m})>_n$ denotes the approximate value of the polymer volume fraction at the node n of the element $\mathbf{m}$; to calculate this value, we take the average value of $\phi(\mathbf{m}')$ over all the elements $\mathbf{m}'$ adjacent to the node n of the element $\mathbf{m}$, that is, for the internal node, $<\phi(\mathbf{m})>_n = \frac{1}{8}\sum_{\mathbf{m}'}\phi(\mathbf{m}')$

The above expressions allow us to formulate the discretized evolution equations for our model. By calculating the nodal displacements as defined in Equation 3.15, we effectively integrate the first equation of our governing system of equations, (Equations 3.1–3.3) defined above. In addition, we numerically integrate Equations 3.2 and 3.3. We refer the reader to Ref. [18] for the derivations and a complete set of the discretized equations that we use to update the concentrations of the dissolved reagent and the oxidized metal-ion catalyst in our sample. Here, we just note that our approach is based on a combination of a final element and final difference techniques. A detailed description of the 3D numerical approach is provided in Ref. [18], which also contains a discussion of how we validated this approach. In particular, we considered a limiting case that can be solved by using our gLSM and an independent method, and showed that there was an excellent agreement between the results obtained with these two different approaches [18].

We also performed numerous simulations to demonstrate that the gLSM approach is both robust and capable of simulating the dynamics of the BZ gels for a wide range of system parameters and sample sizes, including relatively large samples. The images in Figure 3.1b,c illustrate two examples of using the gLSM to simulate the dynamics of BZ samples of different sizes. Here and below, the colors in the images represent the values of the concentration of the oxidized catalyst, v, according to the color bar given in Figure 3.1b. The physical parameters describing the gel properties and reaction parameters are chosen to be the same in these two examples. The only difference between the examples in Figure 3.1b,c is the size of the sample ($24 \times 24 \times 24$ and $200 \times 200 \times 10$ nodes, respectively). The images in Figure 3.1b,c clearly illustrate that the difference in the sample sizes results in a significant difference in the dynamical patterns that form in those samples.

3.2.3
Model Parameters and Correspondence between Simulations and Experiments

The values of most model parameters were chosen on the basis of available experimental data [19]. Thus, the reference values for the Oregonator parameters q and ε were estimated to be $q = 9.52 \times 10^{-5}$ and $\varepsilon = 0.354$, respectively [19]. No experimental data are available to estimate the value of the stoichiometric parameter f, so this parameter is an adjustable model parameter. We specified the value of the parameter $\chi(\phi)$ based on experiments on neutral poly-N-isopropylacrylamide (NIPA) gels [30]. It was shown [30] that this value depends on temperature, T, and polymer volume fraction, ϕ, as follows: $\chi(\phi, T) = \chi_0(T) + \chi_1\phi$, where χ_0 is a function of the temperature T, and χ_1 is a constant. At the temperature of 20 °C the polymer–solvent interactions are described by the function $\chi(\phi) = 0.338 + 0.518\phi$ [30]. The interaction parameter χ^*, which mimics the hydrating effect of the oxidized metal ions, is an adjustable parameter of the model since no experimental data are currently available to guide our choice of χ^*.

The volume fraction of the undeformed gel, ϕ_0, and the dimensionless crosslink density in the gel, c_0v_0, characterize the as-prepared gel. The experimental studies on the chemoresponsive gel reported in Ref. [15] provided the values of $\phi_0 = 0.139$ and $c_0v_0 = 1.3 \times 10^{-3}$. Finally, the transport processes in the gel are characterized by the diffusion coefficient D_u of the dissolved reactant u in the solution and by the dimensionless mobility coefficient of the gel, Λ_0. The value of $D_u = 2 \times 10^{-5}$ cm^2 s^{-1} can be used for the diffusion coefficient [26, 33]. The numerical value of Λ_0 is much greater than one at low values of the volume fraction of polymer [34]. We used $\Lambda_0 = 100$ in all the numerical simulations below. With the above choice of parameters, the characteristic time and length scales in our simulations are $T_0 \sim 1$ s and $L_0 \sim 40$ μm, respectively [19, 23].

Using the above gLSM approach, we obtained qualitative agreement between our results and various experimental findings. In particular, for small samples, we observed the in-phase synchronization of the chemical and mechanical oscillations as illustrated in Figure 3.2a. Here, the image on the left shows the small cubic sample in the least swollen state (and, correspondingly, with the concentration of the oxidized catalyst, v, close to zero). The image on the right in Figure 3.2a shows the sample in the most swollen state; here, the concentration of the oxidized

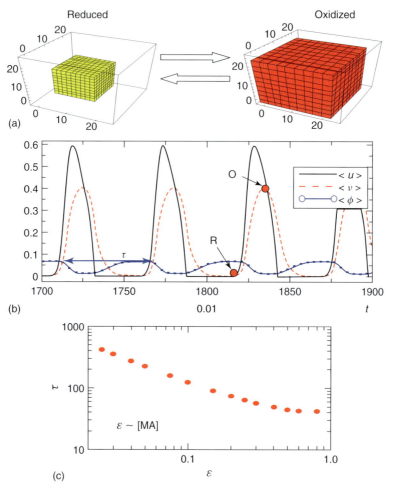

Figure 3.2 (a) Regular oscillations in BZ gel. The size of the sample is 12 × 12 × 12 nodes and the stoichiometric factor in BZ reaction is $f = 0.68$. The sample deswells when the Ru catalyst is in reduced state (the image on the left) and swells when its in the oxidized state (the image on the right). The minimum and maximum values for the color bar are $v_{min} = 8 \times 10^{-4}$ and $v_{max} = 0.4166$, respectively. (b) Evolution of $<u>$, $<v>$, and $<\phi>$ for the sample is shown in (a). Here, τ marks the period of oscillations, and the points marked by R and O correspond to the snapshots in (a) with the reduced and oxidized catalyst, respectively. (c) Dependence of the period of oscillations, τ, on ε, for the sample with $L = 12$ and $f = 0.8$.

catalyst, v, attains its highest value for the given parameters. This is similar to the in-phase synchronization of chemical and mechanical oscillations observed experimentally by Yoshida *et al.* in cubic gel pieces that were smaller than the characteristic length scale of the chemical wave [14]. In Figure 3.2b, we plot the evolution of $<u>$, $<v>$, and $<\phi>$, which are the respective average values of the concentration of the reaction intermediate, oxidized catalyst, and the polymer volume fraction, for the sample shown in Figure 3.2a. The average values are taken over all the elements within the sample at each moment of time. The dots marked R and O correspond to the images in Figure 3.2a for the gel sample in which the catalyst is in the reduced and oxidized states, respectively. The plot clearly shows that the lowest values of the average polymer volume fraction (blue line) correspond to the highest values of the oxidized catalyst (red dashed line) and vice versa.

We also made further qualitative comparisons between our findings and experimental results by investigating the effect of ε on the behavior of the BZ gels [13, 14, 35, 36]. The value of ε is proportional to the concentration of malonic acid (MA) in the experiments. In the following simulations, we fix $f = 0.8$ and $L = 12$ and vary ε. The plot in Figure 3.2c shows that the period of oscillations dramatically decreases with an increase in ε. The observed decrease in the period of oscillations with an increase in ε is in qualitative agreement with the experimental results of Yoshida et al. [13, 14, 35, 36]. Finally, if we further increase the value of ε (i.e., $\varepsilon \geq 0.9$ for the scenario presented above), then we observe a transition to the nonoscillatory regime. Again, this observation agrees qualitatively with experimental studies [13] where researchers observed a transition between the oscillatory and nonoscillatory regimes with an increase in the concentration of MA.

3.3
Results and Discussions

3.3.1
Effect of Confinement on the Dynamics of the BZ Gels

3.3.1.1 Linear Stability Analysis in Limiting Cases

As we illustrate below, our simulations show that the dynamics of BZ gels critically depends on the confinement of the sample. To understand the physical origin of this dependence and to isolate the parameter regions where the effect of confinement is most significant, we begin with a theoretical analysis of the behavior of the BZ gels in two limiting cases [20]. Then, we extend our studies to understand the dynamic behavior in a number of more complex scenarios. We first define the two limiting cases, *Case A* and *Case B*.

Case A: For sufficiently small sample size and high polymer mobility, we can neglect the contributions from diffusion in Equation 3.3 and assume that the evolution of ϕ follows the changes in the reactant concentrations. The latter assumption means that the elastic stress is instantaneously equilibrated with the osmotic pressure (i.e., the first expression in Equation 3.11 is valid at any moment in time). From this assumption, we obtain the concentration of the oxidized catalyst in this limiting case as a function of the polymer volume fraction as

$$v_{\lim}(\phi) = (\phi \chi^*)^{-1} \left(c_0 v_0 \left[(\phi/\phi_0)^{1/3} - \phi/2\phi_0 \right] + \ln(1-\phi) + \phi + (\chi_0 + \chi_1 \phi)\phi^2 \right)$$

(3.17)

Therefore, in this limiting case, the dynamics of the system can be described by the two independent variables, u and ϕ. We find these variables by solving the following system of equations:

$$\frac{d\phi}{dt} = R_\phi(u, \phi) \tag{3.18}$$

$$\frac{du}{dt} = R_u(u, \phi) \tag{3.19}$$

The right-hand sides of Equations 3.18 and 3.19 can be written as [20]

$$R_\phi = \frac{c(\phi, u)}{b(\phi)} \tag{3.20}$$

$$R_u = u((1-\phi)^2 - u) + f \frac{a(\phi)(1-\phi)}{2\phi\phi_0 \chi *} \frac{(1-\phi)^2 q - u}{(1-\phi)^2 q + u} - \frac{u}{1-\phi} \frac{c(\phi, u)}{b(\phi)} \tag{3.21}$$

Above, we have introduced the following definitions:

$$a(\phi) = c_0\left[-\phi + 2\phi^{1/3}\phi_0^{2/3}\right] + 2\phi_0\left[\phi + \phi^2(\chi_0 + \chi_1\phi) + \ln(1-\phi)\right] \tag{3.22}$$

$$b(\phi) = c_0(1-\phi)\left[10\phi^{1/3}\phi_0^{2/3} - 3\phi\right]$$
$$+ 6\phi_0\left[2\phi - \phi^2(1 + \chi_1(1-\phi)\phi) + 2(1-\phi)\ln(1-\phi)\right] \tag{3.23}$$

$$c(\phi, u) = 3\varepsilon(1-\phi)^2\phi(a(\phi) - 2(1-\phi)\phi\phi_0 u\chi *) \tag{3.24}$$

We find the stationary solution $\{u_{st}, \phi_{st}\}$ of Equations 3.18 and 3.19 by solving $G|_{v=v_{\lim}(\phi)} = 0$, $F|_{v=v_{\lim}(\phi)} = 0$. To analyze the stability of the solution $\{u_{st}, \phi_{st}\}$, we linearize Equations 3.18 and 3.19 with respect to small fluctuations and find the growth rate of the fluctuations as $p = (T \pm \sqrt{T^2 - 4D})/2$ (here, $T = [\partial_u R_u + \partial_\varphi R_\varphi]|_{\{u_{st}, \phi_{st}\}}$, $D = [\partial_u R_u \partial_\varphi R_\varphi - \partial_\varphi R_u \partial_u R_\varphi]|_{\{u_{st}, \phi_{st}\}}$, and $\partial_u \equiv \partial/\partial u$).

Case B: Another limiting case is the one where the entire sample is constrained, that is, where all the faces of the 3D sample are attached to hard walls so that the volume of the sample remains constant. Again, we consider a small sample and neglect diffusion in Equation 3.3. Here, we fix $\phi \equiv \phi_{st}$ (determined from Equation 3.11), so that the dynamics is described solely by Equations 3.2 and 3.3 with $\phi \equiv \phi_{st}$.

Figure 3.3 shows the stability map for Cases A (solid lines) and B (dashed lines) in terms of the important variables of the BZ reaction, f and ε; it may be recalled that f characterizes the stoichiometry of the reaction and ε is proportional to the concentration of MA. This map indicates the critical values of $f \equiv f^c$, where the stationary solution loses its stability in Cases A and B. The curves are plotted for three values of the parameter $\chi*$, which characterizes the responsiveness of the gel ($\chi* = 0.01$, $\chi* = 0.05$, and $\chi* = 0.105$ are shown in black, blue, and red, respectively). If the values of $\{f, \varepsilon\}$ lie below the respective solid curves, the sample remains in the steady state, whether or not it is attached to hard walls. If, however, the $\{f, \varepsilon\}$ values are above the respective dashed curves, the sample undergoes oscillations, which are chemomechanical in Case A and purely chemical in Case B since the mechanical oscillations are suppressed by fixing the sample's size. Note that for all the values of $\chi*$, f^c increases with increasing ε.

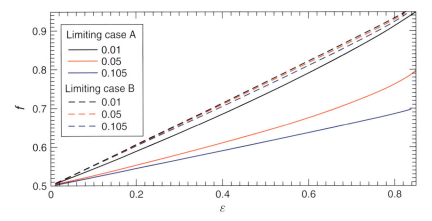

Figure 3.3 Dependence of f^c on ε. For $f > f^c$, the steady-state solution is unstable. Curves for f^c for the limiting Cases A are shown in solid lines and for the limiting Cases B in dashed lines. The values of χ^* in each of the cases are given in the legend.

The most interesting behavior, however, is observed when the parameters are located between the solid and dashed curves. Here, the behavior of the sample significantly depends on whether it is confined: the sample oscillates when it is free and remains in the stationary state when it is attached to the bounding surfaces.

Figure 3.3 also reveals that the parameter region where the oscillations are sensitive to confinement (i.e., the region between the dashed and solid lines in all cases) increases with an increase in ε and is larger for the more responsive sample. It is seen in Figure 3.3 that this region is very small for $\chi^* = 0.01$ and is relatively large for $\chi^* = 0.105$. Figure 3.4 illustrates that the effect is robust with respect to variations in the crosslink density (see the figure legend); however, the area between the solid and dashed curves depends weakly on the value of c_0. Figure 3.4 also shows that the region of parameters where the behavior of the sample significantly depends on whether or not it is confined strongly increases with the increase in the responsiveness of the sample χ^*.

In the limit of small sample sizes for both Cases A and B, our simulations are in excellent agreement with the linear stability analysis. However, for larger samples, where diffusion of the reagent can no longer be neglected and the mobility of the polymer network has a finite value, the analytical treatment is prohibitive. Consequently, we must turn to the computer simulations to study the properties of the system. We ran simulations for a sample of size $L \times L \times L$ with $L = 16$ nodes (i.e., a dimensionless linear size of $(L-1)\lambda_{st}$), $\chi^* = 0.105$, and a range of f values (taken at intervals of $\Delta f = 2.5 \times 10^{-2}$). In Figure 3.5, we plot the critical values of f that correspond to the transitions between the stationary and the oscillatory states for this larger sample when it is free (solid squares) or confined on all sides (solid circles). The interconnecting dotted lines define the corresponding regions in the phase maps; we mark these lines as f_A^c and f_B^c for the cases of the free and constrained samples, respectively. From Figure 3.5, it can be seen that the region

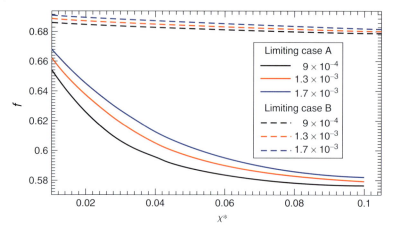

Figure 3.4 Dependence of f^c on χ^*. For $f > f^c$, the steady-state solution is unstable. Curves for f^c for the limiting Cases A are shown in solid lines and for the limiting Cease B in dashed lines. The values of cross-link density, c_0, in each of the cases are given in the legend.

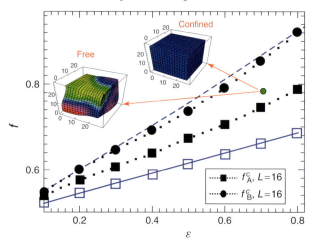

Figure 3.5 Dependence of f^c on ε for the samples of size $L \times L \times L$ nodes with $L = 2$ (blue lines) and $L = 16$ (black lines, with the simulation points marked by the filled squares for the free sample (f_A^c) and by the filled circles for the constrained sample (f_B^c). The images in the inset correspond to the sample size is $16 \times 16 \times 16$ nodes, $f = 0.76$ and $\varepsilon = 0.7$.

where the effects of the confinement are most significant ($f_A^c < f < f_B^c$) is narrower for these large samples than for the smaller gels (marked in blue).

3.3.1.2 Oscillations Induced by the Release of Confinement

To further illustrate the effects of confinement on the behavior of BZ gels, we now choose the parameters in the middle of the range defined above, $f_A^c < f < f_B^c$. We numerically simulate the behavior of the cubic gel sample, which is initially bound by hard walls on all sides. In the simulations, such confinement is implemented by fixing the positions of all the surface nodes. For the chosen parameters, the confined BZ gel is in the nonoscillatory, stationary state (see left inset image in Figure 3.6). The plot in Figure 3.6 shows the evolution of the average concentration of the oxidized catalyst, v, in this sample. The sample remains stationary as long as it is confined. At time $t = 5 \times 10^2$ (at the beginning of the x-coordinate in Figure 3.6), we released the sample from its confinement. This release initiated spontaneous oscillations; immediately after the release, the amplitude of the oscillations is small. Then, the amplitude increases and we observe regular oscillations in both the size of the sample and the concentrations of the chemical reagents, as shown in Figure 3.6. This is exactly the dynamic behavior we expected to observe according to the diagram in Figure 3.5.

In general, the exact values of f_B^c, f_A^c that define the parameters region $f_A^c < f < f_B^c$ in which confinement stops the oscillations depend on the actual dimensions of the sample. However, the results of the linear stability analysis for the two simplest examples and the simulation results for the larger cubic samples provide us with approximate guidelines for predicting the behavior of various cases. If we select an f in the middle of the region $[f_A^c, f_B^c]$, we anticipate that the free sample of *any size* will oscillate, but a bounded sample will remain stationary. We emphasize that this statement is valid for samples where *all* the faces are either free or confined. We demonstrate below that the information in Figures 3.3–3.5 can be used to design the dynamic properties of BZ gels by partially confining of the samples; that is, by bounding only some portions of the sample's surfaces.

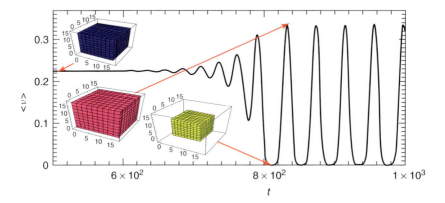

Figure 3.6 Evolution of $<v>$ for the sample of size $10 \times 10 \times 10$ nodes. The color indicates the concentration of oxidized catalyst, v, according to the color bar in Figure 3.1b. Here, we set $v_{min} = 2 \times 10^{-4}$ and $v_{max} = 0.35$, and keep these values fixed throughout this section. Here, $f = 0.76$ and $\varepsilon = 0.7$. The sample remains confined until $t = 500$. Release of the sample's confinement (at $t = 500$) induces chemomechanical oscillations.

3.3.1.3 Behavior of Partially Confined Samples

To illustrate how one can use the diagram in Figure 3.5 to design BZ gels with the desired dynamic properties, we first consider the scenario where only the top and bottom faces are held fixed and the rest of the sample is unrestricted (see left image in Figure 3.7a). In addition, we consider the effect of removing the top confining wall in the same cases (see schematic in Figure 3.7a). For the examples provided below, we choose different dimensions of the sample, but keep the same physical parameters of the system. We set $f = 0.76$ and $\varepsilon = 0.7$ (in the middle of the $[f_A^c, f_B^c]$ region in Figure 3.5). In the limit of small sample thickness, one expects this gel film to behave similar to Case B where the entire sample was constrained. Indeed, the sample in Figure 3.7b remains in the stationary state as long as its top and bottom faces are held fixed (left image). We now remove the top bounding surface, that is, allow the nodes on the top surface of the gel to move. In this case, the sample is no longer constrained as the top of the sample can move relatively freely. Hence, removing the top confinement in this case induces oscillations in the sample, as shown in Figure 3.7b.

In the following simulations, we consider a gel sample that is thin and long in the z-direction (Figure 3.8). In contrast to the case in Figure 3.7a, the majority of the sample is now unconstrained, even if the both top and bottom surfaces are constrained (Figure 3.8a). Hence, we anticipate that the behavior of the sample would be similar to that of a nonconfined sample. Figure 3.8a confirms this prediction, showing two traveling waves that are generated at the top and bottom walls; these waves propagate and collide at the center of the sample. The arrows on

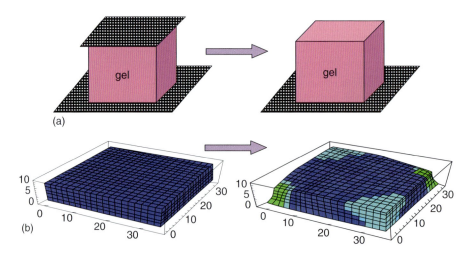

Figure 3.7 Oscillations induced by release of the top confinement of the same. (a) Schematic of the sample: the top and bottom faces are held confined (image on the left) and the top surface is released (image on the right). (b) Snapshots before and after release of the top confinement for the sample size $20 \times 20 \times 6$ nodes, $f = 0.76$ and $\varepsilon = 0.7$.

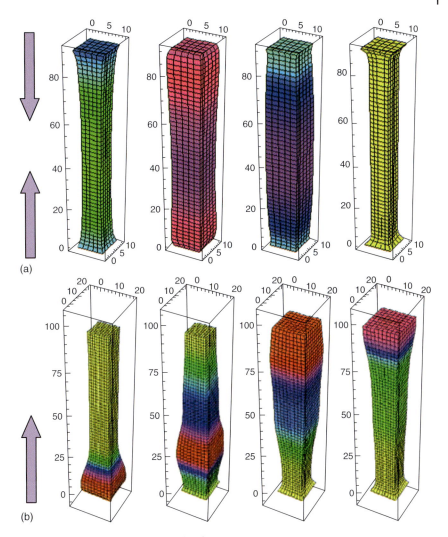

Figure 3.8 Oscillations of the sample of size $6 \times 6 \times 50$ nodes. (a) Top and bottom surfaces are confined (as in the left image in schematic in Figure 3.7a) (b) Top surfaces is resealed (as in the right image in schematic in Figure 3.7a). The snapshots in the top and bottom rows were taken during one period of oscillation. Here, $f = 0.76$ and $\varepsilon = 0.7$.

the left indicate the direction of propagation of the traveling waves. If we remove the top confining wall, we observe that at late times the traveling wave is generated at the fixed end and propagates to the free end (See Figure 3.8b); this observation is consistent with the experimental observations for similar samples [17], and with simulations of 2D BZ gel films [19].

Finally, we consider the examples of nonuniform confinement. We consider thin samples that are attached to hard walls everywhere except at two specific regions, which are marked by a circle of radius R_0 and a rectangle with width W_0 (Figure 3.9a). We choose the value of f higher than f_B^c for the case in Figure 3.9b and within the region $[f_A^c, f_B^c]$ for the case in Figure 3.9c. In Figure 3.9b, a traveling wave propagates throughout the sample, with only small distortions caused by the presence of the free regions. The situation is significantly different, however, in Figure 3.9c, where the oscillations are observed only within the free regions, while the regions that are held fixed remain in the stationary state. Thus, one film contains two well-defined regions, which exhibit qualitatively different dynamics,

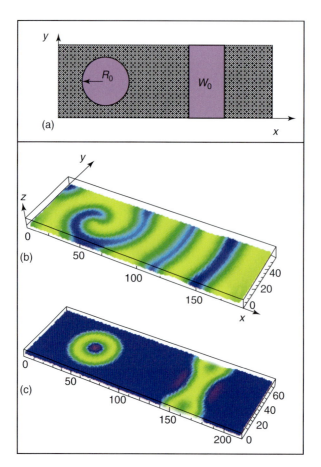

Figure 3.9 Schematic of the sample (top view). White areas are unbounded within the top and bottom surfaces. Sample size is $120 \times 40 \times 2$ nodes; $R_0 = 17$ and $W_0 = 20$ nodes and the separation between the centers is 60 nodes. Here, we fix $\varepsilon = 0.7$ and set the values of f at $f = 0.88$ in (b) and $f = 0.78$ in (c), respectively.

that is, oscillatory and nonoscillatory behavior. We note here that to observe such oscillations within the free regions, the sizes R_0 and W_0 must exceed some critical values. Additional simulations show that for the case in Figure 3.9c, these critical values are $R_0 = 5$ and $W_0 = 7$ nodes. These values also depend on the sample thickness, as well as the actual values within the $[f_A^c, f_B^c]$ region. For example, as one might anticipate, it becomes more difficult to induce oscillations if we choose parameters closer to the lower curve, f_A^c.

Figure 3.9c reveals that "cutouts" in the confining surfaces provide an effective means of controlling the oscillations of active BZ gels and, hence, the functionality of the material. Different arrangements or shapes of these cutouts can be utilized to produce different dynamic behavior. For this reason, the phase map in Figure 3.5 is particularly valuable since it identifies useful parameters for creating films with distinct pulsations in an otherwise stationary layer.

3.3.2
Response of the BZ Gels to Nonuniform Illumination

3.3.2.1 Modeling the Photosensitivity of the BZ Gels

Recent experiments on BZ gels have shown that these gels are sensitive to light of a particular wavelength [37, 38]. Thus, the dynamic behavior of the BZ gels can be modulated by illuminating the sample with this light. The effect of this illumination is to suppress the oscillations [37, 38] and, as we show below, this phenomenon can be exploited to design BZ gel "worms" that exhibit autonomous, directed motion away from the light source.

To model the sensitivity of the BZ gels to light, we modify Equation 3.5 as follows:

$$F(u, v, \phi) = (1 - \phi)^2 u - u^2 - (1 - \phi)\left[f v + \Phi\right] \frac{u - q(1 - \phi)^2}{u + q(1 - \phi)^2} \quad (3.25)$$

The dimensionless variable Φ accounts for the additional flux of bromide ions due to the light [39] and is assumed to be proportional to the light intensity. It is known that this additional flux of bromide ions completely suppresses oscillations within the sample if the light intensity exceeds some critical value, which can be experimentally determined [39]. In our simulations below, we define Φ_c as a critical value of the dimensionless flux of bromide ions (that corresponds to the respective critical value of the light intensity), above which the chemical oscillations are completely suppressed [39]. Generally, Φ_c depends on the reaction parameters and the physical properties of the gel. Using the simulations, we determined that for a system with the above parameters and the size provided below this value is $\Phi_c = 3.2 \times 10^{-4}$. Finally, we note that if we set $\Phi = 0$ in the above equation, we recover our expression for the system in the absence of light, that is, Equation 3.5. Our simulation box is $100 \times 10 \times 10$ nodes in the x-, y-, and z-directions, respectively, unless specified otherwise. We assume no-flux boundary conditions for u at the surfaces of the gel [18].

We consider below the dynamics of BZ gels under different spatial arrangements of the nonuniform illumination. For each element of the sample located in the dark region, we set $\Phi = 0$ and for each element located within the illuminated region, we set a nonzero value of Φ as specified in each of the cases. We initially set the swelling of the sample to its stationary value λ_{st} (taken at $\Phi = 0$) and set the values of u and v to have small random fluctuations around their respective stationary values, u_{st} and v_{st} defined in Equation 3.11. In addition, we note that here we neglect the attenuation of light within the gel and assume that the temperature is constant.

3.3.2.2 Autonomous Motion towards the Dark Region

In our first example involving nonuniform illumination, we consider the light intensity to be a step function, so that one region is kept dark and the other region is illuminated with a constant intensity (Figure 3.10a). Initially, we place the gel sample (of size $30 \times 10 \times 10$) along the x-direction so that one-third of the gel (on the left) is located in the dark region and two-thirds (on the right) is within the illuminated region (Figure 3.10a). Within this illuminated region, we set $\Phi = \Phi_c$, which is the critical value of the dimensionless flux of bromide ions above which the oscillations are completely suppressed [38].

For the chosen parameters, the sample is in the oscillatory regime at $\Phi = 0$ and, thus, a wave of swelling and deswelling is generated at the nonilluminated, left end and travels toward the right, ultimately giving rise to a net motion of the whole sample in the negative x-direction. As the sample moves away from the stationary light source, we set $\Phi = 0$ within those elements that are no longer illuminated. Figure 3.10b shows the time evolution of the BZ gel at the early and late times; the gel is drawn within a larger box to more clearly illustrate the path of its motion; and the nonilluminated region ($\Phi = 0$) is shaded in gray. Correspondingly, Figure 3.10c shows the trajectory of gel's center; the systematic decrease of this point's x-coordinate, x_c, reveals the net motion of the gel along the negative x-direction.

We note that even in the absence of light, traveling waves appear in this sample and can cause a macroscopic displacement of the gel; however, there is no preferred direction for the wave propagation, so that the probabilities of the wave traveling to the left or to the right are equal. Hence, averaged over a large number of samples, the net displacement of this gel is approximately zero. The role of the spatially nonuniform illumination is to break the symmetry in the system. Now, oscillations always originate from the nonilluminated region, and thus, the traveling waves continually propagate from left to right. As a chemical wave travels through a particular element within the sample, it causes the element to swell when the value of v is high and deswell when the value of v is low. The profile of the resulting oscillations within the gel is highly nonlinear and gives rise to a complex dynamics, where smaller amplitude motions to the right alternate with larger amplitude motions to the left. Therefore, we observe that the direction of the net displacement of the gel is opposite to the direction of propagation of the traveling waves; this is consistent with our earlier observations for 2D samples of BZ gels for similar reaction parameters [19, 40]. In other words, traveling waves that originate from

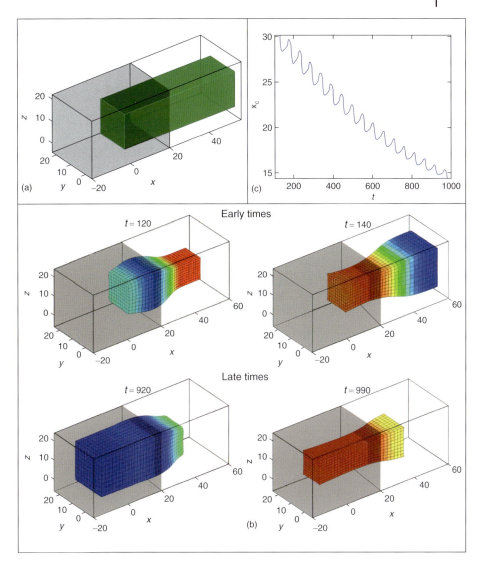

Figure 3.10 Motion of the BZ gel under nonuniform light intensity. (a) Intensity profile of the incident illumination. (b) Evolution of the sample during early ($t = 120$ and 40) and late ($t = 920$ and 990) stages. (c) Trajectory of the x-coordinate of gel's center. Here, we set $\varepsilon = 0.354$, $q = 9.52 \times 10^{-5}$, $f = 0.7$, $\phi_0 = 0.139$, $c_0 = 1.3 \times 10^{-3}$, $\chi_0 = 0.338$, $\chi_1 = 0.518$, and $\chi^* = 0.105$.

the left ultimately push the solvent to the right; hence, through the polymer-solvent interdiffusion, the gel is driven towards the left.

Figure 3.10 illustrates the directed motion of a relatively small sample. We observed that the directed motion remains robust for the longer samples with the same cross section. A large number of simulations on samples of size $100 \times 10 \times 10$

nodes revealed that as the gel travels along the negative x-direction, it also bends in the direction perpendicular to the direction of motion [21]. This bending can occur in any direction, depending on the initial random seed. We note, however, that for such directed motion toward the dark region to take place, it is essential to choose the sample size such that the cross-section of the sample is approximately the same or smaller than the characteristic diffusion length scale in the system.

In addition to controlling the direction of the BZ gel's motion, we can tailor the gel's velocity by varying χ^*, the parameter that characterizes the responsiveness of the gel to the BZ reaction. The plots in Figure 3.11 show how the displacement of the samples at three different values of χ^* (but identical nonuniform illumination) depends on time. We note that varying χ^* changes the stationary values u_{st}, v_{st}, and λ_{st}. Since the samples with different values of χ^* exhibit different initial degrees of swelling (and, hence, have different initial sizes), we characterize the directed motion by calculating the change in the x-coordinate of the gel's center, $x_C(t)$, with respect to its initial value, $x_C(0)$. Figure 3.11 shows the value of $x_C(0) - x_C(t)$ as a function of time for different χ^* and reveals that the slope of the curves, and hence the velocity, increases with increasing χ^*.

By taking the slope of the curve for the evolution of the x-coordinate at relatively early times (but after the oscillations are fully developed within the sample), we find that gel's velocity for the reference value of $\chi^* = 0.105$ along the x-direction is approximately 0.03 in the dimensionless units. The latter value of velocity

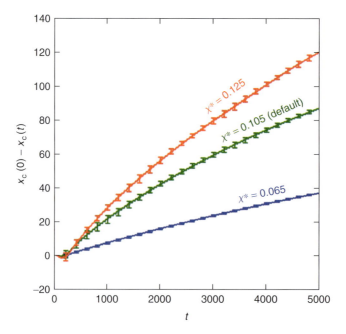

Figure 3.11 Evolution of $x_C(0) - x_C(t)$ at different χ^* for the intensity profile shown in Figure 3.10a.

corresponds to ~1.2 µm s^{-1} at the characteristic length and time scales noted above.

3.3.2.3 Light-Guided Motion along Complex Paths

In the previous section, we showed that one can use the nonuniform illumination to initiate a robust, *unidirectional* motion of BZ gels away from the illuminated region. This ability to control the self-sustained motion of the BZ gels through the nonuniform illumination could potentially open up new possibilities for designing self-regulating soft robots. It might, however, be desirable for such soft robots to not only move along a single direction but also navigate winding routes in order to arrive at a specific location. After the gel arrives at one site, it might be remotely (and noninvasively) given another set of directions. It could also be useful to control the time interval that this object spends at each location before it is sent "on demand" to additional sites.

As we show herein, the BZ gels can autonomously bend and make various turns as they effectively "sense" the variations in the illumination. Hence, these gels are not constrained to move along a sole direction, but can follow a complicated path as they navigate away from the light and, thus, could ultimately perform complex maneuvers and tasks (e.g., carry a payload to specified locations).

In the following studies, we make use of the photosensitivity of the BZ reaction to direct the motion of the gels. We control the arrangement of the light sources to guide the gel's movement along specific paths. Experimentally, the desired intensity profiles could be achieved by using holographic techniques [41–43] or masks. We demonstrate below that under the appropriate illumination, these macroscopic "worms" can make 90° turns, U-turns, travel "up ramps," and even "park" within specified areas [22].

To demonstrate the ability of the BZ gels to follow the complicated paths, we first show how to design the arrangement of the light and dark regions to "instruct" the sample to make a 90° right turn. The intensity profile we use here is shown in the inset of Figure 3.12a; the system consists of two dark regions, A and B, which are shown in gray. The heights of these regions are twice the height of the gel sample. Region A is offset in the *y*-direction so that its edge aligns with the center of region B (Figure 3.12a). Region B extends from the leftmost edge of the gel and initially spans 20% of the gel's length; the *x*-axis of the B region coincides with the long axis of the gel in its initial position. While the simulations are performed in 3D, Figure 3.12 shows only the top view of the simulation box for the sake of clarity.

To mimic variations in the light intensity, in the simulations, we alter the value of Φ (see Equation 3.22 and discussion in Section 3.3.1). Specifically, within the nonilluminated, dark regions, we set $\Phi = 0$. Within the illuminated regions, we set $\Phi = 1.95 \times 10^{-3}$. As the worm starts to move out of this illuminated area, we update the value of Φ, setting it to zero for those gel elements that are now localized in the dark region.

As noted above, for $\Phi = 0$, the system is in the oscillatory regime; therefore, for the example shown in Figure 3.12, the oscillations start from the nonilluminated, left end of the gel and the generated wave of swelling and deswelling travels toward

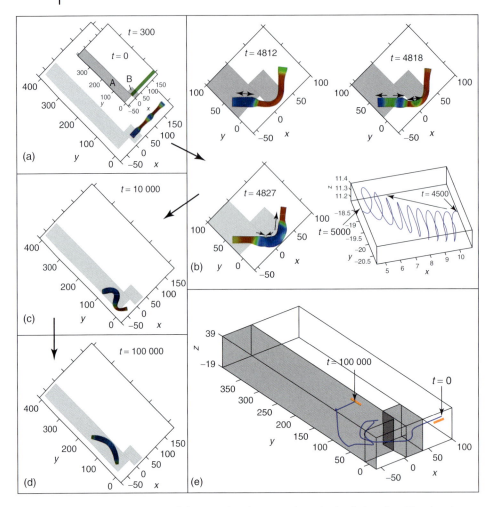

Figure 3.12 Top view of the BZ gel making a 90° turn. (a) The position of the gel at early time ($t = 300$); the inset shows the initial position of the gel (green) with respect to the dark regions (gray) A and B. (b) Details of the dynamic behavior as the gel crosses $x = 0$ threshold. The snapshots show the propagation of the two waves that originate in the dark region. The plot shows the trajectory of the gel's center for a short period of time reveals its net movement (shown by black arrow). (c) and (d) The snapshots demonstrating initial bending and subsequent turning of the gel into region A. (e) Trajectory of the gel's center.

the illuminated, right end. Hence, at early times, as the wave travels in the positive x-direction, it generates the large amplitude motions in the negative x-direction that alternate with smaller amplitude motions in the positive x-direction, so that the net displacement occurs in the direction opposite to the direction of propagation of the traveling wave. In essence, the spatial distribution of light imparts the initial velocity that propels the gel in $-x$-direction.[1] It is in this manner that the gel is driven to move toward the dark region, as explained in more detail in the previous section. Here, however, we build on this concept to further manipulate the motion of the worm. As portions of the gel move leftward through the dark B region and encounter the light, the worm "senses" the contiguous A section and bends to enter this darkened area.

The snapshots in Figure 3.12b detail the dynamic behavior of the sample as it crosses the $x = 0$ threshold and makes a turn from B into A. The figures reveal that oscillations originating in two different dark regions propagate along the sample and interact with each other, causing the worm to twist. The arrows in the snapshots indicate the direction in which these traveling waves propagate in the gel. The plot in Figure 3.12b shows the trajectory for the motion of the center of the gel during a relatively short interval of time. One can clearly see a complex pattern where large amplitude movements toward A alternate with smaller amplitude motions away from A. Hence, the sample undergoes a net displacement that is indicated by the black arrow in Figure 3.12b.

The subsequent snapshots shown in Figure 3.12c,d clearly demonstrate that the gel ultimately makes a 90° right turn and orients along the length of the A region, approximately parallel to y-direction. We also observe that once the gel is completely within the dark region, it straightens out. (The latter structure is less distorted and, hence, has a lower elastic energy than the "bowed" geometry.) We plot the trajectory of the gel's center in Figure 3.12e, which shows that the gel not only turns and reorients by approximately 90° relative to its initial configuration but also becomes trapped within the dark region. We note that in the above case, the cross sections of the A and B regions, as well as the length of the B region, are chosen to be smaller than the length of the gel; therefore, the only possible straight configuration that the gel can attain while still remaining within the dark regions is when it is oriented along the long axis of region A.

In the next simulation study, we design another set of instructions by altering the intensity profile of the illumination in such a way that the gel makes a "U-turn." As shown in Figure 3.13a, the long A region lies on top of the short B domain so that the left edges of these two dark sections are perfectly aligned. The width (y-direction) of these regions is twice the initial width of the gel. The total length (x-direction) of the B region is 20% of the gel's initial length. The gel is placed to the right of region B; initially only 10% of the gel is in the dark region. At the outset, the long axis of the gel lies parallel to the x-axis and the bottom surface of the gel lies five sites lower than the bottom of the B region. As is evident from the snapshots in Figure 3.13b, the gel initially bends and subsequently acquires a straight configuration in the A

---

1) Even in the absence of light, traveling waves will appear in this sample and can cause a macroscopic displacement of the gel; however, there is no preferred direction for the wave propagation, so the probabilities of the wave traveling to the left or to the right are equal. Hence, averged over a large number of samples, the net displacement of this gel is approximately zero.

84 | *3 Self-Oscillating Gels as Stimuli-Responsive Materials*

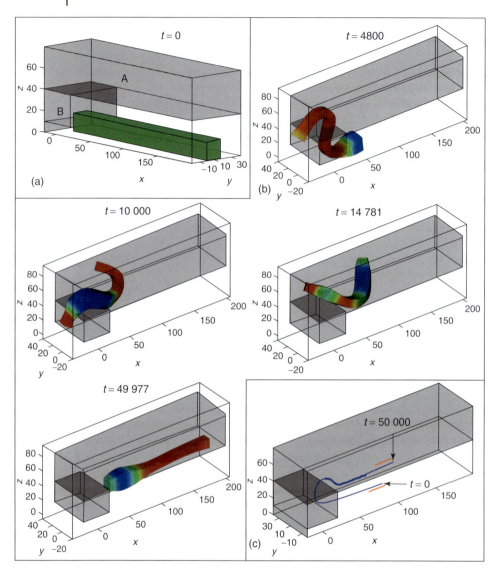

Figure 3.13 Motion of BZ gel when it is initially placed to the right of region A as shown in (a). (b) The snapshots of the key steps as the gel effectively makes a "U-turn." (c) The trajectory of the gel's center.

region. The trajectory of the gel's center (Figure 3.13c) clearly indicates the path the gel underwent to arrive at the late time configuration; the worm effectively makes a "U-turn" in order to become localized within the dark region.

Another of the BZ gel's maneuvers can be seen in Figure 3.14, where we have arranged the dark A and B regions to form a "T-shaped" configuration (Figure 3.14a). The total length (*x*-direction) of region B is 40% of the gel's initial

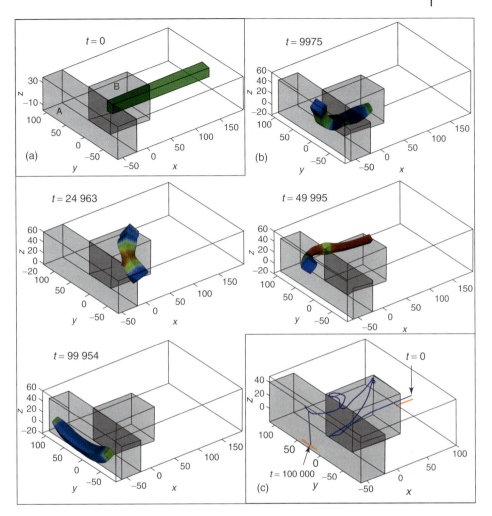

Figure 3.14 Motion of BZ gel subjected to "T-shaped" configuration of regions A and B. (a) Initial position of the gel with respect to regions A and B. (b) Snapshots of the steps that the gel follows to align perpendicular to region B (i.e., within region A). (c) Trajectory of the gel's center revealing the back and forth movement of the gel.

length and the height and width of the B region are twice the gel's initial cross section. The width (x-direction) of region A is 20% of the gel's initial length. The gel is placed in region B so that only 30% of its length is in the dark region (Figure 3.14a). Figure 3.14b shows the steps the gel takes as it follows the darkened path to arrive at its final configuration within A. The plot of the trajectory of the gel's center in Figure 3.14c reveals that the gel shuttles back and forth in B until it ultimately aligns perpendicular to the latter domain ($t = 99\,954$ in Figure 3.14b).

These back and forth movements correspond to images at $t = 24\,963$ and $49\,995$ in Figure 3.14b. Additional simulations indicate that this behavior remains robust even when the long axis of the gel is initially not perfectly aligned along the long axis of the B region. (In particular, we observed similar behavior when we initially set a $5°$ angle between the long axis of the gel and that of the B region.) The ability of the gel to alter its course in order to effectively escape the light strikingly resembles the response of biological systems to adverse environmental conditions [44].

In the final set of simulations, we demonstrate how the nonuniform illumination can be utilized to effectively pin the BZ gels at a specific location. Figure 3.15 ($t = 312$) shows the position of the gel with respect to the dark region at an early time. The size of the gel is $40 \times 10 \times 10$ nodes and initially only 10% of it is in the dark region. The dark region extends to $\pm\infty$ in both y- and z-directions. The asymmetric distribution of light forces the gel to move toward the $-x$-direction such that the two ends of the gel protrude into the bright region ($t = 9954$ in Figure 3.15). As a consequence, the traveling wave is generated at the center of the gel and, thus, the net movement (per cycle) in the $+x$ and $-x$ directions cancels each other. Therefore, the gel becomes pinned in the center of the dark

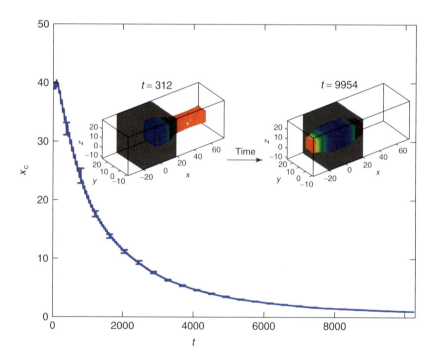

Figure 3.15 Pinning of the gel into the dark region. The top images show the early time ($t = 312$) and the late time ($t = 9954$) positions of the gel. The graph shows the evolution of gel's x-coordinate as a function of time. The vertical lines represent error bars calculated for 10 independent runs with different initial random seeds.

region. The graph in Figure 3.15 shows the evolution of the x-coordinate of the gel's center averaged over 10 independent runs (with different random seeds); the vertical lines represent error bars. These error bars are negligibly small at later times, suggesting that the pinning phenomenon is quite robust. The slope of the curve in Figure 3.15 represents the velocity of the gel in $-x$-direction; this velocity asymptotically approaches zero at late times. (In addition, we observed very small random displacements ($\sim$1%) in y- and z-directions that are dependent on the initial random seed.) We emphasize that once the gel is pinned at the specific location by means of chosen arrangements of the light and dark areas, the sample will remain stationary until we change the light intensity.

Our previous simulations indicate that while long samples will bend, the shorter ones do not (under identical conditions) [22]. In particular, we showed [22] that the bending of the BZ samples is somewhat analogous to a buckling phenomenon where the critical bending force decreases with the length of the sample [45]; that is, the same force can cause the bending of a longer sample but is not sufficient to bend a shorter one. These findings could be utilized to separate the shorter samples from the longer ones in a system that consists of a number of gel samples of different lengths. Depending upon the specific pattern of the light and dark domains, the smaller samples will become pinned at certain locations, while the longer samples will continue to move to a more "favorable" location. We also note that in the samples with large cross sections (greater than the diffusion length), we observed the generation of 3D spiral waves inherent to BZ reaction. Consequently, we found that the bending and reorientation of thicker BZ gels cannot be effectively controlled by light. In other words, to maneuver gel samples along complex paths using light, one has to choose samples with a cross section that is smaller than the characteristic length scale associated with the diffusion of the reagents.

To summarize, these synthetic BZ "worms" exhibit striking biomimetic behavior in their ability to move away from an adverse environmental condition, which in the context of the BZ reaction is the presence of light. It is noteworthy that the latter action mimics the adaptive behavior of the slime mold *Physarum*, which responds to nonuniform light illumination by moving toward the dark region [44].

3.4
Conclusions

We used computational modeling to lay out "blue prints" for designing active gels whose dynamic behavior can be tailored by external stimuli. The approach we developed–the gLSM–provided a much needed method for determining the 3D response of these materials to external cues, and allowed us to analyze both the volume and shape changes that the gels undergo in the presence of different stimuli. In these studies, we focused on the intriguing BZ gels because of their novel biomimetic behavior. The polymer network transduces the chemical energy from the BZ reaction into mechanical energy and, thus, the system mimics biological systems that take in energy to "fuel" their motion. Because of this "lifelike" behavior, the BZ gels could potentially be used in a broad range of applications, from autonomously functioning components in larger scale devices to millimeter-sized soft robots that can be directed to move and potentially transport small-scale objects.

To integrate these soft materials into larger scale devices and still maintain their unique self-oscillating behavior, we must understand the effect that confining surfaces have on the system. Additionally, it is important to determine whether we can obtain "extra value" or unprecedented properties through the judicious placement of the surfaces. In the studies described above, we examined the behavior of BZ gels that were sandwiched between two walls and calculated a phase map that indicates necessary conditions for maintaining the desired functionality of the confined layers. The studies also allowed us to design systems that display novel pattern formation by effectively introducing "cutouts" within the confining walls. With the removal of sections of the walls, the now free regions can exhibit oscillatory behavior. By tailoring the arrangement of the cutouts, one can not only create a stunning array of patterns but also control the propagation of waves within the sample. The ability to induce oscillations and direct the wave propagation via the architecture of the surfaces opens up new opportunities for utilizing these materials. For example, the oscillations induced by the unintentional removal of a surface could be used as an indicator, signaling that the structural integrity of a system has been compromised. Furthermore, the intentional release of a confined sample—or just a portion of the sample—provides a means of tuning the mechanical action of the system.

On a more general level, the ability to generate localized oscillations of soft matter within a hard matrix brings us closer to creating hierarchically structured materials that mimic the structure of biological systems, such as bone. Hence, we can devise new functionalities for these materials that would not be possible with the use of nonactive components.

From considering confining walls as a stimulus for regulating the behavior of hybrid systems, we turned our attention to using light to direct the motion of BZ gel "worms." Here too, we introduced new guidelines for controlling the wave propagation and, hence, the movement of millimeter-sized gel samples. Taking advantage of the photosensitivity of the BZ reaction and nonuniform illumination, we could impart seemingly intelligent behavior to these active worms, so that they appeared to navigate complex paths in a rather straightforward manner. In future studies, we will exploit this ability to create BZ gel "transporters" that respond to changes in illumination by carrying cargo to specified locations. An advantage of these transporters is that they can be made to localize at a particular "station" (by flooding the entire sample with light), where the cargo could be removed. With the reapplication of a nonuniform illumination, the BZ worm could be made to move on and potentially transport other cargo.

The novelty of the behavior described above brings up a key challenge in designing systems that involve responsive, active materials. There still remain few models that allow us to design systems that integrate the unique energy transducing capabilities of these active components. As we attempt to develop more energy efficient, biomimetic systems that can perform self-sustained work by converting one form of energy into another, we will need both new models and new design rules that allow us to optimize the utility and efficiency not only of individual processes but also of the entire system. The studies presented here constitute a step in meeting these important challenges.

Acknowledgments

Financial support from the Army Research Office is gratefully acknowledged.

References

1. Pojman, J.A. and Tran-Cong-Miyata, Q. (2004) *Nonlinear Dynamics in Polymeric Systems*, ACS Symposia Series, Vol. 869, American Chemical Society, Washington, DC.
2. Shibayama, M. and Tanaka, T. (1993) Volume phase-transition and related phenomena of polymer gels. *Adv. Polym. Sci.*, **109**, 1–62.
3. Dhanarajan, A.P., Misra, G.P., and Siegel, R.A. (2002) Autonomous chemomechanical oscillations in a hydrogel/enzyme system driven by glucose. *J. Phys. Chem. A*, **106** (38), 8835–8838.
4. Szanto, T.G. and Rabai, G. (2005) PH oscillations in the $BrO_3^- - SO_3^{2-}/HSO_3^-$ reaction in a CSTR. *J. Phys. Chem. A*, **109** (24), 5398–5402.
5. Labrot, V., De Kepper, P., Boissonade, J., Szalai, I., and Gauffre, F. (2005) Wave patterns driven by chemomechanical instabilities in responsive gels. *J. Phys. Chem. B*, **109** (46), 21476–21480.
6. Howse, J.R., Topham, P., Crook, C.J., Gleeson, A.J., Bras, W., Jones, R.A.L., and Ryan, A.J. (2006) Reciprocating power generation in a chemically driven synthetic muscle. *Nano Lett.*, **6** (1), 73–77.
7. Mujumdar, S.K., Bhalla, A.S., and Siegel, R.A. (2007) Novel hydrogels for rhythmic pulsatile drug delivery. *Macromol. Symp.*, **254** (1), 338–344.
8. Belousov, B.P. (1959) *Collection of Short Papers on Radiation Medicine*, Medgiz, Moscow.
9. Zaikin, A.N. and Zhabotinsky, A.M. (1970) Concentration wave propagation in two-dimensional liquid-phase self-oscillating system. *Nature*, **225**, 535–537.
10. Epstein, I.R. and Pojman, J.A. (1998) *An Introduction to Nonlinear Chemical Dynamics: Oscillations, Waves, Patterns, and Chaos*, Oxford University Press, New York.
11. Sakurai, T., Mihaliuk, E., Chirila, F., and Showalter, K. (2002) Design and control of wave propagation patterns in excitable media. *Science*, **296** (5575), 2009–2012.
12. Yoshida, R., Takahashi, T., Yamaguchi, T., and Ichijo, H. (1996) Self-oscillating gel. *J. Am. Chem. Soc.*, **118** (21), 5134–5135.
13. Yoshida, R., Onodera, S., Yamaguchi, T., and Kokufuta, E. (1999) Aspects of the Belousov-Zhabotinsky reaction in polymer gels. *J. Phys. Chem. A*, **103** (43), 8573–8578.
14. Yoshida, R., Tanaka, M., Onodera, S., Yamaguchi, T., and Kokufuta, E. (2000) In-phase synchronization of chemical and mechanical oscillations in self-oscillating gels. *J. Phys. Chem. A*, **104** (32), 7549–7555.
15. Sasaki, S., Koga, S., Yoshida, R., and Yamaguchi, T. (2003) Mechanical oscillation coupled with the Belousov-Zhabotinsky reaction in gel. *Langmuir*, **19** (14), 5595–5600.
16. Yoshida, R. (2008) Self-oscillating polymer and gels as novel biomimetic materials. *B. Chem. Soc. Jpn.*, **81** (6), 676–688.

17. Yoshida, R., Kokufuta, E., and Yamaguchi, T. (1999) Beating polymer gels coupled with a nonlinear chemical reaction. *Chaos*, **9** (2), 260–266.
18. Kuksenok, O., Yashin, V.V., and Balazs, A.C. (2008) Three-dimensional model for chemoresponsive polymer gels undergoing the Belousov-Zhabotinsky reaction. *Phys. Rev. E*, **78** (4), 041406.1–041406.16.
19. Yashin, V.V. and Balazs, A.C. (2007) Theoretical and computational modeling of self-oscillating polymer gels. *J. Chem. Phys.*, **126** (12), 124707.
20. Kuksenok, O., Yashin, V.V., and Balazs, A.C. (2009) Spatial confinement controls self-oscillations in polymer gels undergoing the Belousov-Zhabotinsky reaction. *Phys. Rev. E*, **80** (5), 056208.1–056208.5.
21. Dayal, P., Kuksenok, O., and Balazs, A.C. (2009) Using light to guide the self-sustained motion of active gels. *Langmuir*, **25** (8), 4298–4301.
22. Dayal, P., Kuksenok, O., and Balazs, A.C. (2010) Designing autonomously motile gels that follow complex paths. *Soft Matter*, **6** (4), 768–773.
23. Yashin, V.V. and Balazs, A.C. (2006) Pattern formation and shape changes in self-oscillating polymer gels. *Science*, **314** (5800), 798–801.
24. Tyson, J.J. and Fife, P.C. (1980) Target patterns in a realistic model of the Belousov-Zhabotinskii reaction. *J. Chem. Phys.*, **73**, 2224–2237.
25. Yashin, V.V. and Balazs, A.C. (2006) Modeling polymer gels exhibiting self-oscillations due to the Belousov-Zhabotinsky reaction. *Macromolecules*, **39** (6), 2024–2026.
26. Scott, S.K. (1994) *Oscillations, Waves, and Chaos in Chemical Kinetics*, Oxford University Press, New York.
27. Barriere, B. and Leibler, L. (2003) Kinetics of solvent absorption and permeation through a highly swellable elastomeric network. *J. Polym. Sci. Part B-Polym. Phys.*, **41** (2), 166–182.
28. Atkin, R.J. and Fox, N. (1980) *An Introduction to the Theory of Elasticity*, Longman, New York, NY.
29. Hill, T.L. (1960) *An Introduction to Statistical Thermodynamics*, Addison-Weley, Reading, MA.
30. Hirotsu, S. (1991) Softening of bulk modulus and negative Poisson's ratio near the volume phase transition of polymer gels. *J. Chem. Phys.*, **94**, 3949.
31. Smith, I.M. and Griffiths, D.V. (2004) *Programming the Finite Element Method*, John Wiley & Sons, Ltd, Chichester.
32. Zienkiewicz, O.C. and Taylor, R.L. (2000) *The Finite Element Method*, vol. 1, Butterworth-Heinemann, Oxford.
33. Tyson, J.J. (1985) in *Oscillations and Traveling Waves in Chemical Systems* (eds R.J. Field and M. Burger), John Wiley & Sons, Inc., New York, NY, pp. 93–144.
34. Achilleos, E.C., Christodoulou, K.N., and Kevrekidis, I.G. (2001) A transport model for swelling of polyelectrolyte gels in simple and complex geometries. *Comput. Theor. Polym. Sci.*, **11**, 63–80.
35. Sakai, T. and Yoshida, R. (2004) Self-oscillating nanogel particles. *Langmuir*, **20** (4), 1036–1038.
36. Suzuki, D. and Yoshida, R. (2008) Effect of initial substrate concentration of the Belousov-Zhabotinsky reaction on self-oscillation for microgel system. *J. Phys. Chem. B*, **112** (40), 12618–12624.
37. Shinohara, S., Seki, T., Sakai, T., Yoshida, R., and Takeoka, Y. (2008) Chemical and optical control of peristaltic actuator based on self-oscillating porous gel. *Chem. Commun.*, (39), 4735–4737.
38. Shinohara, S., Seki, T., Sakai, T., Yoshida, R., and Takeoka, Y. (2008) Photoregulated wormlike motion of a gel. *Angew. Chem. Int. Ed.*, **47** (47), 9039–9043.
39. Krug, H.J., Pohlmann, L., and Kuhnert, L. (1990) Analysis of the modified complete Oregonator accounting for oxygen sensitivity and photosensitivity of Belousov-Zhabotinskii systems. *J. Chem. Phys.*, **94**, 4862–4866.
40. Kuksenok, O., Yashin, V.V., and Balazs, A.C. (2007) Mechanically induced chemical oscillations and motion in responsive gels. *Soft Matter*, **3** (9), 1138–1144.
41. Gabor, D. (1948) A new microscopic principle. *Nature*, **161**, 777–778.

42. Leith, E.N. and Upatnieks, J. (1965) Photography by laser. *Sci. Am.*, **212** (6), 24–35.
43. Tay, S. *et al.* (2008) An updatable holographic three-dimensional display. *Nature*, **451** (7179), 694–698.
44. Nakagaki, T., Iima, M., Ueda, T., Nishiura, Y., Saigusa, T., Tero, A., Kobayashi, R., and Showalter, K. (2007) Minimum-risk path finding by an adaptive amoebal network. *Phys. Rev. Lett.*, **99** (6), 068104.
45. Timoshenko, S.P. and Gere, J.M. (1961) *Theory of Elastic Stability*, Engineering Societies Monographs, McGraw-Hill, New York.

4
Self-Repairing Polymeric Materials
Biswajit Ghosh, Cathrin C. Corten, and Marek W. Urban

4.1
Introduction

This chapter outlines recent advances in self-repairing of polymers that are related to autonomous healing processes mastered by nature. Damages and repairs ranging from angstrom to millimeter levels are discussed in terms of chemical reactions and physical processes governing the damage repairs.

Damages occurring in biological systems typically trigger multilevel chemicophysical responses leading to autonomous repairs. Depending upon the size and the degree of damages in a given biological system, repair mechanisms will vary. As an example, Figure 4.1a illustrates a collage of damages and corresponding repair mechanisms in the context of their dimensions and biological environments. Going from an angstrom (Å) level of the bond cleavage in DNA to millimeter (mm) size bone fractures, each damage and repair mechanisms exhibit their own attributes. For example, endogenous and exogenous sources of DNA damages lead to structural alterations that may result in destructions with potentially dire cellular consequences. These damages, if interfere with the "normal" DNA-template processes of transcription and replication, may cause a permanent mutation of genomes and defect formations [1]. Damages of nucleobases are the most common, and account for thousands of damaged bases in human cells every day. Fortunately, majority of lesions are repaired. Moving to a nanometer scale of skeletal muscle fibers, mysogenic precursors are activated in the site of an injured fiber, leading to myoblasts prolification that repairs damages by differentiating into multinucleated myocytes [2]. Satellite and myelomonocytic cells also contribute to repairing processes in which dysferlin acts as a Ca^{+2} sensor of the muscle membrane damages, triggering vesicle fusion and directing Ca^{+2} ions along the membrane to seal the lesion by sending messages to neutrophils for repair [3, 4]. At micrometer and millimeter scales, skin injury wounds penetrate into the dermis layer of skin, upon which the red blood cells transfer as a result of rupture of blood vessels. At the same time, platelets and inflammatory cells or cytokines arrive at the site of the injury and signal the event. These signals obtained by a specific receptor activate fibroblasts and other connective

Handbook of Stimuli-Responsive Materials. Edited by Marek W. Urban
Copyright © 2011 WILEY-VCH Verlag GmbH & Co. KGaA, Weinheim
ISBN: 978-3-527-32700-3

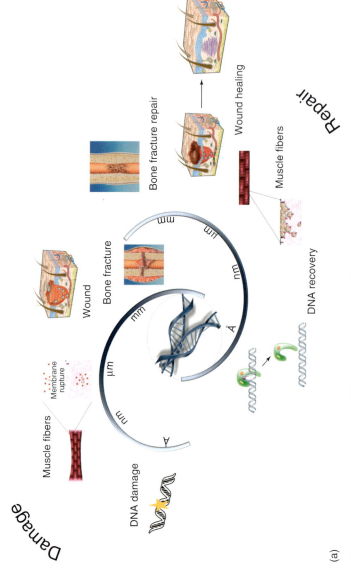

Figure 4.1 (a) Selected examples of damages and repair mechanisms occurring in mammals as a function of time. (b) Selected examples of damages and repair mechanisms occurring in plants as a function of time.

4.1 Introduction | 95

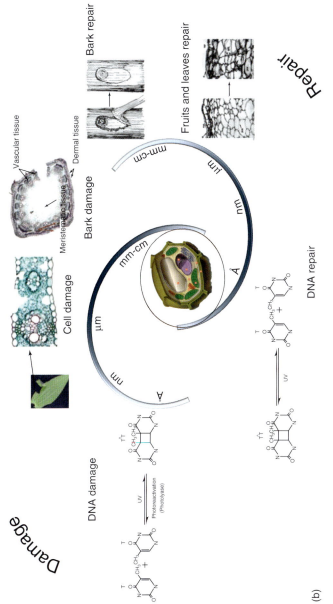

Figure 4.1 continued.

tissue cells to deposit collagen, resulting in a new tissue at the injury site and wound healing [5]. In contrast, the bone fracture repairs involve reactions of inorganic tissues comprised of a cascade of molecular events. Upon the damage, repairs are initiated by immediate inflammatory responses that lead to the recruitment of mesenchymal stem cells, followed by subsequent differentiation into chondocytes that produce bone-forming cartilage and osteoblastes. Upon cartilage matrix formation initiated by the resorption of mineralized cartilage, a transition from mineralized cartilage to bone occurs, and the primary bone formation is followed by remodeling, in which the initial bony callus is reshaped by secondary bone formation and resorption to restore anatomical structures capable of supporting mechanical loads [3, 4].

Unlike in mammals, damages occurring in plants are primarily due to attacks of predators as well as destructive environmental factors. Figure 4.1b schematically illustrates the levels of damage in plants. Again, considering the size, the damages may be occurring from an Ångstroms in DNA [6] to centimeter levels of bark rupture. Because of the similarities of the basic leaf [7, 8], fruit [9, 10], root [11], stem, and bark [12] structural features, micrometer-level damages in plants typically involve cell as well as tissue damages. In contrast to mammals, DNA damages in plants are repaired by photoreactivation [6]. Cell and tissue damage repairs are achieved by intercellular fusion or expansion as well as cell wall thickening with lignification or suberization. At a micrometer-scale damages, release of extrudate in fruits protects from further dehydration upon the damages and the repair mechanism is initiated by sclerified parenchyma cells development, followed by periderm formation beneath the wound in the presence of suberin and lignin [9, 10]. Similarly, damages in leaves can be signaled by concentration changes of jasmonic and salicylic acids, nitric oxide, or ethylene, depending upon the type of the plant, followed by the deposition of suberin and lignin, resulting in periderm formation healing the damage [7, 8]. For larger dimensional damages in millimeter-to-centimeter range, similar mechanism appears to be responsible for self-healing, although molecular-level processes are far from being understood.

While damages and autonomous repairs in mammals and plants occur in a broad spectrum of chemical and morphological environments, one common feature of these processes regardless of their size is that they are macroscopically manifested by highly orchestrated events initiated by signaling molecules. In mammals, they are typically categorized into (i) proinflammatory cytokines, (ii) transforming growth factor (TGF)-β superfamily members, and (iii) angiogenic factors. Each of these groups of cytokines and morphogens exhibit overlapping processes such as hemostasis, inflammation, proliferation, and remodeling, and orchestrated interactions among these processes is essential for successful repairs [5]. Similarly, structurally different plant macromolecules such as oligopeptides and oligosaccharides released from damaged cell walls or molecules inducing hormonal changes such as abscisic, jasmonic, and ascorbic acids, ethylene, and nitric oxide signal the damage, followed by sequences of chemical events [7–11].

Length scales of damages and repairs are inherently related to the time required for healing. Larger dimensions require longer transport and laborious diffusion paths, thus longer healing times. To mimic these complex processes that are not well understood at the molecular level, the development of materials capable of self-repairing by internal or external stimuli, multilevel chemical reactions, and transport requirements is a challenge. Furthermore, to understand superimposed chemical–physical interactions of individual processes at various scale lengths that will lead to autonomous repair is another aspect of self-healing. Although recent attempts and advances to create self-repairable polymeric systems reflect some of these steadily growing trends, the sequence of orchestrated and overlapping signaling, detection, and repairing events are yet to be explored.

4.2
Damage and Repair Mechanisms in Polymers

An ideal self-repairing material ought to continually sense and respond to damages over its lifetime, and restore or possibly enhance original structural features without adverse effects on other properties. Figure 4.2 is a schematic diagram of a potentially ideal self-repairable network, which consists of functional cross-linked macromolecular chains containing sensing and reactive components. While sensor groups are anticipated to remain inactive until activated by the damage, reactive, repairable groups are silent until triggered by sensing groups. This 1D repair will occur usually in one spatially confined environment, and as we recall, the interplay of many cascading events is necessary for repairs in biological systems. Although simplified, Figure 4.2 also illustrates that when the damage occurs, two main events occur: a bond cleavage (path A) and a chain slippage (path B). When the bond cleavage (path A) takes place, sensing groups will trigger the formation of reactive bond-forming species, which may include the formation of free radicals or other reactive groups, for example, –COOH, –NH$_2$, –OH, –SH, –C=O, and –C=C. At a chemical bond level, self-healing processes can be considered as chemical and/or physical adhesion without an adhesive, in which covalent, H-bonding, electrostatic, and/or ionic interactions, as well as Van der Waals forces, may individually or collectively contribute to the repair. Thus in designing polymeric self-repairing reactive networks, directional bonding is essential.

When the chain slippage occurs at the interface/surface of the damage (path B), loose chain ends may form, which may or may not exhibit reforming efficiency, which depends upon the Gibbs free energy (ΔG) of the system. If $\Delta G < 0$, favorable recombination conditions primarily driven by entropic contributions may take place, but self-repair via diffusion may not be favorable when an enthalpic component of the ΔG dominates the process. When slippage of polymer chains occurs during the damage, diffusion at nano- or higher scales, which relates adhesion to interpenetrating processes at two interfaces, is the major driving force. To achieve the repair state, temperature and pressure changes, concentration level changes, or internal stress gradients, along with other stimuli (electromagnetic

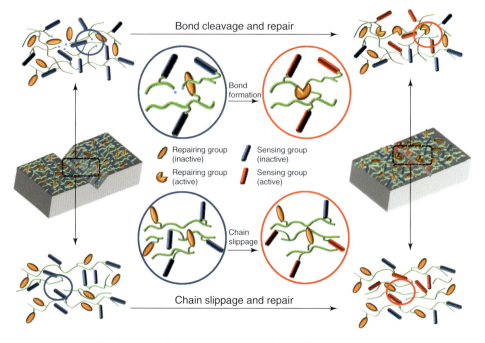

Figure 4.2 Schematic diagram illustrating an ideal self-healing polymeric network.

radiation, chemical potential), may be necessary. To reach an equilibrium state at which the Gibbs free energy will be minimized, diffusion rates are in the range of 10^{-5} m min^{-1} and may vary from system to system, but for liquids they are in the range of 10^{-3} m min^{-1} [13]. Thus, when a liquid is dispersed in a solid matrix, the kinetics of repair will favor the process. Since diffusion rates in solids are also reflected in the free volume of a polymer network, and the free volume depends upon the glass transition temperature (T_g), below the T_g, polymer chains will exhibit limited mobility as well as diminished transport properties. Above the T_g, Fickians diffusion is significantly enhanced [14, 15], which in the context of the self-repairing processes will enhance mobility and self-diffusion, primarily driven by entanglement coupling and reptation [16]. In non-cross-linked and cross-linked polymers, the presence of trapped entanglements affects the elastic modulus during irradiation or chain scission due to diffusion caused by entanglement slippage, which will be significant above the T_g [17]. Similarly, the reptation model in polymeric network solutions attributes topological constraints imposed by the surrounding polymer chains on entanglements to move along tubelike paths that follow their own contour and result in interdiffusion [16]. Although the T_g of the network and its free volume typically represent bulk properties, it should be realized that damage formation increases the T_g at the surface created by the damage. Since the mobility of loose ends further away from the surface is higher compared to the anchoring points, the very top surface entities will have lower T_g. Spectroscopic studies confirmed that T_g varies with the distance from the surface and the thickness of polymer films [18, 19].

By analogy to biological damage and repair mechanisms, Figure 4.3 depicts damages and repairs occurring in materials from angstrom to nano-, micro-, and millimeter ranges, where covalent and H-bonding is primarily attributed to the angstrom-level repairs, whereas metal–ligand and–ionic interactions require longer nanoscale ranges. Similarly, at a micrometer-scale level, controlled diffusion and relaxation processes may occur, whereas larger millimeter repairs may require insertions of the secondary phases. As damages occur at different length scales, healing mechanisms responding to these structural disruptions occur in a synchronous manner. While several studies conducted on each of the individual repairing processes are depicted in Figure 4.3, the fundamental difference between repairs in synthetic and biological materials is the ability of biological systems to superimpose individual events, thus resulting in orchestrated repairs across larger length scales. While there are many opportunities and challenges to further advance materials chemistry of autonomous repairs, studies conducted during the first decade have resulted in significant progress in this field. The remaining sections of this chapter outline the damage and repair mechanisms in polymeric materials at the scale lengths that parallel biological systems depicted in Figure 4.1a,b.

4.2.1
Dimensions of Damages and Repairs

The damages and repairs along with chemical and physical processes associated with them shown in Figure 4.3 exhibit similarities to biological systems, but as pointed out above the main differences are orchestrated, superimposed events of biological processes that occur at different scales. This is not to say that the damages and repairs at nanometer or higher scales do not involve angstrom-level covalent and H-bonding. On the contrary, at microscopic levels the overlap of various length scales is essential for repairs to occur. The main problem is their measurements, namely, what molecular events are associated with a millimeter scale repair. Thus, there are many opportunities for exploring these uncharted waters in synthetic materials involving covalent or H-bonding, ionic and metal–ligand coordination, diffusion and relaxation, or a combination of thereof.

4.2.1.1 Angstrom-Level Repairs

Covalent Bonding As stated above, an ideal self-healing network should contain reactive species that are activated upon damage. The most commonly occurring reactive species in polymer networks are free radicals. Their affinity for reacting with other species is high in solutions, but lower in solid polymer networks. In spite of relatively longer lifetimes of free radicals in polymer networks, for two

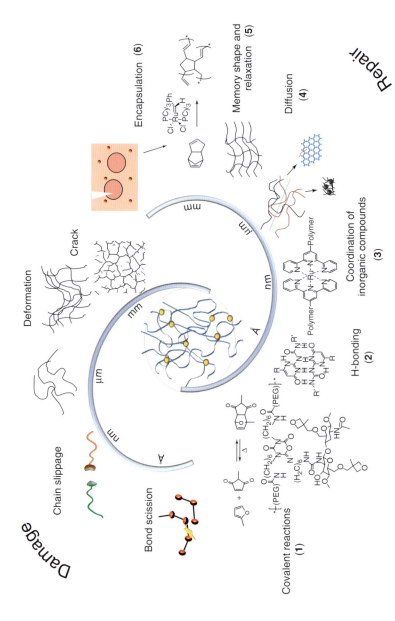

Figure 4.3 Schematic diagram of examples of damages and repairs in synthetic materials.

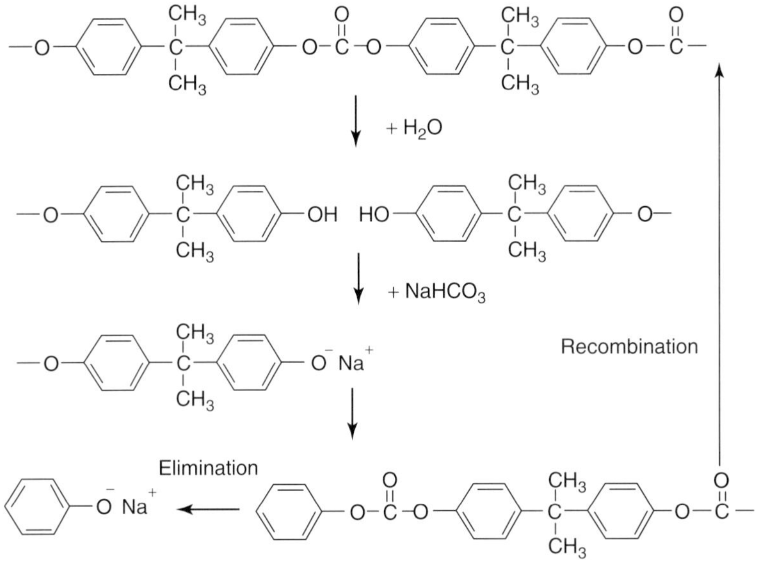

Figure 4.4 Ester exchange reaction in polycarbonate leading to healing [20].

cleaved reactive end chains to react, both ends must diffuse toward each other prior and react before other competing processes that may intercept these reactions. Since the most common process is oxidation of free radicals, one approach to prevent from undesirable reactions is to facilitate suitable environments. For example, polycarbonate (PC) synthesis capable of self-repairing by ester exchange method can be accomplished using a steam pressure at 120 °C [20]. This is illustrated in Figure 4.4, which shows reactions of the first carbonate group upon cleavage by hydrolysis giving rise to higher concentration levels of the end-capped phenoxy functionalities. The addition of a base (e.g., $NaHCO_3$) forces additional ester reactions, thus facilitating substitution reactions between phenoxide and phenyl–carbonyl chain ends, followed by recombination with CO_2 resulting in the reformation of the PC to self-heal the network.

The chain-end recombination was also utilized in poly(phenylene-ether)s (PPEs), in which the repairing agent is generated by oxygen. As shown in Figure 4.5, if a PPE backbone is damaged by heat, light, or external mechanical forces, free radicals are produced as a result of chain scission. However, their stabilization by a hydrogen donor and the presence of Cu (II) secures reactions with each end of the scission chains to form a complex, thus preventing further oxidation. When two end chains are recombined eliminating two protons from both ends, Cu (II) is reduced to Cu (I), and further reacted with molecular O_2 to oxidize again to Cu (II) [21–24]. Although very promising, these approaches at this point exhibit limited applicability for self-healing composites because the recovery of the cleaved bonds requires elevated temperatures. Alternatively, higher concentration levels of effective catalysts should be utilized to activate the recombination of degraded

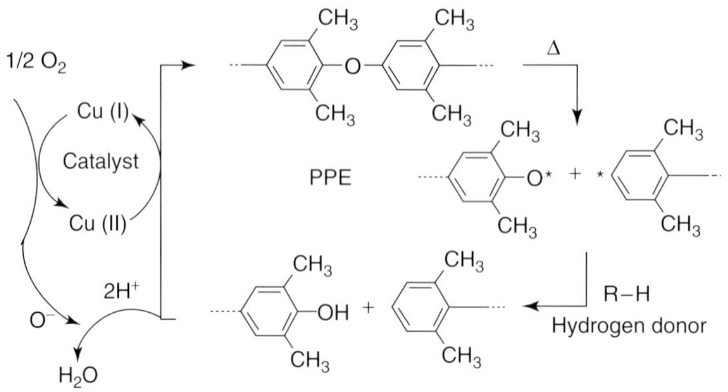

Figure 4.5 Schematic representation of covalent bond formation between scission chain ends in the presence of Cu catalyst leading to healing in polypropylene ethers [21].

oligomers under ambient conditions, which may adversely affect polymer matrix properties. To overcome these issues, the presence of photoactive species such as cinnamate monomer (1,1,1-tris-(cinnamoyloxymethyl)ethane (TCE)) may alleviate high temperature requirements [25]. As shown in Figure 4.6, crack formation results in the C–C bond cleavage of cyclobutane rings between TCE monomers, resulting in the formation of original cinnamoyl groups, and the crack healing occurs due to reversion of cyclobutane cross-links of TCE via [2+2] photocycloaddition upon UV (>280 nm) exposure [26–29].

Furan-maleimide-, dicyclopentadiene (DCPD)-, and anthracene-based polymers exhibit thermal reversibility [30–32]. The concept of reversible cross-linking via Diels–Alder (DA) reaction was utilized in epoxy, acrylate, and polyamide systems, where retro-DA reaction at higher temperature results in disconnection between diene and dienophile, resulting in crack propagation, whereas lower temperature reconstructs the covalent bonds to repair the crack. As illustrated in Figure 4.7, lower temperature results in adduct (exo or endo) formation between furan (diene) and malemide (dienophile) entities, which break apart at 120 °C [33–36]. The main concerns though are the use of relatively expensive and robust methylene chloride solvent as well as longer time requirements for healing. Although a solvent-free route for the synthesis of the next generations of cross-linked networks was reported, the response time frame is still long, which may adversely affect network cross-linking density building [34, 35].

Chitin, one of the most abundantly available natural polysaccharides, can be readily modified by acetylation reactions to form chitosan (CHI) [37]. CHI was modified with the four-member oxetane (OXE), followed by cross-linking with trifunctional hexamethylene di-isocyanate (HDI) in the presence of polyethylene glycol to form heterogeneous cross-linked polyurethane (PUR) network. When there is mechanical damage of the network, self-repairing occurs upon exposure of the damaged area to UV radiation. While the reactions leading to network formation are shown

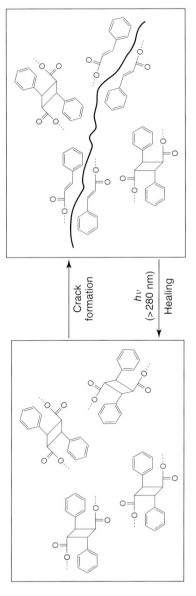

Figure 4.6 [2 + 2] cyclo addition reactions of cinnamoyl groups leading to healing [25].

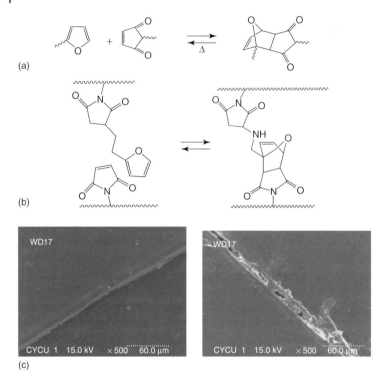

Figure 4.7 Schematic representation of Diels–Alder mechanism (a), specific mechanism of multifuran and multimalemide containing backbone within cross-link polymer system for reversible self-repair (b), and optical images of self-repairing polymer network (c) [34].

in Figure 4.8, the choice of these components was driven by their ability to serve specific functions. PUR in combination with polyurea (PUA) generated as a result of the reaction between isocyanate and amine-functionalized OXE-CHI provides desirable integrity and localized heterogeneity of the network, whereas OXE-CHI provides the cleavage of a constrained four-member ring (OXE) to form free radicals and UV sensitivity (CHI) for self-repair. PUR network repairs occur only when OXE-CHI moieties are incorporated, which upon UV exposure results in the ring opening of OXE along with the PUA–PUR conversion. The primary advantage of these systems is that the OXE–CHI macromolecule can be added to the existing thermosetting polymers, ensuring easy processability and desirable self-healing characteristics [38].

Hydrogen Bonding Although hydrogen bonds between neutral organic molecules are not among the strongest noncovalent interactions, due to their directionality and affinity, they play a significant role in supramolecular chemistry [39]. As

4.2 Damage and Repair Mechanisms in Polymers | 105

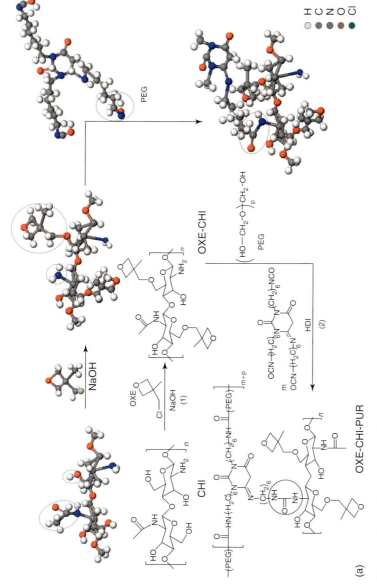

Figure 4.8 Synthetic steps involved in creating self-healing polyurethanes (PUR) containing oxetane (OXE) and chitosan (CHI) (a), and optical images of OXE-CHI-PUR network healed under UV exposure (b) [38].

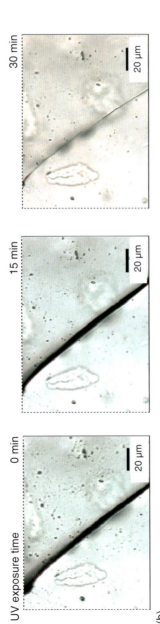

Figure 4.8 *continued.*

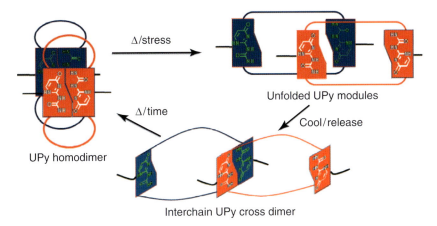

Figure 4.9 Proposed molecular healing mechanism of urea–isopyrimidone (UPy) network via H-bonding.

shown in Figure 4.9, combining four H-bonds in a functional unit of urea isopyrimidone (Upy) resulted in enhanced association strengths between those units and polysiloxane, polyethers, and polyesters [40–42]. Supramolecular polymers based on bifunctional ureidopyrimidinone derivatives behave like conventional polymers, but their mechanical properties exhibit strong temperature dependence. As anticipated, at room temperature, supramolecular polymers exhibit polymerlike viscoelastic behavior in the bulk and in solutions, but at elevated temperatures liquidlike properties are observed [43, 44]. An obvious advantage is the ability to control rheological behavior as a function of temperature, which can be accomplished by combining supramolecular monomers with traditional polymers. When fatty diacids and triacids from renewable resources were utilized in a two-step synthetic route, a self-healing network was obtained. As shown in Figure 4.10, the first step involves condensation of acid groups with an excess of diethylenetriamine, followed by reactions with urea. This material is capable of self-repairing when the two cut ends are brought together at room temperature without external heat and hydrogen bonds are formed between the –C=O groups of amide with amine-functionalized ends. This shows how complexity and chemical tuning may lead to self-healing polymers similar to those in biological systems [45].

4.2.1.2 Nanometer-Level Repairs

Metal–Ligand Coordination Similar to H-bonding, metal–ligand supramolecular interactions can be useful in designing supramolecular polymers [46, 47]. Because of their optical and photophysical properties, metal complexes offer many advantages and the reversibility can be tuned by different metal ion and ligand substitutes. As shown in Figure 4.11a, metal–ligand supramolecules can be disassembled using an excess amount of water-soluble ligands. Under these conditions, temperature

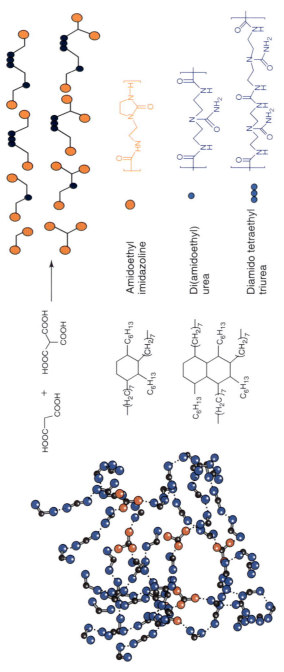

Figure 4.10 Schematic representation of reversible network formed by a mixture of fatty diacid and triacid condensed with di-ethylenetriamine, followed by reaction with urea to produce a mixture of oligomers equipped with a complementary hydrogen bonding [45].

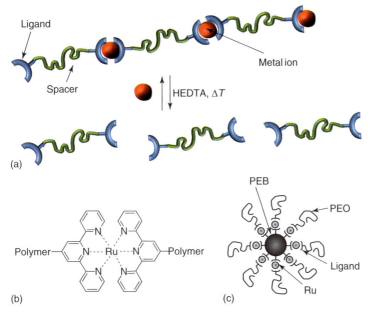

Figure 4.11 Schematic representation of a self-healing supramolecualr polymer network (a) containing coordinated terpyridine–Ru-pincer complex (b), and formation of aqueous micelle (c) by copolymerization with monomers [48].

changes and hydroxyl ethylene diamine triacetic acid (HEDTA) facilitate the reactions to switch back and forth, thus serving as molecular weight or cross-link density controller. Figure 4.11b shows an example of terpyridine-based ligand pincer complexes, which were introduced into a polymer matrix by copolymerization with the ethylene oxide and ethylene-co-butylene monomers [48]. The permanent structural component is the polymer network poly(ethylene-co-butylene) (PEB) and poly(ethylene oxide) (PEO) in Figure 4.11(c), whereas the reversible self-healing component is the metal–ligand coordination complex between a pincer complex and a polymer side-chain pyridine ligand.

Ionic Interactions Ionic interactions including ionomers have been utilized in polymeric materials and selected polymers that may exhibit self-healing attributes include poly(ethylene-co-methacrylic acid) and polyethylene-g-poly(hexylmethacrylate) [49]. Instead of external heat or an alternative stimulus to promote polymer diffusion in a damaged area, repairs occur under ambient and elevated temperatures upon projectile puncture testing [50, 51]. A ballistic puncture in low-density polyethylene (LDPE) does not show healing, whereas puncture in poly(ethylene-co-methacrylic acid) (EMMA) films heals the puncture, leaving a scar on the surface. The proposed healing mechanism is a two-stage process, in which upon projectile impact, ionomeric network is disrupted, and heat generated by the friction during the damage is transferred to the polymer surroundings, generating localized melt state. Elastic responses of the locally molten polymer facilitated by a puncture are known as *puncture reversal*. During the second stage, the molten polymer surfaces fuse together via interdiffusion resulting from ionic interactions between ionic clusters to seal the puncture, followed by rearrangement of the clustered regions and long-term relaxation processes, which continue until ambient temperatures are reached.

4.2.1.3 Micrometer-Level Repairs

Diffusion Diffusion is primarily responsible for self-healing of nonreactive systems or in polymer networks where damages result in a chain slippage (Figure 4.2, path B). In poly(methyl methacrylate) (PMMA), the induced crack healing was achieved by heating it above the T_g under light pressure [52]. A recovery of the fracture stresses was suggested to be proportional to $t^{1/4}$, where t is the heating time. While craze healing occurs at temperatures above and below the T_g, crack healing happens only at or above the T_g [53–56]. To reduce the effective T_g of PMMA, organic solvents were used as plasticizing agents, facilitating enhanced healing [57, 58].

In another approach, small grain thermoplastic epoxy particles adhesive were embedded within glass–epoxy composite matrices, which melted upon heating and, therefore, repaired damages [59]. It appears that the voids left after melting of adhesives have adverse affects on the integrity of epoxy polymer matrix. On the other hand, the higher temperature requirements to melt embedded epoxy particles may cause damage to the surfaces of the polymer. The same concept was utilized by blending thermoplastic poly(bisphenol-A-*co*-epichlorohydrin) with epoxy resin at 80 °C to dissolve it into the thermoset. During damage formation, thermoplastic polymer in the vicinity of the crack may flow to the newly created crack regions and heal the crack [59, 60].

Interfacial diffusion plays a significant role in heterogeneous networks where self-healing is expected to occur. Typically, diffusion is enhanced in a molten state, which enhances diffusion and possibly repairs networks. When γ-Fe_2O_3 nanoparticles were incorporated into emulsion polymerization of *p*-methyl methacrylate/*n*-butylacrylate/heptadecafluorodecyl methacrylate (p-MMA/nBA/HDFMA) matrix and allowed to coalesce, such films exhibited self-repairing when oscillating magnetic field (OMF) was used [61]. The first step in this development was to prepare polymeric films with uniformly dispersed γ-Fe_2O_3 nanoparticles, which was accomplished by *in situ* synthesis of (p-MMA/nBA/HDFMA) colloidal particles in the presence of γ-Fe_2O_3, followed by their coalescence. As shown in Figure 4.12a,b, polymeric films were cut into two pieces and then the broken pieces were physically placed in contact with each other. Upon application of OMF, γ-Fe_2O_3 nanoparticles increase the nanoparticle–polymer interface temperature, thus generating localized melt flow to permanently repair physically separated polymer (Figure 4.12c,d). As shown in Figure 4.12a′–d′, the self-repairing process can be repeated many times. This approach may find many useful applications in

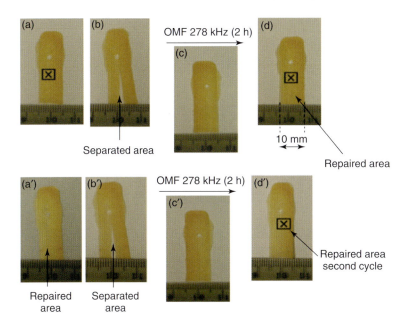

Figure 4.12 Optical images of self-repairing MMA/nBA/HDFMA films containing 14 w/w% of γ-Fe$_2$O$_3$ nanoparticles in the presence of oscillating magnetic field [61].

composites, where damages are not easily detectable, and repairs are necessary to maintain mechanical integrity of a composite.

Organometallic conductive polymer comprising of N-heterocyclic carbenes (NHCs) can be used to control corrosion in coatings and it also exhibits self-healing and conductive properties. As illustrated in Figure 4.13a, conductive and stimuli responsive attributes were achieved by synthesis utilizing NHCs and transition metals, which exhibit conductivity in the order of 10^{-3} S cm^{-1} [62]. When the damaged material is heated at 200 °C without solvent vapor or at 150 °C in the presence of dimethyl sulfoxide (DMSO) vapor, the damage disappears due to the dynamic equilibrium between the metal and the polymer, resulting in the flow of the material into microcracks. This is shown in Figure 4.13b,c [62].

Relaxations and Shape Memories Mechanical deformations of synthetic materials can also be repaired by stress relaxation processes. Shape memory polymers (SMPs) are a group of responsive materials that have the ability to return from a deformed temporary shape to their original permanent shape induced by an external stimulus, such as temperature, electrical or magnetic field [63], solution [64], or light [65]. Similar to diffusion-induced self-healing, the T_g plays a significant role. At temperatures above the T_g, polymer chains display increased mobility, enabling the bulk material to flow, and, therefore, behave as viscous liquid with irreversible mechanical responses. At temperatures below the T_g, mobility is restricted and polymer chains behave like a glassy, elastic solid with reversible

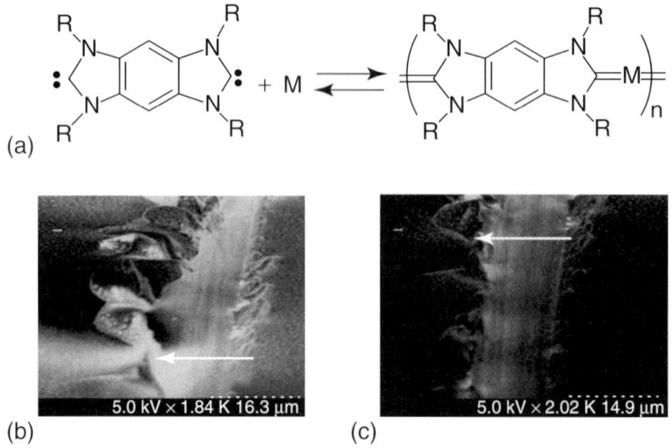

(a)

(b)   (c)

Figure 4.13 Schematic representation of dynamic equilibrium between a monomer species and organometallic polymer leading to healing (a) and scanning electron microscopy (SEM) images of the damaged polymeric film before (b), and after (c) exposure to 200° C, respectively [62].

mechanical responses. Translating macroscopic responses to micro- and molecular deformations is challenging and the mobility of polymer chains in the presence of external stimuli depends on the proximity of the T_g and the degree of cross-linking as well as entanglements present in the network [66]. These competing processes within polymeric networks resulting in chain deformation as a response to external stresses as well as chain contractions as an entropic response result in mass flow [67]. One example is the light-activated shape memory polymers (LASMPs), which utilize one wavelength of light (ν_1) for photo-cross-linking, while the second wavelength (ν_2) of light cleaves the photo-cross-linked bonds, which helps the material to reversibly switch between an elastomer and a rigid polymer by changing the T_g [65].

4.2.1.4 Millimeter-Level Repairs

Multiphase Systems The concept of self-healing thermosetting polymers as well as fiber-reinforced thermosets was adopted from civil engineering, where cementation composites were healed by release of chemicals within the crack, leading to curing in the presence of air [68, 69]. The concept inspired materials scientists to develop polymer matrices containing vessels (fibers or microcapsules) filled with repair chemicals. For example, as shown in Figure 4.14a, to construct polymer composites epoxy-amine or epoxy-vinyl ester, hollow fibers containing repair chemicals were embedded. When damage occurs, and when a crack propagates through the polymer matrix and breaks the fibers, repair chemicals are released in the form of low-viscosity adhesives or two-component resins or monomers dicyclopentadiene (DCPD) or 5-ethylidene-2-norboroene (ENB), which fill up the crack. In the presence of catalysts, cessation of the crack occurs via cross-linking reaction or ring-opening

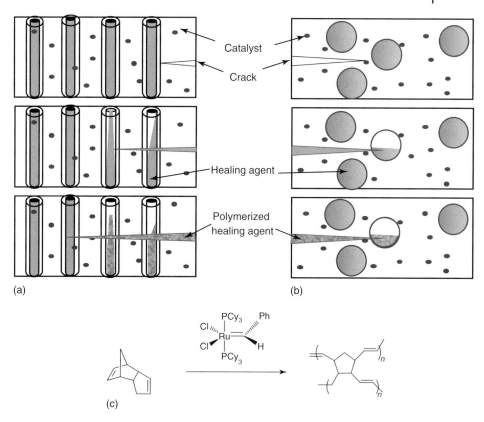

Figure 4.14 Schematic representation of self-repairing in polymer composites containing encapsulated hollow fibers (a) and microparticles (b), by ring-opening metathesis reaction of DCPD (c) [68–72].

metathesis polymerization (ROMP). The same concept was utilized to encapsulate microparticles filled with cross-linking agents. As shown in Figure 4.14b, when crack propagates and breaks the particle, cross-linking agents are released to heal the matrix. Again, the presence of catalysts in the matrix accelerates the process by ROMP reaction as shown in Figure 4.14c. The limitations of this concept are the mechanical integrity of the matrix, size, and diameter dependency of encapsulated vessels, challenges in manufacturing and dispersion, as well as kinetics of deactivation of the catalysts [70–72].

4.3 Summary

This chapter outlined recent advances in the repair of polymeric networks. Recent studies show that the main challenge is to develop new individual self-repairing

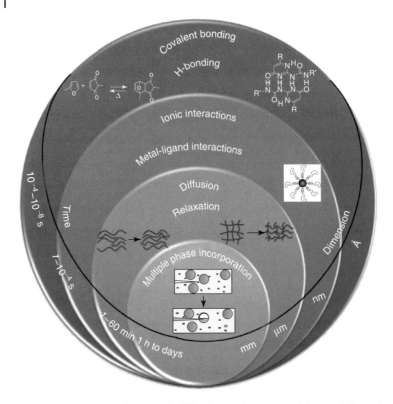

Figure 4.15 Schematic diagram of self-healing in the context of time and dimensions.

events that respond in an orchestrated manner to damages as well as are capable of superimposing with other events. Figure 4.15 depicts several overlapping events occurring at molecular and macroscopic levels, involving interplay between the environment (network) and repairing species. The snapshot of an ideal network shown in Figure 4.2 illustrates that the presence of signaling and repairing species built into the network is only one event, but to mimic biological species, signaling and repairs should occur as sequential events. There are many unexplored areas that will require systematic and sophisticated steps to develop multilevel self-repairing networks. To achieve these goals, analytical tools capable of measuring molecular events as well as macroscopic processes will be necessary.

References

1. Lindahl, T. (1993) *Nature*, **362**, 709.
2. Taylor, D. (2007) *J. Mater. Sci*, **42**, 8911.
3. Long, M.A., Stephane, Y., Crobel, Y., and Rossi, F.M.V. (2005) *Semin. Cell Dev. Biol.*, **16**, 632.
4. Han, R. and Campbell, K.P. (2007) *Curr. Opin. Cell Biol.*, **19**, 409.
5. Diegelmann, R.F. and Evans, C.M. (2004) *Front. Biosci.*, **9**, 283.
6. Sinha, P.R. and Hader, P.D. (2002) *Photochem. Photobiol. Sci.*, **1**, 225.

7. Paris, R., Lamattina, L., and Casalongue, C.A. (2007) *Plant Physiol. Biochem.*, **45**, 80.
8. Leon, J., Rojo, E., and Sanchez-Serrano, J.J. (2001) *J. Exp. Bot.*, **52**, 1.
9. Walter, M.W., Schadel-Randall, B., and Schadel, E.M. (1990) *J. Am. Soc. Hortic. Sci.*, **115**, 444.
10. Artschwager, E. (1927) *J. Agric. Res.*, **35**, 995.
11. Ibrahim, L., Spackman, V.M.T., and Cobb, A.H. (2001) *Ann. Bot.*, **88**, 313.
12. Biggs, A.R. (1985) *Phytopathology*, **75**, 1191.
13. Cussler, E.L. (1997) *Diffusion, Mass Transfer in Fluids Systems*, 2nd edn, Cambridge University Press.
14. Grinsted, R.A., Clark, L., and Koemnig, J.L. (1992) *Macromolecules*, **25**, 1235.
15. Qin, W., Shen, Y., and Fei, L. (1993) *J. Polym. Sci.*, **11**, 358.
16. De Gennes, P.G. (1971) *J. Chem. Phys.*, **55**, 572.
17. Ngai, K.L. and Plazek, D.J. (1990) *Macromolecules*, **23**, 4282.
18. Priestley, R.D., Ellison, C.J., Broadbelt, L.J., and Torkelson, J.M. (2005) *Science*, **309**, 456.
19. O'Connell, P.A. and McKenna, G.B. (2005) *Science*, **307**, 1760.
20. Jang, B.N. and Wilkie, C.A. (2005) *Thermochim. Acta*, **426**, 73.
21. Takeda, K., Unno, H., and Zhang, M. (2004) *J. Appl. Polym. Sci.*, **93**, 920.
22. Hay, A.S. (1967) *Adv. Polym. Sci.*, **4**, 496.
23. White, D.M. and Klopfer, H.J. (1970) *J. Polym. Sci., Part A*, **1**, 1427.
24. Yang, H. and Hay, A.S. (1993) *J. Polym. Sci.: Part A: Polym. Chem.*, **31**, 261.
25. Chung, C.M., Roh, Y.S., Cho, S.Y., and Kim, J.G. (2004) *Chem. Mater.*, **16**, 3982.
26. Paczkowski, J. (1996) in *Polymeric Materials Encyclopedia* (ed. J.C. Salamone), CRC Press, Boca Raton, FL, p. 5142.
27. Ramamurthy, V. and Venkatesan, K. (1987) *Chem. Rev.*, **87**, 433.
28. Egerton, P.L., Hyde, E.M., Trigg, J., Payne, A., Beynon, P., Mijovic, M.V., and Reiser, A. (1981) *J. Am. Chem. Soc.*, **103**, 3859.
29. Hasegawa, M., Katsumata, T., Ito, Y., Saigo, K., and Iitaka, Y. (1988) *Macromolecules*, **21**, 3134.
30. Craven, J.M. (1969) US Patent 3 435 003.
31. Chen, X., Wudl, F., Mal, A.K., Shen, H., and Nutt, S.R. (2003) *Macromolecules*, **36**, 1802.
32. Liu, Y.L. and Hseih, C.Y. (2006) *J. Poly. Sci. Part A*, **44**, 905.
33. Stevens, M. and Jenkins, A. (1979) *J. Polym. Sci.*, **17**, 3675.
34. Liu, Y.L., Hseih, C.Y., and Chen, Y.W. (2006) *Polymer*, **47**, 2581.
35. Liu, Y.L., Hseih, C.Y., and Chen, Y.W. (2007) *Macromolecular Chem. Phys.*, **208**, 224.
36. Liu, Y.L. and Hseih, C.Y. (2005) *J. Polym. Sci. Part A*, **44**, 905.
37. Burrows, F., Louime, C., Abazinge, M., and Onokpise, O. (2007) *Am-Euras. J. Agric. Environ. Sci.*, **2**, 103.
38. Ghosh, B. and Urban, M.W. (2009) *Science*, **323**, 1458.
39. Fredericks, J.R. and Hamilton, A.D. (1996) in *Comprehensive Supramolecular Chemistry*, Chapter 16 (ed. J.-M. Lehn), Pergamon, New York.
40. Beijer, F.H., Kooijman, H., Spek, A.L., Sijbesma, R.P., and Mejer, E.W. (1998) *Angew. Chem. Int. Ed. Engl.*, **37**, 75.
41. Sijsbesma, R.P., Bejer, F.H., Brunsveld, L., Fomer, B.J.B., Hirschberg, J.H.K., Lange, R.F.M., Lowe, J.K.L., and Mejer, E.W. (1997) *Science*, **278**, 1601.
42. Soentjes, S.H.M., Sijgesma, R.P., van Genderson, M.H.P., and Mejer, E.W. (2000) *J. Am. Chem. Soc.*, **122**, 7487.
43. Bosman, A.W., Sijbesma, R.P., and Meijer, E.W. (2004) *Mater. Today*, **7**, 34.
44. Feldmann, K.E., Mathew, J.K., De Greef, T.K.A., Mijer, E.W., Kramer, E.J., and Hawker, C.J. (2008) *Macromoleucles*, **41**, 4694.
45. Cordier, P., Tournilhac, F., Soulié-Ziakovic, C., and Leibler, L. (2008) *Nature*, **451**, 977.
46. Schubert, U.S., Eschbaumer, C., Hien, O., Andres, P.R., and Schubert, U.S. (2001) *Tetrahedron Lett.*, **42**, 4705.
47. Kersey, F.R., Loveless, D.M., and Craig, S.L. (2007) *J. R. Soc. Interface*, **4**, 373.

48. Gohy, J.-F., Lohmeijer, B.G.G., and Schubert, U.S. (2002) *Macromol. Rapid Commun.*, **23**, 555.
49. Valery, R. (2007) in *Ionomers as Self Healing Polymers in Self-Healing Materials: An alternative Approach to 20 Centuries of Materials Science* (ed. S. van der Zwaag), Springer, New York, p. 95.
50. Kalista, S. and Ward, T. (2007) *J. R. Soc. Interface*, **4**, 405.
51. Kalista, S.J., Ward, T.C., and Oyetunji, Z. (2007) *Mech. Adv. Mater. Struct.*, **14**, 391.
52. de Gennes, P.-G. (1980) *Hebd. Seances Acad. Sci.,Ser. B*, **291**, 219.
53. Prager, S. and Tirrell, M. (1981) *J. Chem. Phys.*, **75**, 5194.
54. Wool, R.P. and O'Connor, K.M. (1981) *J. Appl. Phys.*, **52**, 5953.
55. Jud, K., Kausch, H., and Williams, J.G. (1981) *J. Mater. Sci.*, **16**, 204.
56. Kim, Y.H. and Wool, R.P. (1983) *Macromolecules*, **16**, 1115.
57. Wang, P.P., Lee, S. and Harmon, J.P. (1994) *J. Polym. Sci., Part B: Polym. Phys.*, **32**, 1217.
58. Lin, C.B., Lee, S., and Liu, K.S. (1990) *Polym. Eng. Sci.*, **30**, 1399.
59. Zako, M. and Takano, N. (1999) *J. Intel. Mat. Sys. Str.*, **10**, 836.
60. Bleay, S.M., Loader, C.B., Hawyes, V.J., Humberstone, L., and Curtis, P.A. (2001) *Compos. A*, **32**, 1767.
61. Corten, C.C. and Urban, M.W. (2009) *Adv. Mater.*, **21**, 5011.
62. Williams, A., Boydston, J., and Bielawski, W. (2007) *J. R. Soc. Interface*, **4**, 359.
63. Mohr, R., Kratz, K., Weigel, T., Luca-Gabor, M., Moneke, M., and Lendlein, A. (2006) *Proc. Natl. Acad. Sci. U.S.A.*, **103**, 3540.
64. Leng, J., Haibao, L.V., Liu, Y., and Du, S. (2008) *Appl. Phys. Lett.*, **92**, 206105.
65. Lendlein, A. (2009) *Nature*, **434**, 872.
66. Porter, R.S. and Johnson, J.F. (1966) *Chem. Rev.*, **66**, 1.
67. Outwater, J.O. and Gerry, D.J. (1969) *J. Adhes.*, **1**, 290.
68. Dry, C. (1996) *Compos. Struct.*, **35**, 263.
69. Li, C.V., Lim, Y.M., and Chan, Y.W. (1998) *Compos. Part B*, **29B**, 819.
70. Trask, R.S., Williams, G.J., and Bond, I.P. (2007) *J. R. Soc. Interface*, **4**, 363.
71. Motuku, M., Vaidya, U.K., and Janowski, G.M. (1999) *Smart Mater. Struct.*, **8**, 623.
72. Kessler, M.R. and White, S.R. (2001) *Compos. Part A*, **32**, 683.

5
Stimuli-Driven Assembly of Chromogenic Dye Molecules: a Versatile Approach for the Design of Responsive Polymers

Brian Makowski, Jill Kunzelman, and Christoph Weder

5.1
Introduction

Polymers that change their properties in response to stimuli such as chemicals, heat, light, or electricity are of great interest because of their potential use in applications ranging from drug delivery to camouflage systems to artificial muscles [1]. Chromogenic polymers, which alter their absorption and/or fluorescence characteristics in response to an external stimulus, are particularly useful for sensor applications and perhaps represent the largest subgroup of stimuli-responsive materials [2–4]. Many different types of chromophores have been used in this context, ranging from pH-sensitive dyes [5] to fluorescent ligands, which change their properties upon binding to metals [6]. Recently, a versatile new concept to stimuli-responsive fluorescent polymers has emerged [7, 8], which exploits the fact that ordered molecular assemblies sometimes have optical properties that are vastly different from those of their molecular constituents. The approach exploits the stimulus-driven self-assembly or dissociation of nanoscale aggregates of excimer-forming dyes in a host polymer. This transition results in a pronounced change of the material's fluorescence color (Figure 5.1). The approach was further extended to "aggregachromic" dyes, which change their absorption color upon self-assembly. Since comparably few chromogenic polymers are available that respond in a useful way to mechanical stress [2, 9], the framework was initially exploited to create a new family of mechanochromic materials. The absorption or fluorescence color of these materials changes upon deformation, thus providing visible warning signs before mechanical failure occurs [10]. In the meantime, the framework has been used by several research groups for the design of a broad range of sensor materials that are useful for the detection of temperature history, chemical exposure, as well as more complex combinations of stimuli, such as seen in shape-memory materials. This chapter provides an overview of the rapid development of this new class of responsive polymers.

Handbook of Stimuli-Responsive Materials. Edited by Marek W. Urban
Copyright © 2011 WILEY-VCH Verlag GmbH & Co. KGaA, Weinheim
ISBN: 978-3-527-32700-3

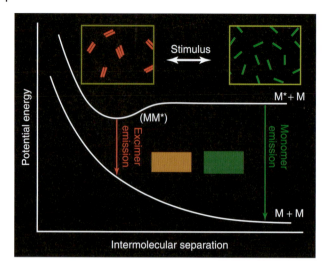

Figure 5.1 Schematic representation of the potential energy level of a pair of molecules M (electronic ground state) and M* (electronically excited singlet state), illustrating the formation of an energetically favorable excimer state (MM*) at appropriate intermolecular separation. The drawings (top) and pictures of polyethylene films (bottom) comprising a cyano-OPV dye in the dispersed (right, green) and aggregated (left, red) state illustrate the operating principle of responsive polymers that exploit the stimuli-driven assembly or disassembly of such molecules.

5.2
Excimer-Forming Sensor Molecules

Excimers, first reported more than 50 years ago by Förster and Kasper [11, 12], are complexes in which a fluorescent molecule in an electronically excited state (M*) shares the absorbed energy with a juxtaposed dye molecule in the electronic ground state, also referred to as the *monomer* (M). A schematic representation of this process, shown in Figure 5.1, reveals that excimer (MM*) formation depends on the intermolecular distance between M* and M; at the appropriate intermolecular distance, the potential energy of the excimer complex MM* is lower than the energy of the separated molecules, causing the excimer to emit light of lower energy (longer wavelength) than the solitary species.

Excimer formation has been found to be most prominent for small organic aromatic molecules such as benzene [13], *p*-xylene [14], naphthalene [15], anthracene [16], pyrene [17], perylene [18], stilbene [19], and others [20]. Some polymers that contain aromatic groups, such as polystyrene [21] and poly(ethylene terephthalate) (PET) [22], and polynucleotides such as cytosine and thymine [23] have also been shown to exhibit excimer emission. Excimer emission is, in fact, widely observed in aromatic hydrocarbons [20].

One of the most intensively studied examples of excimer-forming molecules is pyrene. In dilute cyclohexane solution (10^{-4} M), the emission spectrum contains a

	R1	R2
C1-YB	H	OMe
C1-RG	OMe	OMe
C12-YB	H	$OC_{12}H_{25}$
C18-YB	H	$OC_{18}H_{37}$
C18-RG	OMe	$OC_{18}H_{37}$
C12OH-RG	OMe	$OC_{12}H_{24}OH$

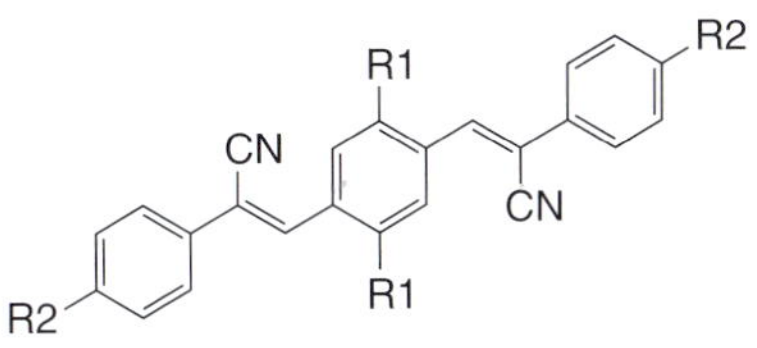

(a)

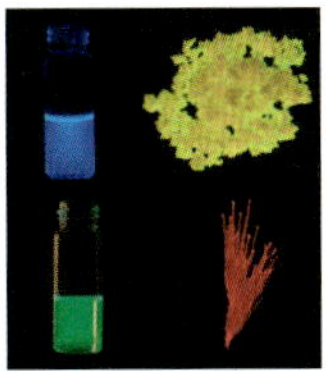

(b)

Figure 5.2 (a) Chemical structures of some of the cyano-OPVs investigated as sensor molecules [24, 35, 36]. (b) Pictures of solutions and crystals of C1-YB (top) and C1-RG (bottom). The pictures are reproduced with permission from Ref. [34].

well-resolved vibronic structure with maxima between 350 and 450 nm corresponding to the emission of individual pyrene molecules [17]. As the concentration of the dye is increased to concentrations above 10^{-3} M, a broad red-shifted emission band with maximum around 475 nm appears, which increases in intensity as a function of dye concentration. This change is due to the increased likelihood of an encounter between an excited molecule and a ground-state molecule at high dye concentration. The excited state lifetime of the excimer species is significantly longer than that of the monomer, and therefore, fluorescence lifetime measurements are typically conducted to confirm the presence of excimers [17].

While many different types of excimer-forming dyes are useful in the context of the here-discussed sensor polymers (vide infra), much of the published work on the subject has been based on cyano-substituted oligo(*p*-phenylene vinylene)s (cyano-OPVs, Figure 5.2), which exhibit particularly attractive features [24–33]. Cyano-OPVs often exhibit a strong tendency toward excimer formation and display remarkably large differences (up to 140 nm) between the emission maxima of molecular solutions and solid state [24]. Their good thermal stability allows one to incorporate them into polymers by conventional melt-processing techniques [34–36]. Further, their optical characteristics and solubility in different host polymers are readily controlled via the nature of substituents attached to the aromatic rings. Figure 5.2 shows six selected examples, which serve to illustrate this aspect. If the central benzene ring is void of any substituents (R1=H) and the peripheral benzene rings carry alkyloxy groups (R2=OR), the emission color changes from blue to yellow upon aggregation – hence, simple dyes from this series have been dubbed CX–YB, where X denotes the length of the peripheral alkyl groups. If two alkyloxy groups are introduced to the central benzene ring, the HOMO–LUMO (highest occupied molecular orbital–lowest unoccupied molecular orbital) gap is reduced and the color scheme changes to green (monomer) and red (excimer); concomitantly, the acronyms used for these dyes are C1-RG, C18-RG, and so forth (RG for red-green). Depending upon the length of the alkyloxy group (R2), the solubility and aggregation behavior in a polymer host can be tailored.

Many members of the family of cyano-OPVs form excimers due to the specific packing of the planarized molecules in a cofacial arrangement (Figure 5.3) [37].

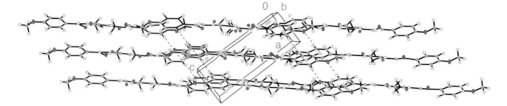

Figure 5.3 Crystal structure of C1-RG. The electron-deficient cyano-vinylene moieties of one molecule overlay with the electron-rich central ring of its neighbor. Reproduced with permission from the supplemental information to Ref. [37].

The crystal structure of C1-RG (cf. Figure 5.2) is characterized by a parallel arrangement of neighboring molecules, which assemble so that the electron-deficient cyano-vinylene moiety of one molecule is overlaid with the electron-rich central ring of its neighbor (Figure 5.3). Such assemblies display excimer emission, which is red shifted by as much as 140 nm relative to monomer emission. Similar observations were made for other RG dyes, as well as for C1-YB, which lacks the electron-donating alkoxy groups in the central ring. Quite surprisingly, increasing the length of the peripheral alkyloxy groups in the YB series to dodecyl (C12-YB) or octadecyloxy (C18-YB) leads to crystal structures in which blue monomer emission is dominant [37]. This suggests that balancing $\pi-\pi$ and aliphatic interactions in cyano-OPVs is possible through the variation of electron density in the central ring and the length of peripheral aliphatic tails. Some of the cyano-OPV dyes investigated by Weder and coworkers, such as C18-RG [38], also exhibit pronounced optical absorption changes upon self-assembly, due to charge-transfer interactions or conformation changes. Such "aggregachromic dyes" extend the above-discussed framework that involved excimer-forming dyes, in that these molecules allow the design of polymer sensors in which assembly leads to a color (not fluorescence color) change. It should be noted that the solid-state structure of these molecules is somewhat difficult to predict, and that the rational design of new dye molecules has remained a veritable challenge.

In solution, excimer formation is frequently a dynamic, diffusion-controlled process [39]. By contrast, diffusion of the dye molecules in solid polymers is usually slow compared to the lifetime of the excited states, and excimers often predominantly arise from preformed ground-state aggregates [40]. It has long been recognized that the formation of excimers in polymers comprising a fluorescent probe can be used to extract structural information on the molecular (e.g., conformation and dynamics of macromolecules in solution [41]) as well as supramolecular level (e.g., miscibility of polymer blends [42], morphology [43], and distribution of dopant-site sizes [44]). The spectroscopic detection of excimers was also employed to sense the aggregation of guest molecules such as pyrene in polystyrene [45], polyethylene [46], and other host polymers. However, the idea to exploit the stimulus-driven self-assembly or dissociation of excimer-forming dyes in a host polymer as a sensing mechanism has only emerged quite recently [8].

5.3
Fluorescent Mechanochromic Sensors

The above description reflects that altering the molecular packing of organic compounds is one possibility to impart dynamic fluorescent behavior to such materials [10]. If this can be achieved by applying pressure, the fluorophore is said to display piezochromic luminescent characteristics [47], in analogy to "normal" piezochromic organic compounds, which change their absorption properties upon compression. In the latter case, the pressure-induced color change is often associated with a metastable state that is maintained at ambient conditions but reverts to the original state upon heat- or solvent-induced recrystallization. While many organic compounds that change their absorption characteristics under pressure are known, piezochromic fluorophores are quite rare [10].

Liquid crystalline dendritic structures that contain anthracene or pyrene cores (Figure 5.4) represent one illustrative example of such piezochromic optical sensors [48, 49]. These compounds are known to form micelles due to the flexible alkyl chains that surround the π-conjugated core. The micelles form disordered structures, due to the difference between intermolecular hydrogen bonding interactions (which take place between the amide groups and are approximately 5.0 Å in length) and the intermolecular π-stacking interactions (with an average distance of 3.5 Å between conjugated cores). The phase that naturally occurs in these molecules is that of segmented columns in a cubic lattice and it exhibits yellow emission due to excimer formation (Figure 5.4). The cubic phase is metastable and contains segmented columns approximately 20 molecules in length. Upon shearing, the cubic phase changes from a segmented columnar structure to a nonsegmented columnar phase, which is accompanied by a color change from yellow to blue. The color change is due to the change in molecular arrangement of the luminescent cores, which are no longer able to form excimers (Figure 5.4).

Cyano-OPVs such as C12-YB and C18-YB (Figure 5.2) represent another family of piezochromic, liquid–crystalline fluorophores, whose emission properties can be reversibly and repeatedly switched from monomer to excimer fluorescence upon compression or quenching the compounds from a nematic state. Heating these solids to enter the smectic regime causes full restoration of the monomer emission. In comparison to the properties of other dyes of the CX-RG series, the piezochromic fluorescent characteristics of C18-YB and C12-YB suggest that balancing $\pi-\pi$ and aliphatic interactions in cyano-OPVs is possible through the variation of electron density in the central ring and the length of peripheral aliphatic tails. This approach allows the design of fluorophores, which can adopt different polymorphs and exhibit piezochromic behavior, if one of the polymorphs allows the formation of excimers.

The original idea to exploit excimer-forming dyes as mechanochromic transducers *in polymers* (as opposed to neat low-molecular-weight compounds) was inspired by the work of Trabesinger *et al.* [50] who showed that aggregated poly(phenylene ethynylene) guest molecules in a polymeric matrix could be dispersed by simple deformation of the material. This deformation transformed the material from a

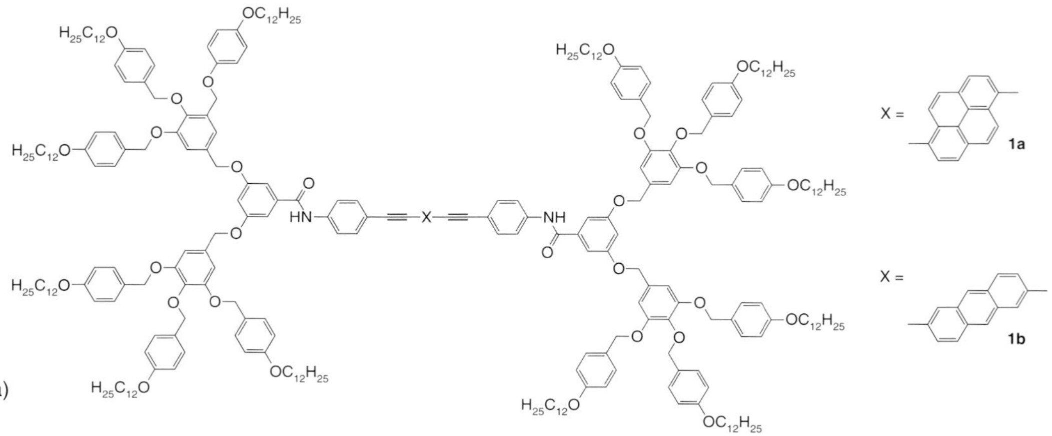

Figure 5.4 (a) Chemical structure of liquid crystalline dendritic compounds that display piezochromic luminescence. (b) Schematic showing the effect of mechanical shearing on the liquid crystal phase of compound **1a**, which is converted from a metastable cubic state to a stable columnar state, concomitant with a change of emission color. (c) Schematic showing the loss of $\pi-\pi$ stacking interactions that occurs upon mechanical shearing and conversion from the cubic to the columnar phase. The figure parts are reproduced with permission from Refs. [48, 49].

5.3 Fluorescent Mechanochromic Sensors | 123

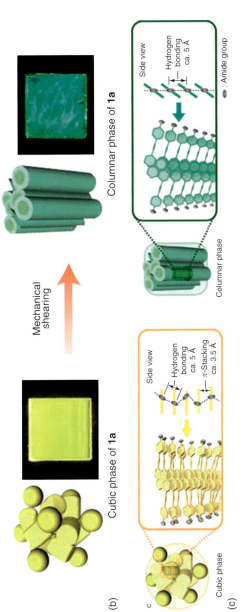

Figure 5.4 *continued*.

nanophase separated blend to a molecularly mixed blend. Statistical analysis was used to show that the cluster size of the conjugated guest molecules decreased as the matrix was deformed, first into smaller clusters and finally into individual molecules. It was therefore speculated that a similar mechanism involving aggregates of excimer-forming dyes in a ductile host polymer could allow the design of sensory polymers with mechanochromic response. The approach was first reduced to practice by Löwe and Weder who integrated C1-RG and C1-YB (Figure 5.2) into linear low-density polyethylene (LLDPE) via a guest-diffusion technique [51]. This process involves the immersion of a polymer film in a dye solution for a given amount of time during which the polymer is impregnated with dye and solvent. Upon removal of the film from the solution, the solvent is evaporated and the dye remains in the polymer host. At sufficiently high concentrations, the dye molecules assemble into nanoscale aggregates, which display excimer emission. It was demonstrated that tensile deformation of blends, which were initially dominated by excimer emission, led to significant changes in their emission properties due to a break up of dye aggregates, and stretched films displayed predominantly monomer emission [51]. Weder's group rapidly expanded this concept by producing blends through melt-processing dye and polymer, creating new cyano-OPVs and investigating numerous polymers as host matrices (Figure 5.5). The polymers explored include a number of different polyethylenes (PEs) [34, 36, 52], poly(ethylene terephthalate) (PET) and poly(ethylene terephthalate glycol) (PETG) [35, 53], fluoropolymers [54], thermoplastic polyurethanes (TPUs) [36], and natural rubber [55]. By employing dyes that form ground-state aggregates (i.e., C18-RG, vide supra), the Weder group was able to extend the concept to materials that change their absorption (not fluorescence) color [38, 53]. Other researchers have adapted this strategy and extended it to a variety of dyes (e.g., cyano-containing poly(phenylene ethynylenes) [56], perylenes [57], CdS nanoparticles [58], and bis(benzoxazolyl)stilbenes [59]) in several polymer matrices, most notably polyesters and polyolefins.

The various studies on excimer-based mechanochromic materials have led to the development of a good understanding of the underlying mechanisms and provide a good picture of how blends with the desired structural design can be reliably produced by conventional fabrication techniques. Some of the important aspects include dye concentration, dye solubility in the host polymer, dye aggregate size, and effective break up of the dye aggregates. The dye concentration in the material must be sufficiently high to cause aggregation of dye molecules and lead to the

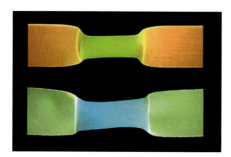

Figure 5.5 Picture of melt-blended films of LLDPE and 0.18% w/w C1-RG (top) and C1-YB (bottom) stretched at room temperature to a draw ratio of 500%. Pictures were taken under excitation with UV light of a wavelength of 365 nm. The melt-processing technique produces films of much higher optical brightness and contrast than guest-diffusion techniques. Reproduced with permission from Ref. [34].

formation of static excimers. Appropriate solubility of the dyes in a particular matrix polymer can be tailored by changing the nature of the substituents attached to the dye core, for example, the length of (aliphatic) tails. Samples based on an LLDPE host were initially produced by rapidly quenching the melt. In the case of blend films comprising C1-RG, the intensity of the excimer emission band increased upon storage under ambient conditions, as exemplarily shown in Figure 5.6a for an LLDPE/C1-RG blend film. As can be seen from the figure, the rate of this process decreases sharply as a function of time and appears to approach zero after a time frame of a few months. Figure 5.6b shows that the speed of this aggregation process can be increased by immersing the quenched blend films in hexane (a poor solvent for the dye, but one that swells LLDPE well). In contrast to the quenched samples, the preparation of samples by slowly cooling the melt results in large-scale phase

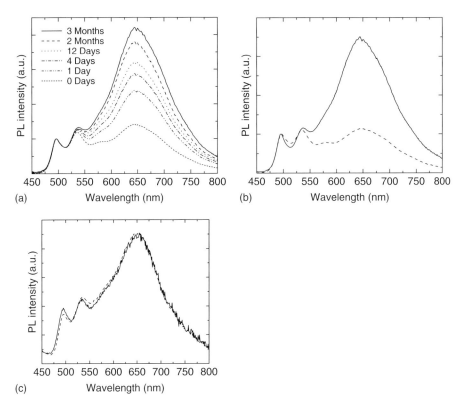

Figure 5.6 (a) Photoluminescence(PL) emission spectra of a 0.18% w/w LLDPE/C1-RG blend film as a function of storage time (as indicated) under ambient conditions. (b) Emission spectra of a 0.18% w/w LLDPE/C1-RG blend film freshly prepared (dashed) and after swelling the film for 15 min in hexane and subsequent drying (solid). (c) Emission spectra of a quenched 0.4% w/w LLDPE/C18-RG blend film initially after quenching (solid line) and after aging under ambient conditions for 50 days (dashed line). All spectra were normalized to the intensity of the monomer peak. Reproduced with permission from Refs. [34, 52].

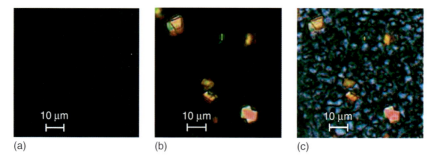

Figure 5.7 Cross-polarized optical micrographs documenting the cooling process of a blend of LLDPE and 0.46% w/w C1-RG at (a) 180 °C, (b) 135 °C, and (c) 100 °C. Reproduced with permission from Ref. [34].

separation between the dye and the PE matrix (Figure 5.7) [34]. This large-scale phase-separated morphology is not suitable, since an efficient break up of the dye aggregates is not possible. The aging phenomenon and dye aggregate size were studied in greater detail by investigating the effect of host polymer crystallinity [52]. The rate at which C1-RG aggregates and the extent of aggregation increase with decreasing polymer crystallinity. This observation is in agreement with the well-established increase of the fractional free volume of the noncrystalline component of PE with decreasing crystallinity and reflects an increase of the dye's translational mobility [60]. Interestingly, dye aggregation appears to be instantaneous in all PE samples comprising C18-RG (Figure 5.6c), leading to the formation of many small aggregates. It was shown that this is due to enhanced nucleation of C18-RG (viz-a-viz C1-RG) [52]. The comparably inefficient nucleation of C1-RG appears to limit the number of dye aggregates formed, leading to the slow growth of larger aggregates via diffusion of the dye molecules. As expected, the smaller dye aggregates in LLDPE/C18-RG blend films are more easily dispersed upon deformation than LLDPE/C1-RG and result in a more substantial emission color change [52]. Thus, besides the length of the solubilizing tails, the dye concentration has an important influence on the aggregate size, which is crucial to creating a mechanochromic response, since the dye aggregates must be small enough to be dispersed during the deformation process.

In situ optomechanical experiments with cyano-OPV containing fluoropolymers and PEs have shown that the emission color change of blends upon deformation matches nicely with the shape of the stress–strain profiles for the samples [52, 54]. Blend films based on several different polyethylene matrices were shown to exhibit a steep increase in color change upon yielding, a moderate increase during neck propagation, and a slightly steeper increase during strain hardening (Figure 5.8) [52]. It was also shown that the magnitude of the emission color change, and therewith the extent of aggregate break up, increases with decreasing strain rate [52]. Investigation of the effect of polymer crystallinity on the mechanochromic response of PE/C18-RG blends revealed a larger extent of the emission color change upon deformation for the higher crystallinity PEs. From a mechanistic aspect, it appears that the ability of the polymer host to

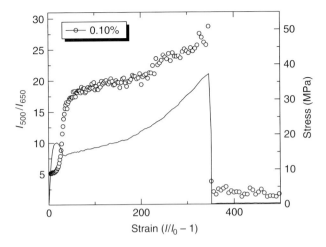

Figure 5.8 Ratio of monomer to excimer emission I_{500}/I_{650} (symbols, measured at 540 and 650 nm), and tensile stress (line) as a function of strain for a blend of LLDPE and 0.1% w/w C18-RG. Adapted with permission from Ref. [52].

disperse dye aggregates upon deformation is primarily related to the plastic deformation process of the PE crystallites. These findings are consistent with the mechanochromic response of several polyesters that have also been investigated [53]. Blends created from semicrystalline PET exhibit a significant color change upon deformation, whereas the mechanochromic response of blends based on the fully amorphous PETG is much less pronounced [35].

A comparative study revealed a surprising difference between the characteristics of cyano-OPVs and fluoropolymers, on the one hand, and polyolefins, on the other [54]. C18-RG, featuring long alkyl tails, provided a superior mechanochromic response in PE matrices in comparison to C1-RG. This was explained by the higher nucleation rate of C18-RG, leading to smaller dye aggregates. Quite surprisingly, the two dyes have essentially opposite behavior in fluoropolymers: all blends containing C18-RG featured large dye aggregates and displayed no mechanochromic response, in contrast to the C1-RG containing fluoropolymers, in which dye aggregates were smaller and the mechanochromic response was pronounced. This suggests that the nucleation of the luminogenic cyano-OPVs investigated is not intrinsic to the dyes, but is largely governed by the polymer host into which they are embedded. The mechanochromic fluoropolymer/C1-RG blends display a pronounced color change upon deformation. *In situ* optomechanical measurements have shown that the mechanochromic effect occurs primarily during plastic deformation and that the mechanically induced dispersion of the dye aggregates becomes more pronounced as the crystallinity of the matrix polymer increases. These findings correlate with previous work in polyolefin matrices and show that the underlying mechanism–dispersion of small aggregates of the excimer-forming sensor molecules upon plastic deformation of polymer crystallites – can be exploited in a range of semicrystalline polymer hosts.

The above-discussed examples of cyano-OPV-based mechanochromic polymer blends involved ductile polymers that undergo irreversible plastic deformation. To explore whether the sensing scheme can also be exploited in elastomers, which reversibly change their fluorescence color as a function of applied strain, thermoplastic polyurethanes(TPUs) that comprise cyano-OPVs were investigated [36]. The study involved both physical blends of a commercial TPU and small amounts of various cyano-OPVs, as well as copolymers produced by the covalent incorporation of a cyano-OPV into TPUs based on poly(tetramethylene glycol) (PTMG), butanediol (BDO), and 4,4'-methylene-bis(phenyl isocyanate) (MDI). Only the latter approach afforded elastomers that in their relaxed state display predominantly excimer emission, but exhibit a significant and largely reversible fluorescence color change upon deformation. Binary blends of Texin 985, an aromatic polyether-based polyurethane based on PTMG, BDO, and MDI, and between 0.05 and 0.4% w/w C1-RG or C18-RG, were prepared by melt mixing. Blends with low concentrations of C1-RG displayed green fluorescence characteristic of monomer emission. Interestingly, when quenched blend films containing C1-RG were strained to a draw ratio of circa 500%, released, and subsequently stored at ambient conditions, the emission spectrum rapidly developed a dominant, unstructured red band centered at 625 nm characteristic of excimer emission within a few hours after stretching. This deformation-triggered aggregation is most intriguing, and such materials may find use in "one-time-stretch" sensing applications. Although the molecular mechanism is still unclear, mechanical deformation appears to move the dye from the hard blocks to the soft-segment-rich domains, where high translational mobility and low solubility lead to rapid phase separation. All blends of C18-RG and Texin 985 (0.05–0.4% w/w) displayed almost exclusively excimer emission due to a lower solubility than C1-RG in polyurethanes. Although blends prepared with the appropriate thermomechanical history displayed phase separation and excimer formation, solid-state tensile deformation of the samples displayed only minor changes upon deformation. This reflects either that the stress experienced by the dye aggregates in these blends is insufficient for their dispersion or that the aggregate size is simply too large to allow for effective dispersion. It appears that large free volume, low crystallinity, and the absence of nucleation sites favor the formation of large sensor dye aggregates and, thus, stifle adequate strain-induced dispersion of these sensor molecules if they are incorporated through physical blending.

Covalent incorporation of a difunctional, ω-hydroxy-functionalized dye from the RG family (1,4-bis(α-cyano-4(12-hydroxydodecyloxy)styryl)-2,5-dimethoxybenzene, C12OH-RG) into the polymer backbone was performed in order to increase the forces experienced by the dye residues during strain and to decrease the chromophore's translational mobility [36]. The hard segment to soft segment (HS : SS) ratio and the dye content were systematically varied. All compression-molded, quenched polymer films containing C12OH-RG were orange and displayed a combination of monomer and excimer emission. The monomer emission decreased with increasing dye concentration and with increased HS : SS ratio as expected. All of the

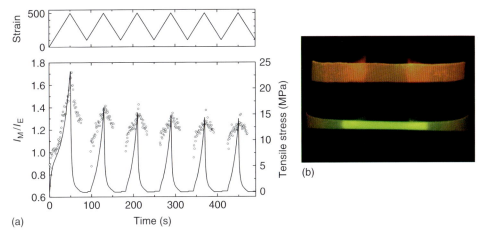

Figure 5.9 (a) Ratio of monomer to excimer emission I_M/I_E (circles, measured at 540 and 650 nm) and tensile stress (solid line) under a triangular strain cycle between 100 and 500% at frequency of 0.0125 Hz. The sample was a mechanochromic elastomer made by integrating C12OH-RG into a TPU backbone. (b) Pictures of the same polymer in unstretched (top) and stretched (bottom) states. Reproduced with permission from Ref. [36].

cyano-OPV-containing TPUs changed their emission color upon mechanical treatment, in the best compositions from an orange-red, excimer-dominated-emission to a green fluorescence that is characteristic of molecularly dispersed chromophores (Figure 5.9). Thus, the covalent incorporation of C12OH-RG into the polyurethane backbone indeed allows for the production of chromogenic elastomers, which change their emission color as a function of applied strain. *In situ* optomechanical studies in which samples were exposed to a cyclic triangular strain pattern showed that the fluorescence color nicely mirrored the reversible stress–strain response of the polymer. Thus, the covalent incorporation of C12OH-RG into the TPU backbone has led to materials that display a very pronounced mechanochromic fluorescence effect in which the emission color is largely coupled to the applied strain.

5.4
Thermochromic Sensors

The phase separation of initially molecularly mixed blends of excimer-forming dyes and glassy host polymers or semicrystalline host polymers with a glass transition in a temperature regime of interest is another versatile sensing mechanism that is useful for the fabrication of threshold temperature sensors or time–temperature indicators (TTIs, Figure 5.10). In this case, the sensing scheme relies on kinetically trapping thermodynamically unstable molecular mixtures of the sensor dyes in the glassy amorphous phase of the polymer by melt-processing and rapid quenching below the glass transition temperature (T_g). Subjecting these materials to temperatures above

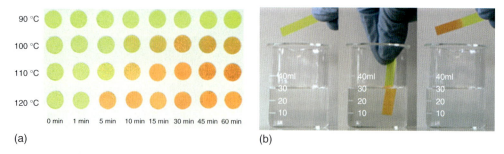

Figure 5.10 (a) Picture of initially quenched 1.1% w/w PETG/C18-RG blends after annealing as indicated. (b) Pictures of an initially quenched 3.1% w/w PETG/C18-RG blend immersed in hot (~100 °C) silicon oil for ~1 min. Reproduced with permission from Ref. [38].

T_g leads to permanent and pronounced changes of their emission spectra, as a result of phase separation and excimer formation. Binary blends of excimer-forming and/or aggregachromic dyes (primarily cyano-OPVs, Figure 5.2, and also other suitable such as 4,4′-bis(2-benzoxazolyl)-stilbene) and a host polymer were produced by melt mixing and subsequently compression molding at the temperature at which they were extruded and rapidly quenching to 0 °C to prevent phase separation between the polymer host and the dye. Initial work investigated poly(methyl methacrylate) (PMMA) and poly(bis-phenol A carbonate) (PC) as prototypes of glassy amorphous matrix materials [61]. The approach was expanded to include PETG as another example of a glassy amorphous polymer and PET as an example of a polar, semicrystalline polymer [35, 38]. The influence of polymer T_g (which is used to define the threshold temperature) was investigated by using a series of poly(alkylmethacrylate) copolymers, in which the composition was varied to vary T_g between 13 and 108 °C [62]. To address the lack of TTIs capable of operation at high temperatures, TTIs that operate in the range of 130–200 °C were recently created by incorporation of dyes into ethylene/norbornene copolymers with high glass transition temperatures [63, 64].

The variety of materials investigated has demonstrated the broad applicability of this approach to create easily tailored threshold temperature sensors and TTIs. It is important to use a dye with appropriate solubility in the host polymer, namely, one that is sufficiently soluble to allow quenching in a dispersed state below T_g, but which will phase-separate upon annealing above T_g. The influence of annealing temperature and dye concentration with respect to aggregation speed has been studied for a number of polymer/dye systems (cf. Figure 5.10 for a visual representation of a system based on the aggregachromic dye C18-RG) [34, 35, 38, 63]. For a quantitative analysis, the relative contributions of monomer (I_M) and excimer (I_E) emission intensities can be extracted from emission spectra as a function of temperature and annealing time. The ratios I_M/I_E from data acquired at different annealing temperatures and different dye concentrations can then be evaluated as a function of annealing time. Data for a series of 0.9% w/w PET/C18-RG blend

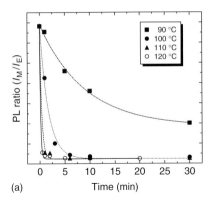 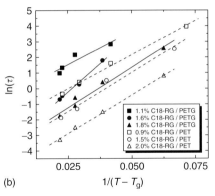

Figure 5.11 (a) I_M/I_E extracted from the emission spectra of a 0.9% w/w C18-RG/PET blend at 100°C as a function of annealing time. (b) Arrhenius plot expressing the temperature dependence of $\ln(\tau)$ of C18-RG/PET and C-18-RF/PETG blends as a function of dye concentration (indicated in the graph in % w/w). Reproduced with permission from Ref. [35].

films annealed between 90 and 120 °C are shown as an example in Figure 5.11. The aggregation kinetics for any given temperature and concentration were shown to be well described by a single exponential function with a characteristic rate constant τ. For any given composition, $\ln(\tau)$ scales linearly with $1/T$ or $1/T - T_g$ (Figure 5.11). Detailed studies have shown that the kinetics can be minutely controlled via the concentration of the dye (more dye = faster aggregation) and the molecular size of the latter (longer dye = slower aggregation). Thus, the aggregation process is predictable and easily controlled via the T_g of the host polymer and the concentration and molecular length of the dye. For any given composition, the fluorescence color of the sample is characteristic of its thermal history.

While the above scheme is uniquely suited for the design of TTIs, it is also feasible to create "real-time" temperature sensors on the basis of controlled assembly of chromogenic systems. One illustrative example is the polyphenylene vinylene poly(2,5-bis[3-N,N-diethylamino)-1-oxapropyl]-1,4-phenylenevinylene) (DAO-PPV) shown in Figure 5.12 [65]. This polymer has been shown to exhibit a significant emission color change when heated in a 1,2-dichlorobenzene solution (Figure 5.12). At low temperatures, the solution fluoresces red, which has been attributed to polymer aggregates being formed at low temperature. As the temperature is increased to 40 °C, the fluorescence color starts changing, turning orange at 70 °C and greenish at 100 °C. Detailed spectroscopic studies have shown that this fully reversible transition is caused by the disassembly of the dye molecules upon heating.

Another interesting example for the use of real-time temperature sensors based on excimer-forming is a multifunctional shape-memory polymer (SMP) with built-in temperature sensing capabilities (Figure 5.13) [66]. This material was prepared by incorporating a fluorescent, chromogenic cyano-OPV dye into a

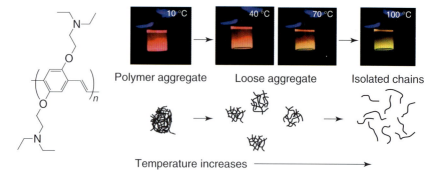

Figure 5.12 Chemical structure of the thermoresponsive DAO-PPV and pictures (taken under illumination with UV light) that document the temperature-induced color changes of this polymer in a solution of 1,2-dichlorobenzene. Reproduced with permission from Ref. [65].

Figure 5.13 Photograph (under illumination by 365 nm UV light) depicting the recovery of a shape-memory sensor with built-in temperature sensor from temporary shape (spiral) to the permanent shape (rod) [66]. The sample was immersed in silicon oil at ~80 °C. Reproduced with permission from Ref. [66].

cross-linked poly(cyclooctene) (PCO) matrix via guest diffusion. The dye concentration was chosen to allow for self-assembly of the dye upon drying, resulting in the formation of excimers. Exposure of the phase-separated blend to temperatures above the melting point (T_m) of the PCO leads to dissolution of the dye molecules and, therefore, causes a pronounced change of their absorption and fluorescence color. The optical changes are reversible, that is, the aggregate absorption and emission are restored upon cooling below T_m. The color is dictated by the phase behavior and is independent of the mechanical state of the SMP. Thus, the effect allows one to monitor reaching of the set/release temperature of the polymer has been reached by observing the color change.

5.5
Chemical Sensing with Excimer-Forming Dyes

Kinetically trapped, thermodynamically unstable mixtures of excimer-forming dyes in glassy amorphous or semicrystalline host polymers cooled below T_g can also be used as the basis for simple chemical detection. In this case, plasticization of the polymer matrix upon exposure to a chemical, that is, lowering T_g, triggers the aggregation of the dye molecules. This general concept was first used for a novel type of chromogenic *humidity-sensing* polymer [67]. Humidity is one of the most commonly measured physical quantities and is of great importance in many contexts [68]. Consequently, a broad range of transduction techniques have been exploited for water sensors. Systems that display reversible color changes include charge-transfer forming dyes [69], solvatochromic dyes [70], chromogenic polymers [71], and inorganic salts [72]. The self-assembly-based moisture sensors described here rely on a different mechanism, which makes their properties fundamentally different from the known systems: they display an irreversible color change upon exposure to moisture, which makes them useful as "moisture history" indicators. The approach combines the above-summarized method of kinetically trapping thermodynamically unstable mixtures of a host polymer and chromogenic sensor molecules below T_g with the well-known concept of plasticizing polymers with chemical vapors to trigger a response [73]. Thus, a new water-sensing polymer system was prepared by quenching molecularly mixed, supersaturated blends of 1,4-bis(α-cyano-4(12-hydroxydodecyloxy)styryl)-2,5-dimethoxybenzene (C12OH-RG), a chromogenic, cyano-substituted OPV dye, and a hygroscopic polyamide to below T_g. Exposure to water indeed plasticized the polyamide, triggered the self-assembly of sensor molecules, and led to the expected irreversible (fluorescence) color changes (Figure 5.14). Initial kinetic experiments reflect a behavior that mirrors the response of the thermally triggered systems discussed

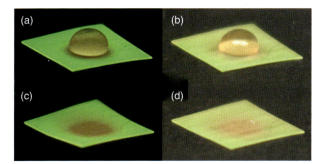

Figure 5.14 Pictures of quenched 1% w/w PA/C12OH-RG blend film exposed to a water droplet taken under illumination with (a) UV light (365 nm) and (b) ambient light. The same sample is shown after removal of water under illumination with (c) UV light (365 nm) and (d) ambient light. Reproduced with permission from Ref. [67].

above: as expected, the switching from monomer to excimer emission is much faster for samples exposed to higher humidity levels and the color changes appear again to be well described by single exponential functions.

The framework was recently also adapted to create epoxy-based, thermosetting epoxy resins, as examples for thermosetting polymers with integrated sensing capabilities [74]. Such resins are widely used for decorative, protective, and/or functional coatings [75]. Protective epoxy coatings are used in many industries due to their excellent chemical resistance, tenacious adhesion, and toughness [76]. The coating not only provides mechanical protection of the underlying substrate but also acts as a chemical barrier. The coating's properties can be compromised by exposure to excessive heat or penetrating chemicals; therefore, the integration of sensing capabilities into such materials is of significant technological interest. Through a systematic study of a relevant model system, it was shown that multifunctional thermoset epoxy coatings with thermal- and chemical-sensing capabilities could be designed on the basis of the aggregachromic sensor approach. The materials investigated here were prepared by dissolving a chromogenic, fluorescent cyano-OPV dye into a cross-linked epoxy resin by reacting dye/monomer mixtures at high temperature and quenching the cured polymer below its glass transition temperature (Figure 5.15). Exposure of these materials to chemical solvents can decrease the materials' T_g through plasticization, which leads to aggregation of monomer to excimer (Figure 5.16). As in the case of the above-described thermal and water sensors, the response is well described by standard kinetic models and can be minutely controlled via the chemical structure and crosslink density of the resin and the dye content. In view of the design simplicity and easy read out, the stimuli-responsive properties exhibited by such coating systems appear to bear significant potential to visualize coating exposure before corrosion occurs. Further, the approach appears to be readily adaptable to other chemistries.

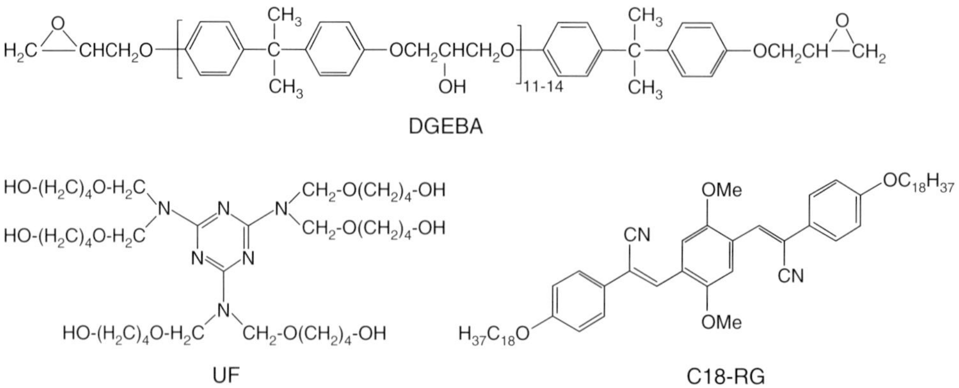

Figure 5.15 Chemical structures of the components used to synthesize chemoresponsive epoxy resins: DGEBA, UF cross-linker, and C18-RG. Reproduced with permission from Ref. [74].

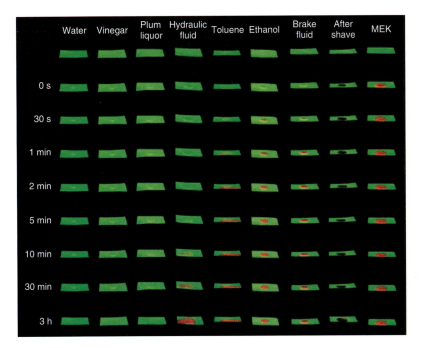

Figure 5.16 Pictures of DGEBA/UF/C18-RG coatings (cured for 1 min at 200 °C) on aluminum substrates as a function of the exposure time to different chemicals that were dropped on the coatings at room temperature. The samples are shown under illumination with UV light of a wavelength of 365 nm. Reproduced with permission from Ref. [74].

5.6
Summary and Outlook

The notion that the stimulus-driven self-assembly or dissociation of (nanoscale) aggregates of small guest molecules in a host polymer can be coupled with the propensity of many organic dyes to exhibit optical properties in ordered molecular assemblies that are vastly different than their molecular constituents has quickly led to a new family of chromogenic polymers, which change their absorption and/or fluorescence characteristics in response to an external stimulus. These optical sensor materials rely on chromophores that display pronounced fluorescence and/or absorption color changes upon self-assembly as a result of charge-transfer interactions and/or conformational changes. Many sensor types including mechanochromic, thermochromic, chemical-sensing, and multistimuli responsive polymers have already been described. Working prototypes based on several different dye classes and polymer families have been developed. A detailed fundamental understanding of the science that enables the technology has been realized and focused industrial interest could result in commercial products in a rather short time frame. Clearly, materials will have to be refined and tailored for specific applications. In this context, the design approach is particularly attractive because it is modular: the dye and the polymer matrix can, in principle, be independently selected so that the sensing and transduction mechanisms can, to a certain extent, be independently addressed. This allows one to readily modify a sensor system of interest.

Acknowledgments

The authors thank Christine Ander, Mark Burnworth, Taekwoong Chung, Brent R. Crenshaw, Mohit Gupta, Devang Khariwala, Maki Kinami, Joseph Lott, Christiane Löwe, Patrick T. Mather, John Protasiewicz, David A. Schiraldi, Robert Simha, Charles E. Sing, and Liming Tang for great collaborations on the stimuli-driven assembly of chromogenic dyes.

References

1. Shahinpoor, M. and Schneider, H.-J. (2008) *Intelligent Materials*, RSC Publishing, Cambridge.
2. Weder, C. (2009) *Nature*, **459**, 45.
3. Jenekhe, S.A. and Kiserow, D.J. (2005) *Chromogenic Phenomena in Polymers: Tunable Optical Properties*, American Chemical Society, Washington, DC, p. 2.
4. Bamfield, P. (1991) *Chromic Phenomena*, RSC, Cambridge.
5. Tomasulo, M., Yildiz, I., and Raymo, F.M. (2006) *J. Phys. Chem. B*, **110**, 3853–3855.
6. Burnworth, M., Rowan, S.J., and Weder, C. (2007) *Chem. Eur. J.*, **13**, 7828.
7. Löwe, C. and Weder, C. (2007) Photoluminescent polymer blends and uses therefore. US Patent 7,223,988 (to Case Western Reserve University).
8. Kinami, M., Crenshaw, B.R., and Weder, C. (2007) Time-temperature indicators. US Patent Application 2007/0158624.
9. Caruso, M.M., Davis, D.A., Shen, Q., Odom, S.A., Sottos, N.R., White, S.R., and Moore, J.S. (2009) *Chem. Rev.*, **109**, 5755.
10. For an excellent perspective on mechanically induced luminescent changes in molecular assemblies see: Sagara, Y. and Kato, T. (2009) *Nat. Chem.*, **1**, 605.
11. Förster, T. and Kasper, K. (1954) *Z., Physik. Chem. NF*, **1**, 275.
12. Förster, T. and Kasper, K.Z. (1955) *Elektrochem. Angew. Physik. Chem.*, **59**, 976.
13. Amicangelo, J.C. (2005) *J. Phys. Chem. A*, **109**, 9174.
14. Birks, J.B., Braga, C.L., and Lumb, M.D. (1965) *Proc. R. Soc. London, Ser. A*, **283**, 83.
15. Kazzaz, A.A. and Munro, I.H. (1966) *Proc. Phys. Soc.*, **87**, 329.
16. Itoh, M., Fuke, K., and Kobayashi, S. (1980) *J. Chem. Phys.*, **72**, 1417.
17. Birks, J.B., Dyson, D.J., and Munro, I.H. (1963) *Proc. R. Soc. London, Ser. A*, **275**, 575.
18. Naciri, J. and Weiss, R.G. (1989) *Macromolecules*, **22**, 3928.
19. Lewis, F.D., Wu, T.F., Burch, E.L., Bassani, D.M., Yang, J.S., Schneider, S., Jager, W., and Letsinger, R.L. (1995) *J. Am. Chem. Soc.*, **117**, 8785.
20. Birks, J.B. (1975) *Rep. Prog. Phys.*, **38**, 903.
21. Phillips, D., Roberts, A.J., and Soutar, I. (1983) *Macromolecules*, **16**, 1593.
22. Chen, L.S., Jin, X.G., Du, J.H., and Qian, R.Y. (1991) *Makromol. Chem.*, **192**, 1399.
23. Crespo-Hernandez, C.E., Cohen, B., and Kohler, B. (2005) *Nature*, **436**, 1141.
24. Löwe, C. and Weder, C. (2002) *Synthesis*, 1185.

25. Gill, R.E., van Hutten, P.F., Meetsma, A., and Hadziioannou, G. (1996) *Chem. Mater.*, **8**, 1341.
26. Döttinger, S.E., Hohloch, M., Hohnholz, D., Segura, J.L., Steinhuber, E., and Hanack, M. (1997) *Synth. Met.*, **84**, 267.
27. Hohloch, M., Maichle-Moessmer, C., and Hanack, M. (1998) *Chem. Mater.*, **10**, 1327.
28. Oelkrug, D., Tompert, A., Gierschner, J., Egelhaaf, H.-J., Hanack, M., Hohloch, M., and Steinhuber, E. (1998) *J. Phys. Chem. B*, **102**, 1902.
29. Henari, F.Z., Manaa, H., Kretsch, K.P., Blau, W.J., Rost, H., Pfeiffer, S., Teuschel, A., Tillmann, H., and Hörhold, H.H. (1999) *Chem. Phys. Lett.*, **307**, 163.
30. Van Hütten, P.F., Krasnikov, V.V., Brouwer, H.J., and Hadziioannou, G. (1999) *Chem. Phys.*, **241**, 139.
31. de Souza, M.M., Rumbles, G., Gould, I.R., Amer, H., Samuel, I.D.W., Moratti, S.C., and Holmes, A.B. (2000) *Synth. Met.*, **111–112**, 539.
32. Martinez-Ruiz, P., Behnisch, B., Schweikart, K.H., Hanack, M., Luer, L., and Oelkrug, D. (2000) *Chem. Eur. J.*, **6**, 1294.
33. Freudmann, R., Behnisch, B., and Hanack, M. (2001) *J. Mater. Chem.*, **11**, 1618.
34. Crenshaw, B.R. and Weder, C. (2003) *Chem. Mater.*, **15**, 4717.
35. Kinami, M., Crenshaw, B.R., and Weder, C. (2006) *Chem. Mater.*, **18**, 946.
36. Crenshaw, B.R. and Weder, C. (2006) *Macromolecules*, **39**, 9581.
37. Kunzelman, J., Kinami, M., Crenshaw, B.R., Protasiewicz, J.D., and Weder, C. (2008) *Adv. Mater.*, **20**, 119–122.
38. Kunzelman, J., Crenshaw, B., Kinami, M., and Weder, C. (2006) *Macromol. Rapid Commun.*, **27**, 1981.
39. (a) Döller, E. and Förster, T. (1962) *Z. Phys. Chem.*, **34**, 132; (b) Birks, J.B., Dyson, D.J., and Munro, I.H. (1963) *Prog. Roy. Soc. Ser. A*, **275**, 575.
40. Spies, C. and Gehrke, R. (2002) *J. Phys. Chem.*, **106**, 5348.
41. (a) Fitzgibbon, P.D. and Frank, C.W. (1981) *Macromolecules*, **14**, 1650; (b) Turro, N.J. and Arora, K.S. (1986) *Polymer*, **27**, 783; (c) Oyama, H.T., Tang, W.T. and Frank, C.W. (1987) *Macromolecules*, **20**, 474; (d) Wang, Y.C. and Morawetz, H. (1989) *Macromolecules*, **22**, 164.
42. (a) Semerak, S.N. and Frank, C.W. (1981) *Macromolecules*, **14**, 443; (b) Gashgari, M.A. and Frank, C.W. (1981) *Macromolecules*, **14**, 1558; (c) Gelles, R. and Frank, C.W. (1982) *Macromolecules*, **15**, 1486; (d) Semerak, S.N. and Frank, C.W. (1984) *Macromolecules*, **17**, 1148.
43. (a) Zimerman, O.E. and Weiss, R.G. (1998) *J. Phys. Chem. A*, **102**, 5364; (b) Vigil, M.R., Bravo, J., Baselga, J., Yamaki, S.B., and Atvars, T.D.Z. (2003) *Curr. Org. Chem.*, **7**, 197.
44. See for example: Naciri, J. and Weiss, R.G. (1989) *Macromolecules*, **22**, 3928.
45. Johnson, G.E. (1980) *Macromolecules*, **13**, 839.
46. Szadkowska-Nicze, M., Wolszczak, M., Kroh, J., and Mayer, J. (1993) *J. Photochem. Photobiol. A: Chem.*, **75**, 125.
47. Sagara, Y., Mutai, S., Yoshikawa, I., and Araki, K. (2007) *J. Am. Chem. Soc.*, **129**, 1520.
48. Sagara, Y. and Kato, T. (2008) *Angew. Chem. Int. Ed.*, **47**, 5175.
49. Sagara, Y., Yamane, S., Mutai, T., Araki, K., and Kato, T. (2009) *Adv. Funct. Mater.*, **19**, 1869.
50. Trabesinger, W., Renn, A., Hecht, B., Wild, U.P., Montali, A., Smith, P., and Weder, C. (2000) *J. Phys. Chem. B*, **104**, 5221.
51. Löwe, C. and Weder, C. (2002) *Adv. Mater.*, **14**, 1625.
52. Crenshaw, B., Burnworth, M., Khariwala, D., Hiltner, P.A., Mather, P.T., Simha, R., and Weder, C. (2007) *Macromolecules*, **40**, 2400.
53. Kunzelman, J., Gupta, M., Crenshaw, B.R., Schiraldi, D.A., and Weder, C. (2009) *Macromol. Mater. Eng.*, **294**, 244.
54. Lott, J. and Weder, C. (2010) *Macromol. Chem. Phys.*, **211**, 28.
55. Anuchai, K., Weder, C., and Magaraphan, R. (2008) *Plast. Rubber Compos.*, **37**, 281.
56. Pucci, A., Biver, T., Ruggeri, G., Meza, L.I., and Pang, Y. (2005) *Polymer*, **46**, 11198.

57. Donati, F., Pucci, A., Cappelli, C., Mennucci, B., and Ruggeri, G. (2008) *J. Phy. Chem. B*, **112**, 3668.
58. Pucci, A., Boccia, M., Galembeck, F., Leite, C., Tirelli, N., and Ruggeri, G. (2008) *React. Funct. Polym.*, **68**, 1144.
59. Pucci, A., Di Cuia, F., Signori, F., and Ruggeri, G. (2007) *J. Mater. Chem.*, **17**, 783.
60. Dlubek, G., Stejny, J., Lupke, T., Bamford, D., Petters, K., Hubner, C., Alam, M.A., and Hill, M.J. (2002) *J. Polym. Sci., Part B: Polym. Phys.*, **40**, 65.
61. Crenshaw, B. and Weder, C. (2005) *Adv. Mater.*, **17**, 1471.
62. Crenshaw, B., Kunzelman, J., Sing, C.E., Ander, C., and Weder, C. (2007) *Macromol. Chem. Phys.*, **208**, 572.
63. Sing, C.E., Kunzelman, J., and Weder, C. (2009) *J. Mater. Chem.*, **19**, 104.
64. Donati, F., Pucci, A., Boggioni, L., Tritto, I., and Ruggeri, G. (2009) *Macromol. Chem. Phys.*, **210**, 728.
65. Wang, C.C., Gao, Y., Shreve, A.P., Zhong, C., Wang, L., Mudalige, K., Wang, H., and Cotlet, M. (2009) *J. Phys. Chem. B*, **113** (50), 16110.
66. Kunzelman, J., Chung, C., Mather, T.M., and Weder, C. (2008) *J. Mater. Chem.*, **18**, 1082.
67. Kunzelman, J., Crenshaw, B.R., and Weder, C. (2007) *J. Mater. Chem.*, **17**, 2989–2991.
68. (a) Rittersma, Z.M. (2002) *Sens. Actuators, A*, **96**, 196;   (b) Lee, C. and Lee, G. (2005) *Sensor Lett.*, **3**, 1;   (c) Chen, Z. and Lu, C. (2005) *Sensor Lett.*, **3**, 274;   (d) Ando, M. (2006) *Trends Anal. Chem.*, **25**, 937.
69. Ando, M. (2006) *Trends Anal. Chem.*, **25**, 937.
70. Kleemann, M., Suisalu, A., and Kikas, J. (2005) *Proc. SPIE-Int. Soc. Opt. Eng.*, **5946**, 54960N–549601.
71. Kondratowicz, B., Narayanaswamy, R., and Persaud, K.C. (2001) *Sens. Actuators, B*, **74**, 138.
72. Kharaz, A. and Jones, B.E. (1995) *Sens. Actuators, A*, **47**, 491.
73. (a) Albert, K.J., Lewis, N.S., Schauer, C.L., Sotzing, G.A., Stitzel, S.E., Vaid, T.P., and Walt, D.R. (2000) *Chem. Rev.*, **100**, 2595;   (b) McQuade, D.T., Pullen, A.E., and Swager, T.M. (2000) *Chem. Rev.*, **100**, 2537;   (c) Wang, L., Fine, D., Sharma, D., Torsi, L., and Dodabalpur, A. (2006) *Anal. Bioanal. Chem.*, **384**, 310;   (d) Potyrailo, A. (2006) *Agnew. Chem. Int. Ed.*, **45**, 702;   (e) Wolfbeis, O.S. (2006) *Anal. Chem.* **78**, 3859.
74. Tang, L., Whalen, J., Schutte, G., and Weder, C. (2009) *Appl. Mater. Interface*, **1** (3), 488–496.
75. Wicks, Z.W., Jones, F.N., Pappas, S.P., and Wicks, D.A. (2007) *Organic Coatings: Science and Technology*, 3rd edn, John Wiley & Sons, Inc., Hoboken, NJ.
76. May, C.A. (ed.) (1976) *Epoxy Resins: Chemistry and Technology*, Dekker, New York, p. 485.

6
Switchable Surface Approaches

Aftin M. Ross, Himabindu Nandivada, and Joerg Lahann

6.1
Introduction

Switchable surfaces can reversibly alter material properties in response to changes in the environment or an external stimulus. Currently, a myriad of stimuli are utilized to control surface properties including electrical potential, light, pH, temperature, and mechanical forces. Exposure to stimuli results in changes in surface conformation, wettability, optical properties, and biocompatibility. Switchable materials exist in nature and their synthetic analogs may find wide-ranging applications. Because the behavior of stimuli-responsive materials can be controlled, potential benefits of their utilization abound, particularly in the biotechnology industry. Exploiting macroscopic changes in these "smart" materials is useful for a variety of applications including tissue engineering, microfluidics, biosensors, molecular electronics, and colorimetric displays.

In this chapter, we highlight diverse material platforms that function as switchable surfaces. Table 6.1 provides a list of the materials discussed herein as well as their respective stimuli. Sections 6.2–6.5 describe the stimuli that induce the material responses and Sections 6.6 and 6.7 the class of molecules that comprise the materials. In each section, we provide a brief description of the materials and their fabrication methods, demonstrate specific examples of their use, and identify potential applications.

6.2
Electroactive Materials

6.2.1
High-Density and Low-Density Self-Assembled Monolayers

The behavior of electroactive materials is controlled by the application of an electrical stimulus. Changes in surface properties as a result of electrical exposure include alterations in wettability, conformation, and protein adsorption. A self-assembled

Handbook of Stimuli-Responsive Materials. Edited by Marek W. Urban
Copyright © 2011 WILEY-VCH Verlag GmbH & Co. KGaA, Weinheim
ISBN: 978-3-527-32700-3

Table 6.1 Switchable surface materials and their stimuli.

	Electrical potential	Light	pH	Temperature
Self-assembled monolayers (SAMs)	X	–	X	–
Azobenzene molecules	–	X	–	–
Spiropyran molecules	–	X	–	–
Shape-memory polymers	–	X	–	X
Brushes	–	–	X	X
Poly(N-iso-propylacrylamide) (PNIPAAm)	–	–	–	X
Rotaxane	X	X	X	–
Catenane	X	–	–	–
DNA monolayers	X	–	X	–
Peptide monolayers	X	–	X	X

monolayer (SAM) is a single layer of amphiphilic molecules that spontaneously organizes onto a substrate as a result of an affinity between the amphiphile and the substrate and is an example of an electroactive material [1]. A gold–alkanethiolate system is commonly utilized owing to the ease of SAM formation and the use of gold as an electrode. Traditional SAMs result in densely packed structures called high-density self-assembled monolayers (HDSAMs). Because HDSAMs have limited steric freedom due to their dense packing, the structure is unable to undergo conformational changes. Thus, low-density self-assembled monolayers (LDSAMs) were created to decrease molecular steric hindrance, allowing for greater conformational freedom of the molecules and the creation of reversibly switchable surfaces. SAMs are created via physioabsorption onto a substrate. Typically, a substrate is immersed in a SAM solution for a sufficient period of time (generally 12–24 h) to form a homogeneous layer. Modifications can be made to the SAM both before and after assembly to augment its responsiveness to various stimuli.

HDSAMs are characteristically made electroactive by changing the electric potential of the gold substrate acting as the electrode. Modulating the electrical potential can result in desorption of thiols from the substrate leading to changes in surface properties, such as the wettability, or hydrophilicity and hydrophobicity, of the material [2–4]. Several methods exist for creating LDSAMs that are modulated by electric potentials [5–7]. In one method, SAMs with bulky end groups are assembled onto a surface. These SAMs are densely packed with respect to the end groups that are then cleaved, resulting in a monolayer that is loosely packed with respect to the alkyl chains. Loosely packed chains are capable of undergoing reversible conformational transitions in response to an external electrical stimulus [8–10]. Lahann et al. utilized this concept to reversibly modulate the wettability of LDSAMs [11]. Upon the application of an electric potential, the top of the molecule, which is negatively charged, is attracted to the positively charged surface and bends into a loop shape as shown in Figure 6.1. Thus the surface changes from hydrophilic in

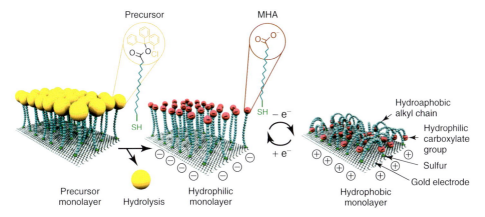

Figure 6.1 Low-density self-assembled monolayer on gold – response to an external electric potential. Lahann et al. Science. 2003 [11].

the "straight" conformation to hydrophobic in the "bent" conformation due to the exposure of the hydrophobic alkyl chains. This conformational alteration also results in a change in impedance, or resistance to current flow [9]. Potential applications for this technology abound and include diagnostics, cell adhesion/motility studies, and tissue engineering.

An alternative method for creating an LDSAM is the assembly of preformed inclusion complexes (ICs) comprising cyclodextrin and alkanethiol [7]. The space-filling cyclodextrin is subsequently unwrapped via dissociation in an appropriate solvent, generating an LDSAM. Liu et al. found that tuning the applied electrical potential resulted in reversible changes in surface wettability as well as protein adsorption. This system was extended, when low-density acids and amino-terminated SAMs were used to coat microfluidic devices in order to reversibly and selectively adsorb proteins of varying isoelectric points in a mixture. When a negative potential was applied, acid-terminated LDSAMs were able to adsorb the positively charged protein and released the protein upon the application of a positive potential. Thus, this method could play a role in protein separation studies for applications in proteomics and sensing technologies.

6.2.2
Self-Assembled Monolayers with Hydroquinone Incorporation

The incorporation of electroactive hydroquinone moieties into SAMs has also been utilized to generate electroactive monolayers. Hydroquinone moieties can be reduced by electrochemical oxidation to generate benzoquinone and this has been used to release ligands attached to a substrate surface [12]. For example, Yeo et al. detached Arginine–Glycine–Aspartic acid (RGD) peptides, which function as a cell adhesive, from the substrate surface; this triggered the release of cells adhered to the peptides [13]. Furthermore, benzoquinone units resulting from this reduction were then utilized (by means of a Diels–Alder reaction) to selectively immobilize diene-functionalized peptides. This reaction resulted in cellular reattachment and migration of cells.

6.3
Photoresponsive Materials

In addition to electric potential, light can also be used to trigger switching properties of surfaces and polymers. Application of ultraviolet (UV) light to these materials may result in reversible changes in characteristics such as hydrophilicity/hydrophobicity, structural arrangement, and shape. Commonly utilized photoresponsive materials include azobenzene molecules, spiropyran molecules, and shape-memory polymers.

6.3.1
Molecules Containing Azobenzene Units

Upon the application of a certain wavelength of light, azobenzene units switch their structure from the trans (straight) to the cis (bent) isomerization. This conformational change corresponds to different dipole moments, with the cis-isomer having a higher dipole moment. Alterations in molecular spatial arrangement can then be translated into macroscopic changes in surface wettability [14, 15]. For example, azobenzene treated surfaces have demonstrated the ability to control liquid droplet and liquid crystals (LCs) alignment [16, 17]. Ichimura *et al.* formed azobenzene monolayers on quartz substrates and established that exposure of UV light resulted in a reversible conformational change that led to the parallel alignment of LCs in contact with the monolayer. Changes in the molecular shape of azobenzene molecules also result in mechanical actuation. Ji *et al.* demonstrated the efficacy of azobenzenes in actuation by coating a microcantilever with a thiol-terminated azobenzene derivative [18]. Application of UV light caused the downward deflection of the coated cantilever, because alterations in azobenzene arrangement from the trans- to cis- conformation resulted in the repulsion of molecules within the monolayer.

The photoresponsiveness of azobenzenes have been exploited in biology *via* the incorporation of azobenzene units into peptides. Hyashi *et al.* synthesized a peptide with an azobenzene backbone and then utilized this peptide and its corresponding RNA-binding aptamer as an *in vitro* selection tool for RNA–ligand pairing [19]. The peptide exhibited reversible photoresponsive binding to target RNA that was turned "on" in the presence of visible light and turned "off" upon application of UV light. Auenheimer *et al.* utilized RGD peptides in conjunction with azobenzene to influence cellular adhesion [20]. Adding an azobenzene derivative to the peptide sequence enabled spatial control as the trans-conformation of azobenzene is 3 Å longer than the cis-conformation. After coating a substrate with the photoresponsive peptide, a reduction in spacing between peptides and substrates in the cis-conformation resulted in lower cell adhesion while the converse was true for molecules in the trans-conformation.

6.3.2
Molecules Containing Spiropyran Units

Spiropyrans are another class of photoresponsive materials with reversible switching capabilities. In this instance, switching is modulated by the photochemical cleavage of a C–O bond in the presence of UV light which changes it from a closed nonpolar form to an open polar form [21–26]. Because the closed form of the molecule is hydrophobic and the merocyanine form is hydrophilic, exposure to UV light alters wettability. Thus incorporating spiropyrans into a substrate or using them as a surface coating, allows for the control of substrate wettability [27]. Athanassiou et al. exploited the ability of spiropyrans to modulate wettability in conjunction with nanopatterning to control volumetric changes [28,29]. In this work, a nanopatterned poly(ethyl methacrylate)-co-poly(methyl acrylate), P(EMA)-co-P(MA) was doped with spiropyran and then exposed to UV illumination. The contact angle of the doped poly(EMA-co-MA) was reduced as compared to the material without the incorporation of spiropyran indicating increased hydrophilicity. Upon exposure of green laser pulses, the doped material returned to a hydrophobic state and the contact angle increased. Alterations in wettability are ascribed to dimensional changes in the nanopatterning due to light irradiation.

Control of wettability is also imperative in modulating cell behavior, particularly cell attachment. Higuchi et al. utilized a copolymer of nitrobenzospiropyran and methyl methacrylate, poly(NSP-co-MMA) to control platelet and mesenchymal stem cell adhesion to substrates [30]. After exposure to UV irradiation, cells or platelets that were previously attached to the copolymer-coated substrate were detached. As protein adsorption, specifically fibrinogen adsorption, is known to play a significant role in platelet adhesion, the impact of UV exposure on fibrinogen adhesion was measured. Because no substantial differences in fibrinogen adhesion were observed in the presence of UV light, it was concluded that cell and platelet detachment resulted from a change in surface energy and/or an alteration in the surface conformation derived from the switch of spiropyran from the closed nonpolar state to the polar merocyanine state upon UV light exposure. Edahiro et al. exploited the wettability control afforded by spiropyrans to create a photoresponsive cell culture surface [31]. In this instance, spiropyrans were incorporated as side chains for a poly(N-iso-propylacrylamide) polymer. Conformational transitions of spiropyran from the nonpolar to polar states after UV exposure modulated the adhesion of living CHO-K1 cells.

A unique application of spiropyrans is in molecular gating. Aznar et al. utilized spiropyrans to control nanoarchitecture through the creation of a gating system consisting of mesoporous MCM-41 incorporated with spiropyran moieties (which functioned as a 3D support) and 1.5 poly(amidoamine) dendrimers, which served as nanomechanical stoppers [32]. UV light exposure resulted in the attraction of the

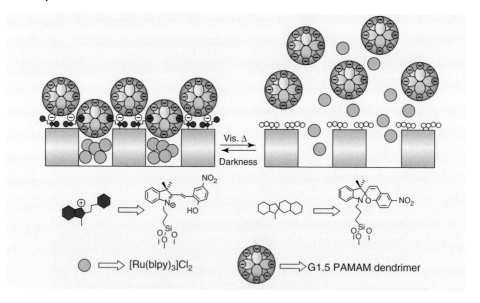

Figure 6.2 Schematic of a gating system that consisted of a mesoporous framework containing a photoswitchable anchored spiropyran, a dye entrapped in the inner pores, and carboxylate-terminated dendrimers as molecular caps. Aznar et al. Adv. Mat. 2007 [32].

negatively charged dendrimers to the positively charged spiropyrans in the polar conformation. This led to the entrapment of the dye molecule, Ru(bipy)$_3$Cl$_2$ as seen in Figure 6.2. However, the dendrimers and therefore the dye molecules were released to the bulk solution upon the application of visible light as spiropyran reverted to its neutrally charged nonpolar conformation.

6.3.3
Photoresponsive Shape-Memory Polymers

Shape-memory polymers can undergo reversible structural transitions in response to UV light [33–37]. These polymers typically comprise elastic polymeric networks and molecular switches. The polymeric network provides the support structure and determines the permanent shape, while the molecular switch undergoes reversible cross-linking in the presence of a photostimulus [38]. Shape-memory polymers modulated by temperature, thermally induced shape-memory polymers, also exist. However, polymers that are light actuated are advantageous in that structural changes in these materials occur at room temperature, thereby diversifying the application scope. For example, external heating effects may damage tissue if the material was utilized in an implant. Applications for shape-memory polymers include optical sensors and actuators, microrobots, optical microtweezers, and photoresponsive medical technologies [36, 39].

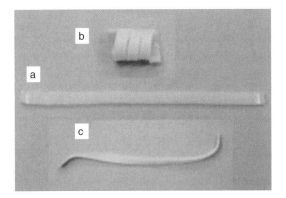

Figure 6.3 Polymer film doped with CAA molecules where (a) is the permanent shape, (b) is the temporary shape, and (c) is the recovered shape. Yu and Ikeda. *Macromol. Chem. Phys.* 2005 [34].

Lendlein and coworkers created two types of photoresponsive shape-memory polymers. The first is a grafted polymer containing ethylene-glycol-1-acrylate-2-cinnamic acid (HEA-CA) molecules and the second is a doped star-poly(ethylene glycol) polymer network containing cinnamylidene acetic acid (CAA) molecules. Both these materials are modulated by the application of an external force and UV light greater than 260 nm [33]. Materials were stretched from some original length and then exposed to UV irradiation. After irradiation, the polymers were fixed in an elongated state because of the [2+2] cycloaddition reaction that occurred in response to the UV exposure. This elongated state was maintained upon removal of the external force but could be reversed upon the application of UV light less than 260 nm due to cross-link cleavage. In addition, the authors found that if UV irradiation greater than 260 nm was applied to only one side of the stretched polymer, the resulting temporary shape was that of a corkscrew spiral. The corkscrew shape occurred because two layers were formed, a top layer with a fixed elongation and a bottom layer that retained its elasticity and thus could contract upon the release of the external stress. The reversible polymer conformations are displayed in Figure 6.3.

6.4
pH-Responsive Materials

Unlike several of the other stimuli considered in this chapter, pH is an internal stimulus and thus requires alteration of the chemical environment. As a result, the use of pH triggers is limited, particularly for basic cell studies that are confined to a narrow pH range. The complexity of reversibility is another drawback that arises with the use of pH-responsive materials, as the solution must be removed or extracted before reuse and this may also alter the physical environment.

Nonetheless, pH-responsive materials have shown promise in fields such as drug delivery and microprocessing [40].

6.4.1
pH-Switchable Surfaces Based on Self-Assembled Monolayers (SAMs)

As manufactured products become miniaturized, there is a need to enhance assembly processes. One of the primary issues is the interaction of the assembly tools with the surface forces of the object (e.g., van der Waals forces) that result in adhesion of the object to the assembly tool(s), thereby preventing object release [41]. Recently, this problem has been addressed through the use of pH-modulated SAMs [42]. In this work, the microobject (a glass microsphere) and the silica gripper tool (microcantilever) were chemically modified with SAMs consisting of aminosilane-grafted 3-(ethoxydimethylsilyl)propylamine (APDMES) and (3-aminopropyl)triethoxysilane (APTES). Both SAMs are amine-functionalized and thus can be protonated at acidic pH. The authors found that as the solution pH was modulated; so were the attractive and repulsive forces between the tool and the microobject. In particular, as the liquid pH was increased from acidic to basic conditions, the repulsive forces increased enabling object release. This concept could be utilized to create a submerged microhandling tool that could grasp and release microbjects based on the solution environment.

6.4.2
pH-Switchable Surfaces Based on Polymer Brushes

Polymer brushes comprise polymer chains that are tethered on one end to a solid substrate [43, 44]. Two methods are commonly employed in the synthesis of polymer brushes, grafting to and grafting from. In the grafting from approach, brushes are created via surface-initiated polymerization (SIP) of the polymer chains [45], while brushes generated in the grafting to method involve adhering presynthesized polymers to the substrate surface [46]. Polyelectrolyte brushes are pH-responsive materials that undergo structural changes at interfaces when their chains are charged and/or discharged because of the protonation/dissociation of acid/base groups. As a result, upon an alteration in pH, polyelectrolyte brushes transform from the swollen state to a shrunken state in which the polymer chains are collapsed [47, 48]. Tam *et al.* exploited this property of polyelectrolyte brushes to create a responsive interface with tunable/switchable redox properties [49]. In this system, indium tin oxide (ITO) served as the substrate and a modified poly(4-vinyl pyridine) modified with an Os-complex redox unit was grafted to the surface. At an acidic pH (pH = 4), the polymer brush was swollen and thus the redox units were in direct contact with the conducting ITO surface resulting in an active electrode. However, at a neutral pH, the brush was in the shrunken state, which restricted the mobility of the polymer chains and therefore access of the redox units to the conducting surface, yielding a nonactive electrode state as indicated in Figure 6.4.

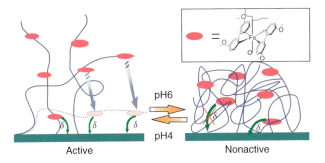

Figure 6.4 Reversible pH control of the redox-poly polymer brush between electrochemically active and inactive states. Tam et al. J. Phys. Chem. 2008 [49].

The pH-responsiveness of polymer brushes has also been exploited as a molecular gate. Motornov et al. created a mixed polyelectrolyte brush consisting of poly(2-vinylpyridine) (P2VP) and poly(acrylic acid) (PAA) that had switchable permeability for anions and cations [50]. In this work, ITO was used as the electrode substrate and the mixed brush was assessed via two soluble redox probes $[Fe(CN)_6]^{4-}$ and $[Ru(NH_3)_6]^{3+}$. When the environment pH was acidic (pH < 3), the P2VP chains (which were positively charged) were permeable to the anionic probe. However, the redox process for the cationic probe was prohibited, resulting in a lack of transport for positively charged ions. The converse was true when the solvent environment was more neutral or basic. This switchable and selective gating technique has applications in drug delivery and could be utilized in biosensors.

6.5
Thermoresponsive Materials

Temperature variations may result in reversible changes in properties such as structural arrangement, size, solubility, and shape. Many materials designed for biomedical or biotechnology applications are confined to a narrow temperature spectrum in order to be effective in a physiological environment. The following thermoresponsive materials are discussed in the next section: poly(N-iso-propylacrylamide) (PNIPAAm), polymer brushes, and shape-memory polymers.

6.5.1
Temperature-Dependent Switching Based on Poly(N-iso-propylacrylamide) (PNIPAAm)

PNIPAAm is a widely studied thermoresponsive polymer that exhibits a lower critical solution temperature (LCST) [51, 52]. Below the LCST, the polymer is expanded and hydrophilic and is thus soluble in water. Above this critical temperature, there is an abrupt phase transition leading to a collapsed and hydrophobic polymeric structure, which renders the polymer insoluble in water [53–55]. The LCST of

PNIPAAm is approximately 32 °C in a solution environment and therefore operates in a physiologically relevant design space [56, 57]. In addition, fine adjustments of the LCST of this material are possible by augmenting the polymer structure. For example, the LCST can be increased by adding ionic comonomers or formulating the polymer with salts or surfactants as pH changes may alter the repulsive forces between the comonomers and thereby influence swelling [58, 59]. Because the LCST of the polymer can be tailored, a wealth of potential applications exist in selective protein adsorption [60], control of cellular adhesion [61], and drug delivery [62–65].

Thermoresponsive polymers have been used in basic cell studies. For example, a PNIPAAm thiol (PNIPAAm-PEG-thiol) was synthesized and used to regulate the adhesion of cells in a microfluidic channel [66]. Fibroblasts were initially cultured at 37 °C to induce cell spreading and adhesion to the microfluidic surface. Then, the temperature of the microfluidic environment was reduced to 25 °C for a period of time. After 20 min, cell monolayers became detached and the morphology of individual cells was transformed to a spherical shape. Cell attachment and detachment as a function of the surface response to changing temperature is provided in Figure 6.5.

The thermoresponsive switching of PNIPAAm is also advantageous in the therapy for cancerous tumors. Chilkolti et al. designed a PNIPAAm polymer with

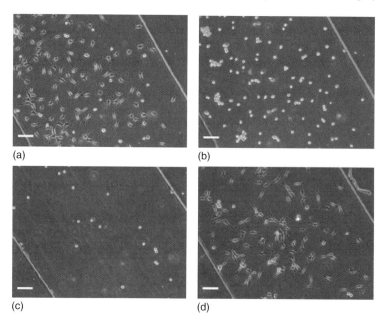

Figure 6.5 Exposure of fibroblasts on PNIPAAm-PEG-coated gold surface to laminar flows in a poly(dimethylsiloxane) (PDMS) fabricated microfluidic channel. (a) Fibroblasts adhered well to polymer surface at 37 °C; (b) cell rounding upon temperature decrease from 37 to 25 °C; (c) cell detachment of rounded cells upon the application of laminar flow at 560 µm s^{-1}; (d) spread cells remain attached to the polymer surface in the presence of 700 µm s^{-1} flow. Ernst et al. Lab Chip. 2007 [66].

an LCST of ∼ 40 °C to be used in combination with the thermal heating of tumors [58]. The LCST was increased by incorporating increased amounts of the acrylamide (AAm) copolymer, as it is hydrophilic. This enabled a polymer loaded with a small fluorescently labeled hydrophobic molecules (used to simulate a chemotherapeutic drug) to remain soluble when injected into a mouse body. However, as tumor permeability was increased by targeted heating to ∼ 42 °C, the polymer underwent a transition to the hydrophobic state and thus aggregated in the tumor. The performance, that is, the degree of polymer localization within the heated tumor, was compared for the thermosensitive polymer and a thermoinsensitive control. Preferential localization of the thermoresponsive polymer was demonstrated while localization was not apparent in the control.

6.5.2
Temperature-Dependent Switching in Polymer Brushes

In addition to pH-responsiveness, the behavior of polymer brushes may also be altered by temperature. For instance, electrochemistry-induced radical polymerization was used to fabricate a PNIPAAm brush [67]. The authors used this PNIPAAm brush in conjunction with ITO to create electrically heated electrodes (HEs). Altered HEs exhibited rapid temperature increases upon heating and this temperature enhancement was reversible. The control and release properties of the HE system were then assessed by means of the model protein hemoglobin. Hemoglobin diluted in phosphate-buffered saline was exposed to the brush at 20 °C when it was in the expanding state. Upon a temperature change above the brush LCST, the brush transitioned to the shrunken state, capturing the hemoglobin molecules. The hemoglobin was released when the temperature was reduced to 20 °C. UV–Vis absorbance indicated that the hemoglobin maintained its native fold as evidenced by a characteristic absorption ban at 408 nm as seen in Figure 6.6.

Zhang et al. controlled the size of nanoparticles by exploiting the pH and temperature responsiveness of a poly(2-(dimethylamino)ethyl methacrylate) (PDMAEMA) brush [68]. Briefly, a PDMAEMA brush was synthesized onto a polystyrene latex nanoparticle via atom transfer radical polymerization (ATRP). Dynamic light-scattering events indicated that pH changes led to temperature changes in solution, which resulted in an alteration in particle size. This approach could be useful in enzyme immobilization or protein separation.

6.5.3
Thermoresponsive Shape-Memory Polymers

Thermally induced shape-memory effects occur in numerous materials including polymers, metallic alloys, and ceramics [69–71]. Lendlein and Langer synthesized degradable thermoplastics that undergo conformational changes in response to a temperature increase [72]. The thermoplastics were comprised of oligo(ε-caprolactone)diol (OCL), which served as the switching segment, and a crystallizable oligo(p-dioxanone diol) (ODX), which provided the physical

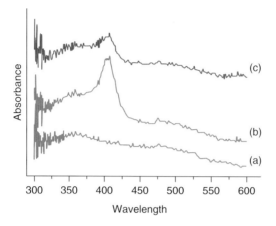

Figure 6.6 UV–vis spectra of hemoglobin incorporated into the PNIPAAm brush (a) without hemoglobin, (b) with hemoglobin at 40 °C, and (c) upon release of hemoglobin at 20 °C. Yin et al. J. Phys. Chem. C. 2009 [67].

cross-links. The material conformational transition from the temporary to the permanent shape was induced by heating the polymers above their transitional temperature (T_{trans}), which was 41 °C in this instance. Shape-memory properties were assessed via a range of cyclic thermomechanical tests and then the polymer was utilized *in vivo*. Sutures were created from the degradable thermoplastics and used to loosely close a rat wound. At 41 °C, the shape-memory effect was induced, which resulted in a tightening of the wound sutures.

Further use of thermoresponsive shape-memory polymers in the biomedical arena is found through the use of main-chain liquid crystalline elastomers (MC-LCEs). The mesophase structure of MC-LCEs may be varied and Rousseau and Mather designed thermosensitive main-chain smectic LCEs [73]. These new materials incorporated two benzoate-based mesogenic groups combine with hydride-terminated poly(dimethylsiloxane) spacers. Dynamic and mechanical analysis was used to investigate the shape-memory effects. It was found that heating the material above the polydomain smectic-C transition temperature led to a loss of smectic ordering and therefore to a large increase in deformability. At this point, a secondary shape could be obtained via mechanical manipulation and this shape could be maintained by cooling below the transition temperature.

6.6
Switchable Surfaces Based on Supramolecular Shuttles

Supramolecular machines contain a discrete number of molecular components capable of converting energy into mechanical work [74]. Like macroscopic machines, users of molecular machines are concerned with energy input, processing time, and output quality. Designers of supramolecular machines need only to look to nature for inspiration as several examples of these natural systems exist. The most significant of these natural motors include adenosine triphosphate (ATP) synthase, myosin, and kinesin [75]. Myosin and kinesin are linear motor proteins that traverse microtubules and actin filaments respectively, converting energy from ATP hydrolysis to molecular motion [76]. Promising synthetic supramolecular assemblies for nanotechnology applications include rotaxanes and catenanes.

6.6.1
Rotaxane Shuttles

As in the case of other materials included in this chapter, rotaxanes can be utilized to control the hydrophilicity and hydrophobicity of surfaces. Rotaxanes are supramolecular systems comprising a dumbbell-shaped axis molecule and a ring molecule identified as a macrocycle [77]. Molecular switching is ideal for rotaxane and pseudorotaxane monolayers because of the interlocking cavities contained within the monolayers [78–82]. Structures that incorporate these supramolecular machines allow for programmable surfaces that can be reversibly switched. One example of the use of rotaxanes in the control of wettability is based on rotaxanes incorporated into SAMs [83]. The SAMs were comprised of rotaxane combined with mercaptoundecanoic acid on gold. This system was used to move liquid droplets in response to light-induced changes, as seen in Figure 6.7.

In addition to light, pH may also modulate the behavior of rotaxanes. Jun *et al.* created a fluorescent rotaxane in which the fluorescent intensity of the molecule was pH responsive [84]. The rotaxane comprised a fluorenyltriamine axle and a cucurbituril macrocycle. Nitrogen atoms in the fluorenyltriamine axle could be protonated at a low pH (e.g., pH = 1) and this resulted in the placement of the macrocyle at the protonated diaminohexane site. Once deprotonation occurred at a high pH such as pH = 8, the macrocycle was transported to the diprotonated diaminobutane site causing a change in color and fluorescence intensity.

Bioelectronics is another application area, in which rotaxanes, particularly redox-active rotaxanes, could make a significant impact. Enzyme electrodes are altered in these applications by direct electron transfer between the electrode surface and the redox enzyme. Electronic communication between the surface and the redox enzyme centers is hindered, because a separation exists. This impediment can be circumvented by aligning the enzyme with the electrode and utilizing the redox relay units as go-betweens. The aforementioned concept has been exploited to associate an apoprotein, apo-glucose oxidase (apo-GOx), onto relay-functionalized materials including flavin adenine dinucleotide (FAD) monolayers, nanoparticles, and carbon nanotubes [85–88]. Katz *et al.* used reversible redox-active rotaxane shuttles in the bioelectrocatalyzed oxidation of glucose [80].

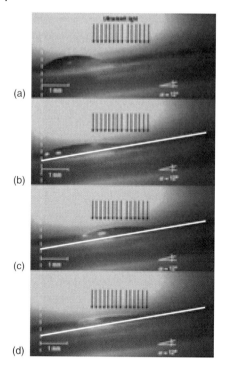

Figure 6.7 Lateral photographs of light-driven transport of a 1.25-μL diiodomethane drop on a E-1.11-MUA Au(111) substrate on mica up a 12° incline. The solid white line indicates the substrate surface. (a) Before irradiation, (b) after 160 s of irradiation (just before transport), (c) after 245 s of irradiation (just after transport), and (d) after 640 s irradiation (at photostationary state). Berna et al. Nat. Mater. 2005 [83].

6.6.2
Switches Based on Catenanes

Like rotaxanes, catenanes are mechanically interlocked molecules. However, instead of interlocking one ring shaped macrocycle and a dumbbell shape, catenanes consist of interlocked macrocycles. The number of macrocycles contained in a catenane is indicated by the numeral that precedes it. Catenanes have bistable and multistable forms and a switchable, bistable [2]catenane is commonly exploited in nanotechnology and molecular electronics because its behavior can be controlled by electrochemical processes [89]. Collier et al. was the first to demonstrate the electroactivity of interlocked catenanes [90]. The authors affixed phospholipid counterions to a monolayer of [2]catenanes and then sandwiched this system between two electrodes. This work resulted in a molecular switching device that opened at a positive potential of 2 V and closed at a negative potential of 2 V.

Recently, Spruell et al. synthesized and characterized a single-station [2]catenane that exhibited fundamental differences from the bi- and multistation catenane switches that currently exist [91]. Tetrathiafulvalene (TTF) is a electrochemically switchable unit that is often incorporated into catenanes to yield architectures that can be controlled electrically. Normally, electrochemically switchable catenanes are comprised of a primary binding station consisting of tetracationic cyclophane cyclobis(paraquate-p-phenylene) ($CBQT^{4+}$), which releases oxidized TTF^{2+} via electrostatic repulsion, and a secondary binding station that provides reversibility following the release of TTF2. However, in this work, a single-station [2]catenane that functions as a reversible switch between two translational states was created, eliminating the need for the secondary station and thus reducing system complexity. Applications for this technology may include solid-state electronic devices.

6.7
Switchable Surfaces Comprising DNA and Peptide Monolayers

The specificity of DNA base pairs is advantageous in biosensing applications such as genetic screening because it results in highly specific binding between targets of interest and probe moieties [92–94]. In addition, DNA can be analyzed with a myriad of surface analysis techniques including those that are label-free, or methods that do not require fluorescent tagging of the DNA. Some of these techniques include atomic force microscopy [95], electrochemical methods [96, 97], and surface plasmon resonance spectroscopy [98, 99]. Mass, conductivity, and electric field measurements may also be utilized to characterize DNA biosensing [100]. Though DNA monolayers have attracted significant attention, peptide monolayers are beginning to gain momentum in biotechnology areas [101]. In particular, researchers are interested in utilizing these building blocks to create materials from a bottom-up approach.

6.7.1
DNA-Based Surface Switches

DNA molecules can be directed by electrical potentials when tethered to electrode surfaces and this interaction is commonly exploited with gold as the substrate [102–104]. Kelley *et al.* was able to modulate the conformation of a thiol-modified oligonucleotide by altering electric potential exposure [105]. Originally, the monolayer was 50 Å in length and tilted at a 45° angle in open-circuit voltage. Application of a negative potential resulted in an increase in thickness of the monolayer to 55 Å and a distinct orientation change. When a positive potential was applied, the thickness was decreased to 20 Å. Electrical modulation has been combined with fluorescence to facilitate real-time optical monitoring of switchable DNA monolayers [106]. Single-stranded oligonucleotides were labeled with a fluorophore and DNA hybridization was initialized by the addition of unlabeled complimentary strands. As a result of the phosphates in the sugar–phosphate backbone of the negatively charged oligonucleotides, the DNA monolayers could be switched from a tilted or flat conformation to a vertical conformation. This orientation difference was indicated by changes in fluorescence intensity because the tilted conformation

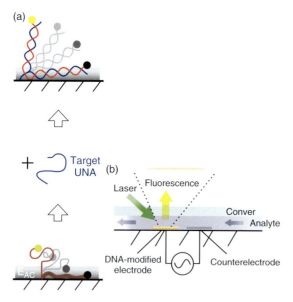

Figure 6.8 (a) Electrically controllable DNA films; (b) schematic describing the measurement setup consisting of gold electrodes modified by ss-DNA molecules and a fluidic channel containing the analyte sequences. Rant et al. Proc. Natl. Assoc. Sci. 2007 [106].

partially quenched the fluorescence signal. The system setup is described in Figure 6.8.

Electric field changes have been used in conjunction with pH changes to alter the conformation of DNA monolayers [107–110]. For example, DNA comprise two distinct domains, one domain consisted of a single strand linked to the gold substrate and the other domain contained four strands called an *i-motif* [107]. The i-motif could be modulated by modifications to the system pH. As these DNA molecules formed a SAM, the i-motif functioned as a bulky headgroup, resulting in a monolayer that was densely packed with respect to the i-motif but loosely packed with respect to the single-stranded DNA. As a result of this configuration, small molecules could not penetrate the monolayer. However, upon an increase in pH, the i-motif transitioned to a single-stranded conformation, which increased the permeability of the system to small molecules.

6.7.2
Switches Based on Aptamers

An aptamer is a short, single-stranded DNA or RNA sequence. These molecules typically demonstrate a high affinity to a multitude of molecules from proteins to cells [111]. The sensitivity and selectively with which aptamers bind make them promising options in diagnostic and therapeutic applications [111, 112].

Because aptamers experience reversible denaturation and regeneration, their use is particularly attractive in reusable sensing technologies [113]. Many of the potential uses for aptamers require that these molecules are immobilized on the surface. This requirement complicates their use because the retaining method employed must sustain the high binding affinity aptamers demonstrate in solution. Another challenge concerns the need to acquire a flexible, ordered receptor molecule while maintaining the appropriate aptamer folding that induces binding.

Numerous techniques are employed to study the reversible binding of aptamers and various biomolecules including cantilever-based sensors, surface plasmon resonance (SPR), and electrochemical impedance [114]. An aptamer that binds thrombin was assembled on a gold nanowire by Huang and Chen [115]. In this work, the aptamer probe was controlled by changes in electric potential and was used in a fluorescent protein test. The probe was affixed to the nanowire and specifically interacted with a biotinylated, fluorescently labeled thrombin. Upon the application of a potential, this complex was either attracted or repelled depending on the sign of the applied potential. Because the complex was negatively charged, a positive potential resulted in a reduction in the fluorescence intensity due to increased distance between the fluorescent label and the nanowire.

6.7.3
Switches Based on Helical Peptides

Helical peptides address an important limitation of alkanethiol monolayers, *i.e.* the existence of a monolayers; that a size mismatch between the functional head group and the alkyl chain exists. With these molecules, a disulfide group acts as an anchor on one side and the functional group is present on the other side. Another unique feature of α-helical peptides is the speed of electron transfer as this speed is faster along the dipole moment of the peptides than against it. SAMs comprised of two peptide sequences, with dipole moments in opposing directions were generated by Yasutomi *et al.* [116]. The sensitizer head groups on the helical peptides could be actuated at a precise excitation wavelength and each head group could be augmented independently. An anodic or cathodic current was produced depending on the excitation wavelength, leading to a directional change in the overall photocurrent. Yasutomi deviated from the two-peptide monolayer system and generated a single-component SAM that was pH responsive [117]. The 16-mer helical peptide was composed of a pH-responsive carboxyl group and a L-3-(3-N-ethylcarbazolyl)alanine at the C-terminus. For this system, changing the solution pH reversed the orientation of the photocurrent.

6.7.4
Switchable Surfaces of Elastinlike Polypeptides (ELP)

Elastin-like polypeptides (ELPs) are thermoresponsive biopolymers with a broad range of biotechnological applications including tissue engineering, drug delivery, biosensing, and protein purification [118, 119]. Like PNIPAAm, ELPs also experience reversible hydrophilic–hydrophobic phase transitions at an LCST. At temperatures below this LCST the peptides are soluble in aqueous environments, but above this temperature, the peptides become dehydrated and aggregate resulting in insolubility. Sequence (VPGXG)$_m$ serves as the common structural motif for ELPs with X being any amino acid but proline and m indicating the number of repeats. It is this structural motif that provides the polypeptides with thermosensitivity. ELP structure can be controlled via genetic engineering and thus augmentations to the peptide sequence and/or molecular weight can alter the LCST of the polypeptide [120].

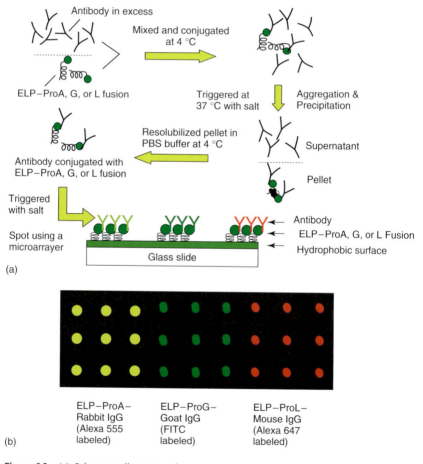

Figure 6.9 (a) Schematic illustrating the formation and purification of ELP–protein fusion–antibody complexes that are spotted onto the glass slide to fabricate antibody microarrays. (b) Fluorescence image of microarrays fabricated by immobilizing antibodies from rabbit, goat, and mouse, in complexes formed with fusions, respectively. Antibodies from rabbit, goat, and mouse are labeled with fluorophores Alexa 555, fluorescein isothiocyanate (FITC), and Alexa 647, respectively. Gao et al. J. Am Chem Soc. 2006 [122].

Dip pen lithography has been utilized to anchor ELPs to patterned monolayer on a gold surface [121]. The surface was used to investigate reversible release and binding of biomolecules of interest upon changes in temperature and ionic strength. Gao *et al.* synthesized ELP fusion proteins that were complexed with antibodies and then immobilized these complexes onto a hydrophobic surface to create an antibody array [122]. Complex immobilization stemmed from hydrophobic interactions that were triggered by temperature changes. Furthermore, this method was extended to fabricate a tumor detection system by immobilizing a capture antibody for the tumor marker of interest utilizing the ELP–antibody complex as depicted in Figure 6.9 [122]. Recently, the thermoreversible-phase properties of ELPs were utilized to generate a stimuli-responsive cell-adhesion system [123]. The polypeptide sequence $[(VPGVG)_4(VPGKG)]_8[(VPGVG)]_{40}$ was synthesized and then immobilized onto aldehyde-derivatized surfaces. These surfaces could then selectively control the attachment and release of fibroblasts to a substrate.

6.8
Summary

In this chapter, the use and means of actuation of various stimuli-responsive materials have been described in the context of creating programmable surfaces. With the ability to modulate a wide range of surface properties from hydrophobicity and hydrophilicity to structural arrangement, dynamic surfaces can be fine-tuned to address needs that span the biotechnology industry. As opportunities for innovation continue to be explored, materials and stimuli will emerge to generate more complex surface architectures with defined responses. Thus the future is bright for "smart" materials research.

References

1. Smith, R.K. Lewis, P.A., and Weiss, P.S. (2004) Patterning self-assembled monolayers. *Prog. Surf. Sci.*, **75**, 1–68.
2. Abbott, N.L., Gorman, C.B., and Whitesides, G.M. (1995) Active control of wetting using applied electrical potentials and self-assembled monolayers. *Langmuir*, **11**, 16–18.
3. Gorman, C.B., Biebuyck, H.A., and Whitesides, G.M. (1995) Control of the shape of liquid lenses on a modified gold surface using an applied electrical potential across a self-assembled monolayer. *Langmuir*, **11**, 2242–2246.
4. Abbott, N.L. and Whitesides, G.M. (1994) Potential-dependent wetting of aqueous-solutions on self-assembled monolayers formed from 15-(ferrocenylcarbonyl)pentadecanethiol on Gold. *Langmuir*, **10**, 1493–1497.
5. Berron, B. and Jennings, G.K. (2006) Loosely packed hydroxyl-terminated SAMs on gold. *Langmuir*, **22**, 7235–7240.
6. Wang, X.M., Kharitonov, A.B., Katz, E., and Willner, I. (2003) Potential-controlled molecular machinery of bipyridinium monolayer-functionalized surfaces: an electrochemical and contact angle analysis. *Chem. Commun.*, 1542–1543.
7. Liu, Y. *et al.* (2004) Controlled protein assembly on a switchable surface. *Chem. Commun.*, 1194–1195.

8. Peng, D.K. and Lahann, J. (2007) Chemical, electrochemical, and structural stability of low-density self-assembled monolayers. *Langmuir*, **23**, 10184–10189.
9. Peng, D.K., Yu, S.T., Alberts, D.J., and Lahann, J. (2007) Switching the electrochemical impedance of low-density self-assembled monolayers†. *Langmuir*, **23**, 297–304.
10. Peng, D.K., Ahmadi, A.A., and Lahann, J. (2008) A synthetic surface that undergoes spatiotemporal remodeling. *Nano Lett.*, **8**, 3336–3340.
11. Lahann, J. et al. (2003) A reversibly switching surface. *Science*, **299**, 371–374.
12. Yeo, W.S. and Mrksich, M. (2006) Electroactive self-assembled monolayers that permit orthogonal control over the adhesion of cells to patterned substrates. *Langmuir*, **22**, 10816–10820.
13. Yeo, W.S., Yousaf, M.N., and Mrksich, M. (2003) Dynamic interfaces between cells and surfaces: electroactive substrates that sequentially release and attach cells. *J. Am. Chem. Soc.*, **125**, 14994–14995.
14. Barrett, C.J., Mamiya, J.I., Yager, K.G., and Ikeda, T. (2007) Photo-mechanical effects in azobenzene-containing soft materials. *Soft Matter*, **3**, 1249–1261.
15. Wan, P.B., Jiang, Y.G., Wang, Y.P., Wang, Z.Q., and Zhang, X. (2008) Tuning surface wettability through photocontrolled reversible molecular shuttle. *Chem. Commun.*, 5710–5712.
16. Ichimura, K., Suzuki, Y., Seki, T., Hosoki, A., and Aoki, K. (1988) Reversible change in alignment mode of nematic liquid crystals regulated photochemically by command surfaces modified with an azobenzene monolayer. *Langmuir*, **4**, 1214–1216.
17. Ichimura, K., Oh, S.K., and Nakagawa, M. (2000) Light-driven motion of liquids on a photoresponsive surface. *Science*, **288**, 1624–1626.
18. Ji, H.F. et al. (2004) Photon-driven nanomechanical cyclic motion. *Chem. Commun.*, 2532–2533.
19. Hayashi, G., Hagihara, M., Dohno, C., and Nakatani, K. (2007) Photoregulation of a peptide-RNA interaction on a gold surface. *J. Am. Chem. Soc.*, **129**, 8678–8679.
20. Auernheimer, J., Dahmen, C., Hersel, U., Bausch, A., and Kessler, H. (2005) Photoswitched cell adhesion on surfaces with RGD peptides. *J. Am. Chem. Soc.*, **127**, 16107–16110.
21. Yoshida, M. and Lahann, J. (2008) Smart nanomaterials. *ACS Nano*, **2**, 1101–1107.
22. Jiang, W.H. et al. (2005) Photo-switched wettability on an electrostatic self-assembly azobenzene monolayer. *Chem. Commun.*, 3550–3552.
23. Siewierski, L.M., Brittain, W.J., Petrash, S., and Foster, M.D. (1996) Photoresponsive monolayers containing in-chain azobenzene. *Langmuir*, **12**, 5838–5844.
24. Wang, S.T., Song, Y.L., and Jiang, L. (2007) Photoresponsive surfaces with controllable wettability. *J. Photochem. Photobiol. C-Photochem. Rev.*, **8**, 18–29.
25. Rosario, R. et al. (2004) Lotus effect amplifies light-induced contact angle switching. *J. Phys. Chem. B*, **108**, 12640–12642.
26. Bunker, B.C. et al. (2003) Direct observation of photo switching in tethered spiropyrans using the interfacial force microscope. *Nano Lett.*, **3**, 1723–1727.
27. Xia, F., Zhu, Y., Feng, L., and Jiang, L. (2009) Smart responsive surfaces switching reversibly between super-hydrophobicity and super-hydrophilicity. *Soft Matter*, **5**, 275–281.
28. Athanassiou, A. et al. (2006) Photo-controlled variations in the wetting capability of photochromic polymers enhanced by surface nanostructuring. *Langmuir*, **22**, 2329–2333.
29. Athanassiou, A., Kalyva, M., Lakiotaki, K., Georgiou, S., and Fotakis, C. (2005) All-optical reversible actuation of photochromic-polymer microsystems. *Adv. Mater.*, **17**, 988–992.
30. Higuchi, A. et al. (2004) Photon-modulated changes of cell attachments on poly(spiropyran-co-methyl methacrylate) membranes. *Biomacromolecules*, **5**, 1770–1774.

31. Edahiro, J. et al. (2005) In situ control of cell adhesion using photoresponsive culture surface. *Biomacromolecules*, **6**, 970–974.
32. Aznar, E., Casasus, R., Garcia-Acosta, B., Marcos, M.D., and Martinez-Manez, R. (2007) Photochemical and chemical two-channel control of functional nanogated hybrid architectures. *Adv. Mater.*, **19**, 2228–222+.
33. Lendlein, A., Jiang, H.Y., Junger, O., and Langer, R. (2005) Light-induced shape-memory polymers. *Nature*, **434**, 879–882.
34. Yu, Y.L. and Ikeda, T. (2005) Photodeformable polymers: a new kind of promising smart material for micro- and nano-applications. *Macromol. Chem. Phys.*, **206**, 1705–1708.
35. Yamato, M. and Okano, T. (2004) Cell sheet engineering. *Mater Today*, **7**, 42–47.
36. Yu, Y.L., Nakano, M., and Ikeda, T. (2003) Directed bending of a polymer film by light – miniaturizing a simple photomechanical system could expand its range of applications. *Nature*, **425**, 145–145.
37. Lendlein, A. and Kelch, S. (2002) Shape-memory polymers. *Angew. Chem. Int. Ed.*, **41**, 2034–2057.
38. Yoshida, M., Langer, R., Lendlein, A., and Lahann, J. (2006) From advanced biomedical coatings to multi-functionalized biomaterials. *Polym. Rev.*, **46**, 347–375.
39. Li, M.H., Keller, P., Li, B., Wang, X.G., and Brunet, M. (2003) Light-driven side-on nematic elastomer actuators. *Adv. Mater.*, **15**, 569–572.
40. Murthy, N., Campbell, J., Fausto, N., Hoffman, A.S., and Stayton, P.S. (2003) Bioinspired pH-responsive polymers for the intracellular delivery of biomolecular drugs. *Bioconjug. Chem.*, **14**, 412–419.
41. Gauthier, M., Re'gnier, S., Rougeot, P., and Chaillet, N. (2006) Analysis of forces for micromanipulations in dry and liquid media. *J. Micromech.*, **3**, 389–413.
42. Dejeu, J.R.M., Gauthier, M.L., Rougeot, P., and Boireau, W. (2009) Adhesion forces controlled by chemical self-assembly and pH: application to robotic microhandling. *ACS Appl. Mater. Interfaces*, **1**, 1966–1973.
43. Kizhakkedathu, J.N., Norris-Jones, R., and Brooks, D.E. (2004) Synthesis of well-defined environmentally responsive polymer brushes by aqueous ATRP. *Macromolecules*, **37**, 734–743.
44. Pyun, J., Kowalewski, T., and Matyjaszewski, K. (2003) Synthesis of polymer brushes using atom transfer radical polymerization. *Macromol. Rapid Commun.*, **24**, 1043–1059.
45. Prucker, O. and Ruhe, J. (1998) Mechanism of radical chain polymerizations initiated by azo compounds covalently bound to the surface of spherical particles. *Macromolecules*, **31**, 602–613.
46. Ducker, R., Garcia, A., Zhang, J.M., Chen, T., and Zauscher, S. (2008) Polymeric and biomacromolecular brush nanostructures: progress in synthesis, patterning and characterization. *Soft Matter*, **4**, 1774–1786.
47. Minko, S. (2006) Responsive polymer brushes. *Polym. Rev.*, **46**, 397–420.
48. Netz, R.R. and Andelman, D. (2003) Neutral and charged polymers at interfaces. *Phys. Rep.: Rev. Sec. Phys. Lett.*, **380**, 1–95.
49. Tam, T.K., Ornatska, M., Pita, M., Minko, S., and Katz, E. (2008) Polymer brush-modified electrode with switchable and tunable redox activity for bioelectronic applications. *J. Phys. Chem. C*, **112**, 8438–8445.
50. Motornov, M. et al. (2009) Switchable selectivity for gating ion transport with mixed polyelectrolyte brushes: approaching 'smart' drug delivery systems. *Nanotechnology*, **20**, 10.
51. Luzinov, I., Minko, S., and Tsukruk, V.V. (2004) Adaptive and responsive surfaces through controlled reorganization of interfacial polymer layers. *Prog. Polym. Sci.*, **29**, 635–698.
52. Sheeney-Haj-Ichia, L., Sharabi, G., and Willner, I. (2002) Control of the electronic properties of thermosensitive poly(N-isopropylacrylamide) and Au-Nano-particle/Poly(N-isopropylacrylamide) composite hydrogels upon phase

transition. *Adv. Funct. Mater.*, **12**, 27–32.
53. Chung, J.E., Yokoyama, M., and Okano, T. (2000) Inner core segment design for drug delivery control of thermo-responsive polymeric micelles. *J. Control. Release*, **65**, 93–103.
54. Dimitrov, I., Trzebicka, B., Muller, A.H.E., Dworak, A., and Tsvetanov, C.B. (2007) Thermosensitive water-soluble copolymers with doubly responsive reversibly interacting entities. *Prog. Polym. Sci.*, **32**, 1275–1343.
55. Dreher, M.R. et al. (2008) Temperature triggered self-assembly of polypeptides into multivalent spherical micelles. *J. Am. Chem. Soc.*, **130**, 687–694.
56. Yamada, N. et al. (1990) Thermoresponsive polymeric surfaces – control of attachment and detachment of cultured-cells. *Makromol. Chem.-Rapid Commun.*, **11**, 571–576.
57. Sun, T.L. et al. (2004) Reversible switching between superhydrophilicity and superhydrophobicity. *Angew. Chem. Int. Ed.*, **43**, 357–360.
58. Chilkoti, A., Dreher, M.R., Meyer, D.E., and Raucher, D. (2002) Targeted drug delivery by thermally responsive polymers. *Adv. Drug Deliv. Rev.*, **54**, 613–630.
59. Eliassaf, J. (1978) Aqueous-solutions of poly(N-Isopropylacrylamide). *J. Appl. Polym. Sci.*, **22**, 873–874.
60. Huber, D.L., Manginell, R.P., Samara, M.A., Kim, B.I., and Bunker, B.C. (2003) Programmed adsorption and release of proteins in a microfluidic device. *Science*, **301**, 352–354.
61. Yamato, M., Konno, C., Utsumi, M., Kikuchi, A., and Okano, T. (2002) Thermally responsive polymer-grafted surfaces facilitate patterned cell seeding and co-culture. *Biomaterials*, **23**, 561–567.
62. Yoshida, R. et al. (1994) Positive thermosensitive pulsatile drug-release using negative thermosensitive hydrogels. *J. Control. Release*, **32**, 97–102.
63. Kavanagh, C.A., Rochev, Y.A., Gallagher, W.A., Dawson, K.A., and Keenan, A.K. (2004) Local drug delivery in restenosis injury: thermoresponsive co-polymers as potential drug delivery systems. *Pharmacol. Ther.*, **102**, 1–15.
64. Kikuchi, A. and Okano, T. (2002) Pulsatile drug release control using hydrogels. *Adv. Drug Deliv. Rev.*, **54**, 53–77.
65. Kost, J. and Langer, R. (2001) Responsive polymeric delivery systems. *Adv. Drug Deliv. Rev.*, **46**, 125–148.
66. Ernst, O., Lieske, A., Jager, M., Lankenau, A., and Duschl, C. (2007) Control of cell detachment in a microfluidic device using a thermo-responsive copolymer on a gold substrate. *Lab Chip*, **7**, 1322–1329.
67. Yin, Z.Z., Zhang, J.J., Jiang, L.P., and Zhu, J.J. (2009) Thermosensitive behavior of poly(N-isopropylacrylamide) and release of incorporated hemoglobin. *J. Phys. Chem. C*, **113**, 16104–16109.
68. Zhang, M. et al. (2006) Double-responsive polymer brushes on the surface of colloid particles. *J. Colloid Interface Sci.*, **301**, 85–91.
69. Kagami, Y., Gong, J.P., and Osada, Y. (1996) Shape memory behaviors of crosslinked copolymers containing stearyl acrylate. *Macromol. Rapid Commun.*, **17**, 539–543.
70. Swain, M.V. (1986) Shape memory behavior in partially-stabilized zirconia ceramics. *Nature*, **322**, 234–236.
71. Lendlein, A., Schmidt, A.M., and Langer, R. (2001) AB-polymer networks based on oligo(epsilon-caprolactone) segments showing shape-memory properties. *Proc. Natl. Acad. Sci. U.S.A.*, **98**, 842–847.
72. Lendlein, A. and Langer, R. (2002) Biodegradable, elastic shape-memory polymers for potential biomedical applications. *Science*, **296**, 1673–1676.
73. Rousseau, I.A. and Mather, P.T. (2003) Shape memory effect exhibited by smectic-c liquid crystalline elastomers. *J. Am. Chem. Soc.*, **125**, 15300–15301.
74. Balzani, V., Credi, A., Ferrer, B., Silvi, S., and Venturi, M. (2005) *Molecular Machines*, Vol. 262, Springer-Verlag, Berlin, pp. 1–27.
75. Schliwa, M. and Woehlke, G. (2003) Molecular motors. *Nature*, **422**, 759–765.

76. Vale, R.D. and Milligan, R.A. (2000) The way things move: looking under the hood of molecular motor proteins. *Science*, **288**, 88–95.
77. Kay, E.R., Leigh, D.A., and Zerbetto, F. (2007) Synthetic molecular motors and mechanical machines. *Angew. Chem. Int. Ed.*, **46**, 72–191.
78. Balzani, V., Credi, A., and Venturi, M. (2008) Molecular machines working on surfaces and at interfaces. *ChemPhysChem*, **9**, 202–220.
79. Hiratani, K., Kaneyama, M., Nagawa, Y., Koyama, E., and Kanesato, M. (2004) Synthesis of [1]rotaxane via covalent bond formation and Its unique fluorescent response by energy transfer in the presence of lithium ion. *J. Am. Chem. Soc.*, **126**, 13568–13569.
80. Katz, E., Lioubashevsky, O. and Willner, I. (2004) Electromechanics of a redox-active rotaxane in a monolayer assembly on an electrode. *J. Am. Chem. Soc.*, **126**, 15520–15532.
81. Fioravanti, G. *et al.* (2008) Three state redox-active molecular shuttle that switches in solution and on a surface. *J. Am. Chem. Soc.*, **130**, 2593–2601.
82. Bunker, B.C. *et al.* (2007) Switching surface chemistry with supramolecular machines. *Langmuir*, **23**, 31–34.
83. Berna, J. *et al.* (2005) Macroscopic transport by synthetic molecular machines. *Nat. Mater.*, **4**, 704–710.
84. Jun, S.I., Lee, J.W., Sakamoto, S., Yamaguchi, K., and Kim, K. (2000) Rotaxane-based molecular switch with fluorescence signaling. *Tetrahedron Lett.*, **41**, 471–475.
85. Willner, I. *et al.* (1996) Electrical wiring of glucose oxidase by reconstitution of FAD-modified monolayers assembled onto Au-electrodes. *J. Am. Chem. Soc.*, **118**, 10321–10322.
86. Katz, E., Riklin, A., Heleg-Shabtai, V., Willner, I., and Bückmann, A.F. (1999) Glucose oxidase electrodes via reconstitution of the apo-enzyme: tailoring of novel glucose biosensors. *Anal. Chim. Acta*, **385**, 45–58.
87. Xiao, Y., Patolsky, F., Katz, E., Hainfield, J.F., and Willner, I. (2003) Plugging into enzymes: nanowiring of redox enzymes by a gold nanoparticle. *Science*, **299**, 1877–1881.
88. Patolsky, F., Weizmann, Y., and Willner, I. (2004) Long-range electrical contacting of redox enzymes by SWCNT connectors. *Angew. Chem. Int. Ed.*, **43**, 2113–2117.
89. Flood, A.H. *et al.* (2004) Meccano on the nanoscale – a blueprint for making some of the world's tiniest machines. *Aust. J. Chem.*, **57**, 301–322.
90. Collier, C.P. *et al.* (2000) A [2]Catenane-based solid state electronically reconfigurable switch. *Science*, **289**, 1172–1175.
91. Spruell, J.M. *et al.* (2009) A push-button molecular switch. *J. Am. Chem. Soc.*, **131**, 11571–11580.
92. Cosnier, S. and Mailley, P. (2008) Recent advances in DNA sensors. *Analyst*, **133**, 984–991.
93. Daniels, J.S. and Pourmand, N. (2007) Label-free impedance biosensors: opportunities and challenges. *Electroanalysis*, **19**, 1239–1257.
94. Odenthal, K.J. and Gooding, J.J. (2007) An introduction to electrochemical DNA biosensors. *Analyst*, **132**, 603–610.
95. Wang, K., Goyer, C., Anne, A., and Demaille, C. (2007) Exploring the motional dynamics of end-grafted DNA oligonucleotides by in situ electrochemical atomic force microscopy. *J. Phys. Chem. B*, **111**, 6051–6058.
96. Drummond, T.G., Hill, M.G., and Barton, J.K. (2003) Electrochemical DNA sensors. *Nat. Biotechnol.*, **21**, 1192–1199.
97. Meng, F.B., Liu, Y.X., Liu, L., and Li, G.X. (2009) Conformational transitions of immobilized DNA chains driven by pH with electrochemical output. *J. Phys. Chem. B*, **113**, 894–896.
98. Georgiadis, R., Peterlinz, K.P., and Peterson, A.W. (2000) Quantitative measurements and modeling of kinetics in nucleic acid monolayer films using SPR spectroscopy. *J. Am. Chem. Soc.*, **122**, 3166–3173.
99. Yang, X.H. *et al.* (2006) Electrical switching of DNA monolayers investigated by surface plasmon resonance. *Langmuir*, **22**, 5654–5659.

100. Immoos, C.E., Lee, S.J., and Grinstaff, M.W. (2004) Conformationally gated electrochemical gene detection. *Chembiochem*, **5**, 1100–1103.
101. Chow, D., Nunalee, M.L., Lim, D.W., Simnick, A.J., and Chilkoti, A. (2008) Peptide-based biopolymers in biomedicine and biotechnology. *Mater. Sci. Eng. R-Rep.*, **62**, 125–155.
102. Rant, U. *et al.* (2006) Dissimilar kinetic behavior of electrically manipulated single- and double-stranded DNA tethered to a gold surface. *Biophys. J.*, **90**, 3666–3671.
103. Rant, U. *et al.* (2006) Electrical manipulation of oligonucleotides grafted to charged surfaces. *Org. Biomol. Chem.*, **4**, 3448–3455.
104. Rant, U. *et al.* (2004) Dynamic electrical switching of DNA layers on a metal surface. *Nano Lett.*, **4**, 2441–2445.
105. Kelley, S.O. *et al.* (1998) Orienting DNA helices on gold using applied electric fields. *Langmuir*, **14**, 6781–6784.
106. Rant, U. *et al.* (2007) Switchable DNA interfaces for the highly sensitive detection of label-free DNA targets. *Proc. Natl. Acad. Sci. U.S.A.*, **104**, 17364–17369.
107. Mao, Y.D. *et al.* (2007) Alternating-electric-field-enhanced reversible switching of DNA nanocontainers with pH. *Nucleic Acids Res.*, **35**, 8.
108. Wang, S.T. *et al.* (2007) Enthalpy-driven three-state switching of a superhydrophilic/superhydrophobic surface. *Angew. Chem. Int. Ed.*, **46**, 3915–3917.
109. Liedl, T., Olapinski, M., and Simmel, F.C. (2006) A surface-bound DNA switch driven by a chemical oscillator. *Angew. Chem. Int.Ed.*, **45**, 5007–5010.
110. Mao, Y.D., Chang, S., Yang, S.X., Ouyang, Q., and Jiang, L. (2007) Tunable non-equilibrium gating of flexible DNA nanochannels in response to transport flux. *Nat. Nanotechnol.*, **2**, 366–371.
111. Balamurugan, S., Obubuafo, A., Soper, S.A., and Spivak, D.A. (2008) Surface immobilization methods for aptamer diagnostic applications. *Anal. Bioanal. Chem.*, **390**, 1009–1021.
112. Mok, W. and Li, Y.F. (2008) Recent progress in nucleic acid aptamer-based biosensors and bioassays. *Sensors*, **8**, 7050–7084.
113. Song, S.P., Wang, L.H., Li, J., Zhao, J.L., and Fan, C.H. (2008) Aptamer-based biosensors. *Trac-Trends Anal. Chem.*, **27**, 108–117.
114. Rodriguez, M.C., Kawde, A.N., and Wang, J. (2005) Aptamer biosensor for label-free impedance spectroscopy detection of proteins based on recognition-induced switching of the surface charge. *Chem. Commun.*, 4267–4269.
115. Huang, S.X. and Chen, Y. (2008) Ultrasensitive fluorescence detection of single protein molecules manipulated electrically on Au nanowire. *Nano Lett.*, **8**, 2829–2833.
116. Yasutomi, S., Morita, T., Imanishi, Y., and Kimura, S. (2004) A molecular photodiode system that can switch photocurrent direction. *Science*, **304**, 1944–1947.
117. Yasutomi, S., Morita, T., and Kimura, S. (2005) pH-Controlled switching of photocurrent direction by self-assembled monolayer of helical peptides. *J. Am. Chem. Soc.*, **127**, 14564–14565.
118. Simnick, A.J., Lim, D.W., Chow, D., and Chilkoti, A. (2007) Biomedical and biotechnological applications of elastin-like polypeptides. *Polym. Rev.*, **47**, 121–154.
119. Chilkoti, A., Christensen, T., and MacKay, J.A. (2006) Stimulus responsive elastin biopolymers: applications in medicine and biotechnology. *Curr. Opin. Chem. Biol.*, **10**, 652–657.
120. Meyer, D.E. and Chilkoti, A. (1999) Purification of recombinant proteins by fusion with thermally-responsive polypeptides. *Nat. Biotechnol.*, **17**, 1112–1115.
121. Hyun, J., Lee, W.K., Nath, N., Chilkoti, A., and Zauscher, S. (2004) Capture and release of proteins on the nanoscale by stimuli-responsive elastin-like polypeptide "switches". *J. Am. Chem. Soc.*, **126**, 7330–7335.

122. Gao, D. *et al.* (2006) Fabrication of antibody arrays using thermally responsive elastin fusion proteins. *J. Am. Chem. Soc.*, **128**, 676–677.

123. Na, K. *et al.* (2008) "Smart" biopolymer for a reversible stimuli-responsive platform in cell-based biochips. *Langmuir*, **24**, 4917–4923.

7
Layer-by-Layer Self-Assembled Multilayer Stimuli-Responsive Polymeric Films

Lei Zhai

7.1
Introduction

Multilayer polymeric films fabricated through the layer-by-layer (LBL) self-assembly technique are able to integrate stimuli-responsive materials into the films through various intermolecular interactions. The ease of the fabrication process and the excellent control of the film composition and morphology, combined with the capability of coating substrates conformally, offers a wide range of multilayer films that change film properties such as permeability, wettability, color, and solubility upon external stimuli. This chapter reviews the methods of building LBL multilayer films and the approaches to control the structures of the films. These smart films provide numerous applications in controlled drug release, sensing, energy generation, and flow regulation with their unique response to external stimuli including pH and temperature change, photo irradiation, application of electrical fields, and specific molecular interactions.

Stimuli-responsive coatings, also known as *smart coatings*, are able to change their properties upon external stimuli including temperature, pH, salt concentration, light, presence of specific molecules, and other factors. The capability of changing surface behavior in an accurate and predictable manner has generated numerous applications ranging from chemical sensing, separations, and drug delivery to various bioapplications. To name a few, smart coatings with tunable hydrophilic/hydrophobic wettability have been used in the development of micro- and nanofluidic devices, self-cleaning and antifog surfaces, and sensor devices [1–3]. Photochemical-responsive coatings have been applied to build a surface that releases nitric oxide [4] over nanometer length scales for photochemotherapeutic applications, and a surface that can be photoactivated for spatiotemporal control of cell adhesion during cell cultivation [5, 6]. The control of the properties of these smart coatings can be achieved through the stimuli-responsive materials forming self-assembled monolayers (SAMs) and polymer films. Among various approaches to fabricate responsive polymer films, multilayered polymer films, generally referred as *polyelectrolyte multilayer* (*PEM*) films, fabricated through the LBL assembly technique, have been proven to be effective smart coatings with

Handbook of Stimuli-Responsive Materials. Edited by Marek W. Urban
Copyright © 2011 WILEY-VCH Verlag GmbH & Co. KGaA, Weinheim
ISBN: 978-3-527-32700-3

tailored chemical composition and architecture in the micro- and nanoscales. First developed by Decher in the early 1990s, the LBL assembly technique has made rapid progress in film-preparative methodologies and functional film materials [7]. Advanced multilayer films with components such as synthetic linear polymers [8], polymeric microgels [9–12], biomacromolecules [13–18], particles [19–21], dendritic molecules [22, 23], organic components [24–28], block copolymers [29–32], and polyelectrolyte stabilized micelles [33] have been successfully fabricated. The driving force for multilayer film fabrication is diverse, including electrostatic interaction [34, 35], hydrogen-bond [36–41], coordination-bond [42–44], charge-transfer interactions [45], biospecific interaction (e.g., sugar–lectin interactions) [46, 47], guest–host interaction [48, 49], cation–dipole interaction [50, 51], and the synergetic interaction of the above forces. The simple fabricating process is depicted in Figure 7.1 with some polymers that have been used to make PEMs. These LBL-deposited multilayer films have applications in areas such as antireflection coatings [52], controlled releasing coatings [9–12, 32, 53–55] biosensors [56–59], nonlinear optics [60–64], solid-state ion-conducting materials [65], solar-energy

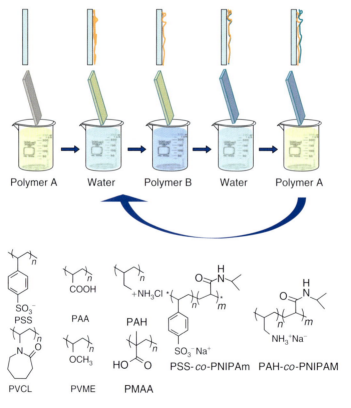

Figure 7.1 A schematic illustration of the fabrication of polyelectrolyte multilayer coatings using the layer-by-layer self-assembly technique.

conversion [66, 67], and separation membranes [68–73]. Besides constructing multilayer films on planar substrates, multilayer films have also been deposited on sacrificial colloid particles. These multilayer film coatings turn into capsules upon the dissolution of the cores, offering versatile vehicles for controlled drug release and delivery [74–77]. Extension of materials and substrates for multilayer film fabrication enrich the structures and functionalities of multilayer films. Polymeric building blocks possessing versatile structures in solution are helpful to obtain polymeric films with well-tailored structures as well as functionalities, as exemplified by the PEMs of weak polyelectrolytes [52, 78–83]. The demonstrated high capacity of PEMs to a wide range of functional molecules, including dyes [28, 83], nanoparticles [21], and carbon nanotubes [84, 85], combined with the engineered response properties of these films, presents an attractive combination for future applications of such films as functional responsive coatings.

7.2
Fabrication of Multilayer Polymer Coatings

The LBL molecular-level adsorption of polymers through various intermolecular interactions is now a well-established methodology for creating conformal thin-film coatings with precisely tuned physical, biochemical, and chemical properties (Figure 7.1). This technique involves sequential adsorption of materials that can form intermolecular interactions [34–51]. The growth of the film is a bottom-up approach and allows the precise control of film composition and dimension in nanoscale. This technique provides a versatile platform for the assembly of materials and nanostructures of interest in surface functionalization to respond to external stimuli. Owing to the conformal nature of the polyelectrolyte adsorption process, such a straightforward technique allows the formation of conformal thin films on virtually any solution-reachable structures.

The electrostatic attraction between oppositely charged molecules has been extensively used as a driving force for multilayer buildup. Strong electrostatic attraction occurs between a charged surface and an oppositely charged molecule in solution. The driving force of the film formation is the energy change of the complexation of polyelectrolytes, which is derived mainly from the release of counter ions. Upon complexation, polyelectrolytes form ions pairs through electrostatic interaction and release their counter ions to maintain the electrostatic neutrality. Since the polymer films are formed through the ion pairing of polyelectrolytes with opposite charges, any environment change such as solution ion concentration that can affect the ion pairs will induce the morphology change. Two categories of polyelectrolytes, strong polyelectrolytes and weak polyelectrolytes, have been used to fabricate multilayer films with different properties. LBL assembly of strong polyelectrolytes usually generates smooth films with low bilayer (polycation/polyanion) thickness. For example, for poly(styrene sulfonate) (PSS)/poly(allylamine hydrochloride) (PAH) films deposited at low ionic strength, an interlayer roughness was found between adjacent layers of 1.2–1.6 nm [86–88]. In the case of weak polyelectrolytes, the dependence of their degree of ionization on the pH of their solutions offers an approach to manipulate both internal and surface composition. For example, when poly(acrylic acid) (PAA) and PAH are assembled from their low pH solutions (pH = 2) where the degree of ionization of PAA chains is low, while the PAH is fully ionized, readily swellable multilayer films with lower ionic cross-link density and high concentration of PAA are formed. When PAA and PAH are assembled at neutral pH (pH = 6.5) where both polyelectrolytes are fully charged, more highly ionically cross-linked films are created with nearly equal amounts of PAA and PAH. This influence of solution pH on film morphology has made weak PEM films interesting smart coating materials for various applications.

Besides electrostatic interactions, hydrogen bonds have been utilized to fabricate multilayer films of polymers with hydrogen bond donors and acceptors [89–91]. Pairs of water-soluble uncharged polymers with intermolecular hydrogen-bonding interactions, such as polycarboxylic acids at low pH values and poly(ethylene oxide) (PEO) or poly(vinylpyrrolidone) (PVP), were self-assembled at surfaces [90, 91]. Since the hydrogen bonds rely on the presence of protons, the formed hydrogen bonds can be destroyed by changing the environmental pH. Utilizing this property of hydrogen bonds, Sukhishvili and Granick have demonstrated that LBL multilayers of poly(methacrylic acid) (PMAA) and PVP can be stable up to pH 6.9, but dissolve when the pH is raised above this point. Therefore, hydrogen-bonded LBL assembly can be used to fabricate layered, erasable, ultrathin polymer films. The stability of hydrogen-bond multilayer films can be improved by cross-linking using thermal and photochemical approaches [90] or carbodiimide chemistry [92]. One of the hydrogen bond-accepting polymers, poly(N-isopropylacrylamide) (PNIPAM), is a thermoresponsive polymer that introduces a variety of thermoresponsive properties to the multilayer films of hydrogen bonds. PNIPAM exhibits its lower critical solution temperature (LCST) in an aqueous solution and shows a sharp phase transition at 32 °C [93–97]. PNIPAM is soluble in water below its LCST (32 °C), while at temperatures above its LCST, it forms intramolecular hydrogen bonds and becomes insoluble. This thermal-responsive polymer has generated interesting applications in the field of flow regulation and drug delivery.

Multilayered polymer films can also be fabricated through other intermolecular connections such as covalent bonds, molecular recognition, and charge transfer interaction [98–101]. The formation of covalent bonds between adjacent polymers is usually achieved through a mild, well-controlled reaction with high yield. Such an approach was first developed by Mallouk and coworkers through a multilayer synthesis based on a sequential adsorption of $ZrOCl_2$ and 1,10-decanebisphosphonic acid, which can form stable zirconium 1,10-decanebisphosphonate crystals [98]. Various reactions and interactions including the Heck reaction [99], click chemistry [100], and disulfide chemistry [101] have also been utilized to build multilayer films. Different fabrication approaches of multilayer films have allowed the incorporation of a variety of functional materials into the films, offering the capability of responding to external stimuli that include pH, ion concentration, temperature, photo irradiation, and specific molecular interaction. Stimuli-induced property changes such as swellability, stability, ion permeability, porosity, and wettability bring about a variety of applications in controlled drug release, tissue engineering, and electronics.

7.3
Response of Multilayer Polymer Coatings to External Stimuli

7.3.1
Salt Concentration Change

For multilayer films fabricated through electrostatic interaction, environment salt concentration or ionic strength change can attenuate intermolecular interactions in polyelectrolytes with permanent charges where polyelectrolyte/polyelectrolyte pairs are replaced by polyelectrolyte/salt ion pairs. The process is similar to a typical ion exchange, except that polyelectrolytes are not free to leave the multilayer films but attach each other through fewer electrostatic interactions. At certain critical salt concentration, the reaming polyelectrolyte/polyelectrolyte ion pairs are no longer capable of holding the film together, and the individual polyelectrolyte dissociates from the film, leading to film dissolution. The decreasing polyelectrolyte/polyelectrolyte interaction with the increment of salt concentration offers a type of coating that is responsive to environment salt concentration. The structure changes include the film thickness, permeability, and decomposition of the film. Sukhishvili has nicely reviewed the research in this field [102]. The work by Ibarz et al. has shown that capsules made from PAH and PSS can reversibly switch their permeability to large molecules from impermeable to permeable by the variation of ionic strength in a range of low concentrations from 10^{-3} to 10^{-2} M [103]. For PEM films deposited onto solid substrates, ionic strength variations can also introduce nanoporosity [104] or result in film smoothing [105, 106]. The seeming contradiction between these results is probably due to the different binding energies of various polyanion/polycation pairs and their different tendency to dissociation in salt solution. For diffusion of small molecules through a polyelectrolyte film, two mechanisms have been proposed: (i) free volume diffusion through relaxing polymer network and (ii) diffusion through small pores whose size was suggested to be on the order of 1 nm [107]. In the past several years, better understanding of the relationship between the sign and the amount of charge introduced within a film by pH variations, ionic strength of solution, and film permeability and swelling has developed. Antipov et al. showed that permeability of PSS/PAH capsules to a fluorescent dye was strongly enhanced in a 0.5 M NaCl solution [108], and that the salt effect on the permeability covers a wide concentration range from 0.01 to 0.5 M [109]. Their work indicates that the diffusion of molecules in PEMs mostly through cavities in the polyelectrolyte complex (PEC) and the increment of salt concentration weaken the interaction between polyelectrolytes, therefore increasing the film permeability.

In the Rubner group, a controlled etching of PAA/PAH PEMs was investigated in NaCl solutions where the PEM thickness decreases to a fixed value depending on the initial film thickness and salt concentration [110]. The dissolution of PEMs

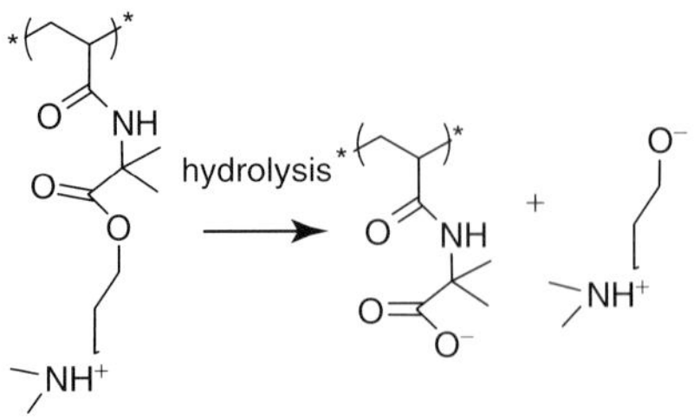

Figure 7.2 Hydrolysis of the "charge-shifting" polymers generates negatively charged polyelectrolytes.

is due to a salt-induced film reorganization that leads to the release of water-soluble PECs. On the basis of the PEC phase diagram [104], the thickness loss at certain NaCl concentration suggests that water-soluble PECs can only form for particular ratios of polycation to polyanion units at a given salt concentration. Increasing the salt concentration widens the ratio range and results in further film dissolution. In addition, the kinetic data support the diffusion of water-soluble PECs throughout the films. Such controlled PEM dissolution responsive to salt concentration can find applications in controlled drug release as well as generate thickness gradient patterns that are useful platforms to control cell attachment and migration. It is also important to note that the salt concentration and pH-induced PEM film disruption is not appropriate for the release of drugs under physiological conditions. To solve this problem, Lynn and Hammond have developed PEMs containing hydrolytically degradable poly(α-amino ester) and calf thymus DNA as a model functional polyanion, and demonstrated that these films eroded slowly upon incubation under physiological conditions [60, 111, 112]. Furthermore, linear poly(ethylenimine) (LPEI) functionalized with methyl ester side chains and plasmid DNA were used to build PEMs where the slow release of DNA was caused by the gradual hydrolysis of the ester functionality under physiologically relevant conditions, resulting in a controlled reduction in cationic charge density and a change in the nature of the electrostatic interactions between the polymers and DNA [113]. Besides using hydrolytically degradable polymers, new "charge-shifting" cationic polymers have been used for controlled release of DNA [114–116]. Upon the hydrolysis of the charge-shifting polymers that generate negative charges on polymers, the PEMs disassembled and released imbedded molecules (Figure 7.2). This approach allows the slow release of DNA up to 90 days [117]. The controlled release of drugs under physiological conditions is discussed in the following topics.

7.3.2
pH Alternation

The degree of ionization of weak polyelectrolytes such as PAA and PAH depends on the pH of their solutions. The linear charge density on the weak polyelectrolyte backbones in the assembled films is a function of pH. Therefore, the electrostatic interaction within weak PEMs can be easily tuned and excess charge can be created by varying the environment pH. Although the bulk weak PEMs fabricated at certain deposition pH does not carry excess charge, the fraction of charged parts in polyelectrolytes is determined by the self-assembly pH and polyelectrolyte pKa values. The fraction of chargeable functional groups introduced during self-assembly determines the film morphology as well as the capacity to accommodate excess charge through postassembly pH variation. The variation of the charges on the polyelectrolytes leads to the rearrangement of the polyelectrolytes in the weak PEMs, generating interesting structures in response to external pH change. This feature was widely exploited to tune the properties of PEM films such as permeability, morphology, or wettability.

Rubner and coworkers have systematically investigated the effect of assembling PAA and PAH at different pH values on the resultant film structures [118]. Figure 7.3 highlights the representative film structures attributed to different assembling pH values of PAA and PAH. When the PAA and PAH layers are assembled from their low pH solutions (pH = 2.0), the degree of ionization of the PAA chains is kept low while PAH is fully charged. This results in a nonstoichiometric pairing of PAA/PAH bilayer composition with more PAA (70%) than PAH (30%). This assembly condition generates a readily swellable film with a relatively low ionic cross-link density and high concentration of carboxylic acid groups. When PAA and PAH are assembled at neutral pH (pH = 6.5) where both polymers are fully charged, a higher degree of ionic cross-linked films is obtained. Since most PAA and PAH are coupled through electrostatic interaction, the film is smooth and not responsive to environmental pH change. The third example illustrates the weak PEM films assembled from partially charged PAA (pH = 3.5) and PAH (pH = 8.5). During the deposition of PAA, the PAA chains with few negative charges absorb on a substrate with fully charged PAH. This step produces a loopy structure with a lot of free acid groups on the surface. In contrast, during the adsorption of PAH, the pH is increased and the remaining acid groups become fully ionized. The PAH chains

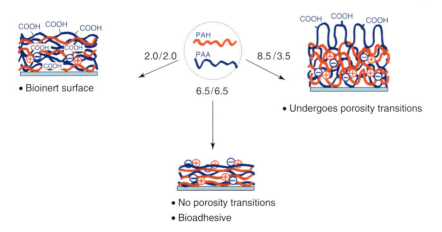

Figure 7.3 Schematic illustration of the representative film structures attributed to different assembling pH values of PAA and PAH.

penetrate into the PAA surface layer and neutralize this extra charge, creating an internal ion-paired bilayer with a PAH-enriched surface. Therefore, the film surface composition depends on the finishing step. The surface contains a significant fraction of free acid groups when PAA is the outermost layer, while there are essentially no free acid groups on the surface when PAH is the outermost layer. These structures represent a less energetically favored arrangement compared with PEM films fabricated from fully charged PAA and PAH and undergo a pH-driven spinodal decomposition to produce porous structures. This is accomplished by exposing a 3.5/8.5 PAA/PAH PEM film to acidic water (pH = 1.8–2.4) for a brief period of time followed by rinsing with neutral water and blow drying with air. The possible mechanism involves protonating carboxylate groups by treatment at low pH, breaking numerous electrostatic PAA/PAH interactions to free the structure, and re-pairing in more favorable stitching arrangements. The level and type of porosity developed depends on parameters such as pH of the acidic water, temperature, and exposure times to the treatment. It is interesting that the size of pores can be tuned from tens of nanometers to tens of microns by varying the pH of the acidic water and the treating procedure. For example, when the film is immersed into a pH 1.8 acidic water followed by neutral water rinse, pores with size from 20 to 40 nm are created. When the films are immersed into a pH 2.4 acidic water followed by neutral water rinse, pores with size of 1 μm are created. Most interestingly, when the film undergoes sequential immersion into pH 1.8 and 2.4 water, followed by a neutral water rinse, a honeycomb-like structure is generated with pores with sizes of tens of microns . Furthermore, the porous structure can be eliminated by immersing in acidic water followed by air drying. Although the mechanism is not fully understood, these porous films have found numerous applications in antireflection coatings, biocompatible membrane for controlled release, and superhydrophobic coatings. By mimicking the Bragg reflector structures of a *Morpho* butterfly wings, Zhai and coworkers have fabricated polymeric multilayer Bragg reflectors by multilayer films containing alternating dense materials and porous materials, generating reflecting color like butterfly wings [119]. The multilayer Bragg reflectors were fabricated from PAA, PAH, and sulfonated poly(styrene) (SPS) in appropriate combinations (Figure 7.4a). The porous zones were developed postassembly via the immersion of the heterostructure film into an aqueous acidic medium followed by rinsing in deionized water. The ability to introduce nanoporosity controllably in select regions of a multilayer heterostructure makes it possible to design and fabricate conformable Bragg reflectors that change the reflection band with different vapors. The shift of the reflection band is attributed to the refractive index change caused by the condensation of different vapors into the nanoporous structures. Figure 7.4b shows the shift of the reflection band of the Bragg reflector upon the exposure of different vapors.

Changes in pH have been used to induce film swelling in PEMs by generating a pH-tunable charge density and counteractive attractive and repulsive forces. The mechanism of such pH responsive properties has been nicely explained by Glinel and coworkers [120]. In PSS/PAH PEMs, PSS is a strong polyanion, whose charge is fixed at any pH, while PAH is a weak polycation and its charge density varies with

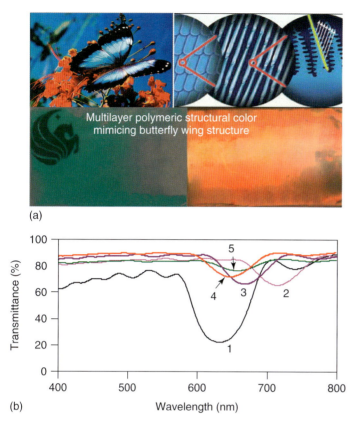

Figure 7.4 A multilayer Bragg reflector with alternating high- and low-refractive index blocks resembling the butterfly wing structures (a). Transmittance spectra of a treated [(PAH/PAA)$_8$–(PAH/SPS)$_{50}$]$_9$ film in air (1),' after exposure to water ($n = 1.33$) (2), ethanol ($n = 1.36$) (3), acetone ($n = 1.39$) (4), and toluene ($n = 1.49$) (5) vapors (b).

the acidity of the aqueous solution. Its apparent pKa in such multilayers has been determined to be about 10.8 [121]. Upon increasing the pH value above the pKa of the PAH, the charge density of the polymer decreases and the backbone becomes almost neutral in more basic solutions. As a result, many charges in PSS/PAH systems are not compensated any more and an excess of negative charges from PSS is present. The accumulation of negative charges causes the polymer chains to repel one another and there is a local increase of osmotic pressure, leading to swelling of the multilayer films. The swelling, induced by environment pH changes, has been utilized in the design of capsules whose shell permeability can be controlled for encapsulation and release procedures. In the case of PSS/PAH capsules templated on various cores, increasing the pH above 11 led to a dramatic swelling of the structures and finally to their dissolution, but this swollen state could be stabilized with the use of organic solvents [122, 123]. The swelling of these microcontainers in alkaline solutions is accompanied by an increase in the wall permeability and has been used for the controlled picogram dose encapsulation and release of macromolecules [124]. More complicated systems with two weak polyelectrolytes have also been developed to be responsive both in acidic and alkaline regions. By replacing the strong polyelectrolyte PSS with a weak polyanion PMAA in combination with PAH, new microcapsules have been fabricated [125, 126]. Similar to the mechanism involved in PSS/PAH capsules, PMAA/PAH shells swell and then dissolve when PMAA becomes less than 10% charged (pH < 2.7) and symmetrical when PAH loses most of its charges (pH > 11.5). In this case, a stable swollen state could be observed in a narrow acidic pH range (0.2 units), where hydrophobic interactions between PMAA chains prevent the structure from dissolution. In order to stabilize the swollen state at basic pH, PAH can be replaced by a much more hydrophobic polycation, poly(4-vinyl pyridine) (PVPy). which becomes insoluble and precipitates in water when the pH decreases below 5. PVPy/PMAA capsules have been assembled and have exhibited pH-sensitivity both in acidic and basic solutions [127]. Since the PEMs are stabilized by both electrostatic interactions and hydrogen bonding from pH 2 to 8, stable swollen states are observed at both the limiting pHs because of counteracting hydrophobic as well as hydrogen-bonding interactions. In summary, pH-responsive microcapsules could be prepared when using at least one weak polyelectrolyte in the LBL assembly. At a pH value that dramatically decreases the charge density of the weak polyion, a drastic swelling of the capsules was observed. Without any stabilization, this swelling irreversibly led to the dissolution of the structure. However, stable swollen states were obtained with the help of hydrophobic interactions and hydrogen bonds. In these particular cases, the swelling could be inversed by tuning back the pH to the initial value. Reversibility of the swelling is a key parameter for these systems in applications such as encapsulation and release, or sensing.

On the other hand, since PEMs fabricated through hydrogen bonds between hydrogen bond donor and acceptor polymers are sensitive to environment pH, the formation and application of pH-responsive polymer multilayers via hydrogen bonds have been extensively investigated. Several types of multilayer films containing PVP/PMAA, PEO/PMAA, or PEO/PAA at low pH have been fabricated and the disintegration of such films at pH levels above 6.9, 4.6, and 3.6, respectively, have been investigated by the Sukhishvili group [128, 129]. It is proposed that the critical pH for the decomposition is controlled by a balance of internal ionization, especially the fraction of carboxylic groups. Caruso and coworkers reported the pH-induced film deconstruction characteristics of hydrogen-bonded layers consisting of PVPy/PAA [130] in the electrostatically assembled layer and showed that the thickness of the hydrogen-bonded layers played an important role in controlling the film dissolution. Ono and Decher used PAA and poly(ethylene glycol) (PEG) to construct the pH-responsive film segment. This film covers the substrate surface and disintegrates by a pH change in order to release the PEM film constructed on top of the pH-responsive multilayer [131].

7.3.3
Temperature Variation

PEM films that contain polymers exhibiting a phase change through a temperature stimulus, such as PNIPAM have been extensively investigated for applications in surface-wetting regulation, controlled cell adhesion, and drug release. Temperature-responsive poly(N-isopropylacrylamide coacrylic acid) (PNIPAM-co-AA) microgels and PAH were alternately deposited on substrates through electrostatic interactions by the Lyon group [9, 132]. These microgel particles in PEM films demonstrated fast response to temperature change when the environment pH is below 4.25 (pK_a of acrylic acid) and were used for pulsed thermal release of insulin and doxorubicin [133]. This approach immobilizes the fast temperature-responsive microgels with high drug-loading capacity in PEMs, promising an efficient drug-release approach. The authors discovered that the swelling–deswelling of the microgel was influenced by solution pH, suggesting that polymer interactions affect the phase change of temperature-responsive polymers. Further studies performed by the Sukhishvili and the Schlenoff groups have shown that the strength of hydrogen bonding in PEMs strongly attenuates the temperature response [133] and the presence of extra charges in PEMs can compensate the deswelling of temperature-responsive polymers by enhancing the film swelling [134]. In the work by the Sukhishvili group, two types of temperature-responsive polymers, poly(N-vinylcaprolactam) (PVCL) and poly(vinyl methyl ether) (PVME), were included in ultrathin films using the LBL method, alternating adsorption of these polymers with PMAA at low pH. It was found that the temperature response of PMAA/PVME and PMAA/PVCL multilayers occurred over a wide range, about 10–15 °C, while phase separation of PVME or PVCL in solution occurred within a narrow temperature region close to the LCST of the polymer solution. Furthermore, an increase of permeation of thymol blue through the films in the range of the LCST of PVCL or PVME reflected loosening of the film structure as a result of PVCL or PVME phase transitions. When polymers were not covalently fixed within the film, deswelling of PVME or PVCL chains at temperatures near and above the LCST was accompanied by the appearance of voids in the multilayer film and an increase in dye permeability. The effect of the strength of intermolecular interactions and the composition of interpolymer complexes was demonstrated by the widening of temperature range of PMAA/PVME and PMAA/PVCL phase change as well as the larger permeability increase of PMAA/PVCL (stronger intermolecular interaction) compared to PMAA/PVME. When thermally responsive PEMs were made from charged PNIPAM copolymers, the film permeability decreased upon heated above the LCST. This phenomenon is attributed to the hydration caused by large amount of charges in the PEMs that compensate the phase change of the PNIPAM copolymer. The dehydration of the PNIPAM copolymer blocks the hydrated opening, leading to reduced permeability [134].

An alternative approach to introduce thermoresponsive polymers into multilayer films was based on chemical reactions between activated polymers (i.e., PAA) during sequential deposition of temperature-responsive copolymers containing amino or carboxylic groups [10, 100] or polymerization grafted from activated polymers [3]. A fully integrated microfluidic valve with a switchable, thermosensitive polymer surface has been fabricated through the combination of the responsive PEMs and microfabrication. To build a thermosensitive valve, a microfluidic patch was conformally coated with PAA/PAH multilayer coatings followed by functionalization with a thermosensitive polymer, PNIPAM. Figure 7.5 shows the change in the water contact angle with surface temperature, which decreases dramatically at the LCST. When the valve was heated above LCST, the valve became hydrophobic and inhibited the flow of water. In contrast, as the valve was cooled down to room temperature, it became hydrophilic and allowed the flow of water. Dyed water was allowed to flow in the microfluidic channels under capillary action only, to test the capability of the valve to control liquid flow. Figure 7.6 shows the operation of the switchable valve at 70 °C. At this temperature, the PNIPAM forms intramolecular hydrogen bonds and becomes hydrophobic. The hydrophobic patch with high contact angle (122°) stops the dyed water at the polymer front. This achieves the "closed" status of the valve at room temperature, where PNIPAM forms intermolecular hydrogen bonds and becomes hydrophilic. The hydrophilic patch with low contact angle (14°) allows the aqueous solution to flow through. This corresponds to the "open" status of the valve (Figure 7.6) [3, 135].

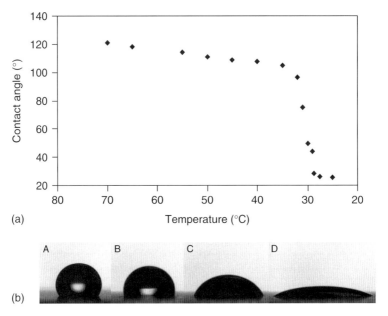

Figure 7.5 (a) Contact angle as a function of temperature (b) a water droplet which changes its contact angle (from 14° at 25 °C to 122° at 65 °C) on a PNIPAM-functionalized surface [3].

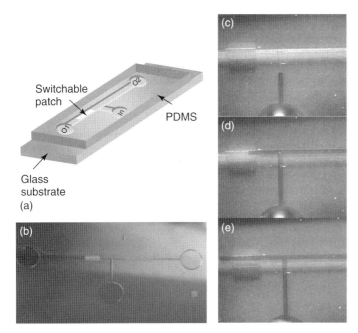

Figure 7.6 (a) Schematic of an LBL-deposited switchable polymer for flow regulation, (b) fabricated device; dyed water (c) approaches T-junction, (d) stops over the polymer surface ($\theta_c > 120°$) at elevated temperature, and (e) passes through at room temperature [135].

Another approach to produce temperature-responsive films is through hydrogen-bonded self-assembly [136]. Quinn and Caruso have demonstrated that the dye absorption and release capability of hydrogen-bonded PNIPAM/PAA multilayers was significantly enhanced at elevated temperatures [137]. In addition, temperature-controlled release of dyes from electrospun PAA/PAH nanofibers was obtained by depositing temperature-sensitive PAA/PNIPAM multilayers onto the fiber surfaces [138].

A hydrogen-bond PEM based on PMAA and PNIPAM that responds to both temperature and pH was developed by Swiston et al. to release the capped functional PEM patches with cells [139]. The pH mechanism depends on the ionization level of PMAA's acid groups that function as hydrogen bond donors. Below a critical pH (6.2), PMAA/PNIPAM PEMs are stable at all biologically useful temperatures. The temperature response depends on the solubility of PNIPAM in water where PNIPAM is insoluble above its LCST (32 °C). Therefore, PMAA/PNIPAM PEMs dissolve in physiological pH conditions (∼7.4) at 4 °C (below LCST, PNIPAM is soluble) but remain stable at 37 °C (above LCST, PNIPAM is insoluble).

7.3.4
Light Radiation

Light radiation is potentially very useful in generating responsive systems, with a variety of mechanisms by which polymers can be switched, including isomerization, elimination, photosensitization, and local heating. A variety of nanoparticles [140] and dyes [141] have been incorporated in PEM films and capsules to generate light-responsive systems. For example, microcapsules composed of PAH/PSS PEMs and light-absorbing gold nanoparticles were prepared by Caruso's group for light-responsive delivery applications. The encapsulated macromolecules (fluorescein isothiocyanate-labeled dextran and lysozyme) were released on demand upto irradiation with short (10-ns) laser pulses in the near-infrared (NIR) (1064 nm) range [142]. The laser-induced release results from the gold nanoparticle-mediated heating of the capsule shell to extreme temperatures, which produces significant thermal stresses that ultimately cause the shell to rupture. On the other hand, since the pulses of NIR laser light required to induce release are short, and the laser light energy is effectively confined to the capsule shell, such releasing approach has no detrimental effect on encapsulated molecules. Compared with other light-responsive PEMs using UV–vis irradiation, NIR light is particularly attractive in biomedical applications due to the weak absorption of NIR radiation by most tissues. This concept has also been used by Volodkin and coworkers to activate the gold nanoparticle aggregates and gold-nanoparticle-integrated microcapsules absorbed on biocompatible hyaluronic acid/poly(L-lysine) films. The laser irradiation is capable of affecting, releasing, or removing the upper coatings of the films depending on the laser power [143]. Corbitt and coworkers fabricated interesting light-activated antimicrobial micro "roach motels," using microcapsules composed of poly(phenylene ethynylene)-type (PPE) conjugated polyelectrolytes (CPEs). These microcapsules are capable of capturing and entrapping bacteria such as *Cobetia marina* or *Pseudomonas aeruginosa* effectively. It was observed that more than 95% of the bacteria were killed after 1 h exposure of capsule and bacterial mixture [135, 144].

Light-responsive PEMs have been used by Jiang *et al.* to fabricate highly sensitive NIR sensing devices [145]. In their studies, nanoscale (with a thickness 20–60 nm) freely suspended composite membranes with encapsulated gold nanoparticles were obtained from spin-assisted LBL polyelectrolyte deposition [146]. The authors built a large array (64 × 64) of microscopic cavities covered by the composite membrane to test their response to ambient temperature change. This free-standing membrane demonstrated an interesting deflection change upon the variation of ambient temperature with the sensitivity of $1\,nm\,mK^{-1}$. The fabricated membrane array also showed a clear local change of optical reflection upon direct NIR radiation without any cross talk between neighboring cells of 80 μm diameter. These membrane arrays have great potential in IR microimaging applications.

Azobenzenes have been integrated into PEMs to introduce light-responsive capabilities using polyelectrolyte grafted azobenzenes. Azobenzenes have two isomeric states: a thermally stable trans configuration, and a metastable cis form.

Under irradiation, a fraction of the trans-azobenzenes will be converted to the cis form, which will thermally revert to the more stable trans configuration on a timescale determined by the molecule's particular substitution pattern [147]. This clean photochemistry gives rise to numerous remarkable photoswitching and photoresponsive behaviors observed in azobenzene PEMs. Besides the nonlinear optical properties associated with azobenzene [67, 148], one of the most interesting effects induced by light irradiation is a surface relief grating (SRG) that was first reported by Batalla and Kim [149, 150]. Specifically, when the azo-PEMs were irradiated with two coherent laser beams (which generate a sinusoidally varying light pattern at the sample surface), the materials spontaneously deformed so as to generate a sinusoidal SRG (Figure 7.7) [151, 152]. This surface-deformation response of azo systems is not only useful for patterning but also allows nanoscale motion to be induced when desired, using light. The change of PEMs surface contact angle cause by the switch of PAA-derived azobenzene was observed by Wu et al. [153]. They also demonstrated that the isomerization rates depended on the azo chromophore structure and the local environments of the condensation states.

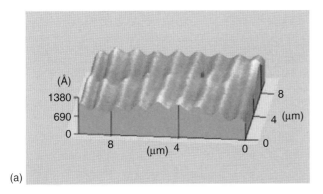

(a)

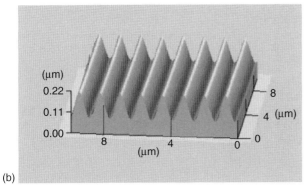

(b)

Figure 7.7 Atomic force microscopy (AFM) images of SRG formation on (a) a multilayer film (10 bilayer of pH of 11/5.5 for poly(diallyldimethylammonium chloride)(PDAC)/PAA-AN) and (b) a spin-coated film [151].

Another important application of light-responsive PEMs is to convert solar energy to electricity, in other words, photovoltaics (PVs). Organic photovoltaic (OPV) devices have emerged as one of the most promising candidates for large-scale economical solar power generation because of the low material cost and ease of processing. The compatibility of the materials with large area, lightweight, and flexible substrates makes OPV devices extremely attractive as mobile devices. With the capability of incorporating light-responsive materials such as CPE, fullerene derivatives, and semiconductive nanoparticles, PEMs have also been extensively investigated for application in organic bulk heterojunction photovoltaics (OBHPVs). An OBHPV film commonly consists of two light-active materials, an electron donor and an electron acceptor. The operating principle is based on exciton generation upon the solar radiation on light-active materials, charge separation from the excited state across an interface, followed by charge transport to the electrodes [154]. The donor/acceptor interface plays a key role in improving the device efficiency, which makes PEMs that provide molecular-level control of the architecture of the active layer of the cell an interesting candidate for OBHPV applications. Fabricating PEM OBHPV devices was pioneered by the collaborative work between Kotov and Prato [155], and Rubner and Tripathy [156], where different approaches were used to make electron donor/acceptor multilayer films. Kotov and Prato covalently bonded fullerenes to electron-donating ruthenium(II)-polypyridyl to form a dyad (efficiency = 1% at 400-nm radiation), while Rubner and Tripathy formed a layered heterojunction by first depositing electron-donating layers poly(phenylene vinylene) (PPV) and PAA, and then a second multilayer of fullerenes and PAH (efficiency = 0.01%). The lower efficiency of the latter devices is possibly due to the insulating PAA and PAH that hinder charge transport. Following these two pioneering works, CPEs and fullerene derivatives have been used to make a series of PEM-based OBHPV devices [157–159], however, with low efficiency (<1%) compared with conjugated polymer/fullerene systems (~7%) [160]. To date, the highest reported photon-to-current efficiency in PEM OBHPV is 5.5% from PPE and a water-soluble fullerene (Figure 7.8). The overall power-conversion efficiency of this system was 0.04%. The challenge to improve the efficiency may depend on how to improve

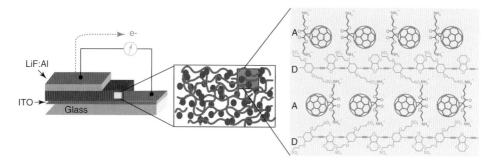

Figure 7.8 Schematic representation of the photovoltaic cell structure showing the alternating donor (D) and acceptor (A) layers, forming the active material [160].

the charge transport across the sandwich structures that are filled with alternating electron donor/acceptor materials.

7.3.5 Electrochemical Stimuli

Electrochemical stimuli, compared with previously mentioned pH, temperature, and salt concentration changes, offer rapid, reverse, and local inducement that avoid the disruption of bulk materials. Electrochemically responsive PEMs have found numerous applications in sensors, electrochromic devices, light-emitting diodes, and controlled drug release. Van Vliet and Hammond have fabricated electroactive PEM composite films containing cationic LPEI and anionic Prussian Blue (PB) nanoparticles [161]. Upon an electrochemical reduction, PB doubles the negative charge on the particles, leading to an influx of water and ions from solution to maintain the charge neutrality in the film. The influx causes the film to swell and change the mechanical properties. It was found that reverse swelling and charge of Young's elastic modulus is up to 10 and 50%, respectively. The electrochemically controlled swelling offers important implications for responsive mechanically tunable surfaces. The electrochemical reaction at electrodes can also cause the film dissolution due to the generation of ions that break the electrostatic interaction between polyelectrolytes. Boulmedais and coworkers studied the release of heparin from poly(L-lysine)/heparin multilayer films. Upon the application of an electric current, a rapid dissolution process occurred via the local decrease in pH near the anode (resulting from the production of H^+ ions). This decrease in pH induced the disruption of electrostatic bonds between the two oppositely charged polyelectrolytes, leading to a local electrodissolution of poly(L-lysine)/heparin multilayers and the formation of nanoporous films [162].

CPEs have unique properties such as water solubility, processability, variable bandgap light absorption, fluorescence, and redox properties, offering a wide range of applications in responsive PEMs including light-emitting diodes, electrochromic devices, and PVs. The Rubner group initiated research on PEM light-emitting thin-film devices using PPV [163] and polymeric tris(bipyridyl) ruthenium(II) containing multilayers [164]. Device with brightness levels of 1000 cd m^{-2} have been achieved for sequentially adsorbed films of PPV [165]. The efficiency of PEM light-emitting devices was later improved by Bazan and coworkers by optimizing the charge transport layers. They fabricated a multilayer heterostructure using a fluorene-based copolymer, poly{[9,9-bis(6 (N,N,N-trimethylammonium)hexyl)-fluorene-2,7-diyl]-*alt*-[2,5-bis(p-phenylene)-1,3,4-oxadiazole]}(PFO-PBD-NMe^{3+}), as an electron transport layer along with PEDOT:PSS (poly(3,4-ethylenedioxythiophene poly(styrenesulfonate)) as a hole transport layer, achieving a brightness of 3450 cd m^{-2} [166].

PEMs with electrochemical materials change color with an applied potential through a redox process. The LBL-based fabrication technique effectively immobilizes electrochromically active polymers or nanoparticles on an electrode to achieve efficiency access of active materials for a maximum response. The first PEM electrochromism phenomenon was reported by Schlenoff and Laurent using PEM films with positively charged polyviologen and negative charge PSS [167] while the first solid-state PEM electrochromic device was fabricated by the Hammond group, utilizing PEDOT colloids and poly(aniline) (PANI). PEDOT, a conjugated polymer, can change from a blue semiconductive polymer to a clear conductive polymer upon oxidation. Using the LBL assembly of poly(3,4-ethylenedioxythiophene) colloids doped with sulfonated poly(styrene) (PEDOT:SPS) and PANI, the cathodically coloring film was assembled from the polycation LPEI and PEDOT:SPS as the polyanion while the anodically coloring film was assembled from the polycation PANI and poly(2-acrylamido-methane-2-propanesulfonic acid) (PAMPS) as the polyanion. The resulting solid electrochromic device gave a maximum transmittance change of 30% within 1 s (Figure 7.9) [168]. The performance of the device was later greatly improved by replacing PANI with poly(hexylviologen) [169]. The combination of PEDOT's electrochromic and conducting capabilities with polyviologen's deep absorption at higher potentials led to a transmittance change of 82.1% with a switching speed of 0.6 s. In order to improve the device performance by removing inactive PSS, Reynolds and coworkers used a functionalized PEDOT and PAH and produced a device with 45% change in transmittance [170]. Other electrochromic materials that can undergo redox reaction including sulfonated poly(thienothiophene) [171], PB [172, 173], and polyoxometalates (POMs) [174–177] have also been used to

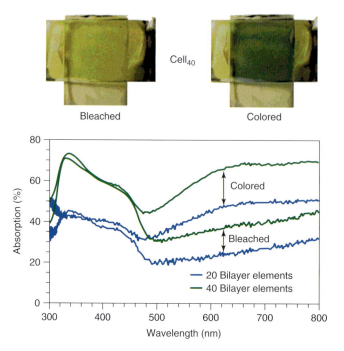

Figure 7.9 Demonstration of electrochromic properties of $(PANI/PAMPS)_{20}|PAMPS·H_2O|(LPEI/PEDOT:SPS)_{20}$ (Cell$_{20}$) and $(PANI/PAMPS)40|PAMPS·H_2O|(LPEI/PEDOT:SPS)_{40}$ (Cell$_{40}$) multilayer devices [168].

make electrochromic devices. The potential applications of these electrochromic devices include tunable windows, display panels, and smart paper.

7.3.6
Specific Molecular Interaction

Polymer coatings that are responsive to specific molecules such as biotins and DNAs have numerous applications in biosensing, controlled drug release, and tissue engineering. The easy and mild aqueous-based fabrication procedure of PEMs allows the incorporation of various environment-sensitive biomolecules into the films to generate PEMs responsive to specific molecules. Anzai and coworkers have designed PEMs that are responsive to specific small molecules by using receptor–ligand interactions. In contrast to electrostatic interaction or hydrogen bond, the interaction between tetravalent glycoprotein avidin and a 2-iminobiotin functionalized polymer is stable in aqueous media but disintegrated rapidly upon the addition of biotin. The dissolution of the film is due to the competitive binding of biotin to avidin which is stronger than that of 2-iminobiotin to avidin [177, 178]. In addition, the authors have fabricated films from glycogen and concanavalin A (a tetravalent protein that reversibly binds sugars) disintegrating upon the addition of small molecule carbohydrates that bind specifically and competitively to concanavalin A [179]. A systematic *in vivo* and *in vitro* study of controlled degradation of polysaccharide (chitosan and hyaluronan) multilayer films was performed by Picart and coworkers [180]. They found that the films can be degraded by enzymes presented in plasma and by plasma *in vitro*. The native films were rapidly degraded, while the cross-linked films are resistant to degradation *in vitro*. The *in vivo* stability of the films was studied by implanting film-coated slides in the mouse peritoneal cavity. Although cross-linked films are more stable than native films (mostly degraded in 6 days), they generated inflammatory response and the formation of fibrous tissue. These polysaccharide PEMs offer potential *in vivo* biodegradable coatings with tunable degradation rate by controlling chitosan molecular weight, degree of cross-linking, and concentration of enzymes. Shen and coworkers have reported a similar strategy to control the rate of enzymatic degradation of polypeptide (poly(L-lysine) and salmon sperm DNA)-based films. It is found that glutaraldehyde cross-linking of multilayered assemblies enhanced the stabilities of these assemblies in the presence of the enzyme trypsin [56].

Immobilization of bioactive enzymes or DNA within PEMs has been investigated to fabricate various biosensing systems. Zhou *et al.* have fabricated PEMs (20–40 nm) containing myoglobin or cytochrome P450cam and DNA grown on electrodes for rapid detection of DNA damage that could serve as a basis for *in vitro* genotoxicity screening [181]. When activated by hydrogen peroxide, the enzyme in the film generated metabolite styrene oxide from styrene that reacted with double-stranded (ds)-DNA in the same film, mimicking metabolism and DNA damage in human liver. The DNA damage was detected by square-wave voltammetry using catalytic oxidation with a Ru-complex and monitoring of the binding of a Co-complex. A novel biosensor that can continuously measure the esterase activity of NEST, a catalytically active fragment of neuropathy target esterase (NTE) has been developed by the Lee group. The biosensor was fabricated by immobilizing NEST on the top of tyrosinase(Tyr)/poly(lysine) PEMs on a gold electrode. NEST's esterase activity was measured by constant potential amperometry with a rapid response time (<10 s). The biosensor's response increased linearly with the concentration of NEST's substrate (phenyl valerate) and was inhibited in a concentration-dependent manner by a known NEST inhibitor (PMSF). This versatile approach can be applicable to other medically relevant esterases, including acetylcholinesterase (AChE) and butyrylcholinesterase (BChE) [182]. Glucose oxidase has been assembled in PEMs with PB nanoparticles [183, 184] and carbon nanotubes [59] for the detection of glucose.

The redox reaction between the polyelectrolytes in PEM films and specific molecules in solutions has been used to induce responsive film swelling. To name a few, multilayer films containing a ferrocene-derivatized polyallylamine hydrochloride (PAH-Fc) [185] and an osmium complex-derivatized polyallylamine hydrochlorides (PAHOs) [186] can swell by 10% of initial film thickness upon oxidation of the Os(II). Multilayer capsules with layered anionic and cationic polyferrocenylsilanes to form multilayer capsules expand and increase their permeability upon chemical oxidation of the ferrocene units [187]. Additionally, a poly(L-glutamic acid)/PAH multilayer film is reported to take up ferrocyanide ions from solutions and can expand and contract by 5–10% in response to electrochemical oxidation and reduction of the ferrocyanide species [188].

7.4
Conclusion and Outlook

Over the last decade, multilayered polymer films fabricated by the LBL self-assembly technique have experienced rapid growth due to the ease of the fabrication process and flexibility in incorporating various active materials into the films. The resulting multilayer films have opened the doors for complex applications including smart coatings. This chapter reviews the development of multilayered polymer films that are responsive to external stimuli and presents the applications of these smart coatings. It is clearly demonstrated that stimuli-responsive multilayer films have great potential in a wide spectrum of applications ranging from controlled drug release to energy generation. Significant research effort will be directed to improving the sensitivity of the films and the productivity of multilayered films. Traditional LBL dipping process apparently is not suitable for large-scale production and the thickness growth of the film (nanometers per bilayer) limits the applications where thick films are required. Both theoretical [189] and experimental [190, 191] studies have been performed to study the linear and exponential growth of multilayer films to obtain thick films using limited cycles of deposition. Besides traditional dipping approach, multilayer films have also been built using spraying, which has the potential of large-scale fabrication [192–194]. Joint effort among chemists, materials scientists, biologists, and engineers is necessary to push the multilayer smart coatings from laboratories to real applications. A lot of nice reviews about the stimuli-responsive polyelectrolyte multilayer films and their applications can provide future understanding of this exciting field.

References

1. Gras, S.L., Mahmud, T., Rosengarten, G., Mitchell, A., and Kalantar-Zadeh, K. (2007) *Chem. Phys. Chem.*, **8**, 2036.
2. Howarter, J.A. and Youngblood, J.R. (2007) *Adv. Mater.*, **19**, 3838.
3. Chunder, A., Etcheverry, K., Londe, G., Cho, H.J., and Zhai, L. (2009) *Colloids. Surf. A*, **333**, 187.
4. Sortino, S., Petralia, S., Compagnini, G., Conoci, S., and Condorelli, G. (2002) *Angew. Chem. Int. Ed.*, **41**, 1914.
5. Nakanishi, J., Kikuchi, Y., Takarada, T., Nakayama, H., Yamaguchi, K., and Maeda, M. (2004) *J. Am. Chem. Soc.*, **126**, 16314.
6. Nakanishi, J., Kikuchi, Y., Inoue, S., Yamaguchi, K., Takarada, T., and Maeda, M. (2007) *J. Am. Chem. Soc.*, **129**, 6694.
7. Decher, G. (1997) *Science*, **277**, 1232.
8. Lvov, Y., Decher, G., and Möehwald, H. (1993) *Langmuir*, **9**, 481.
9. Nolan, C.M., Serp, M.J., and Lyon, L.A. (2004) *Biomacromolecules*, **5**, 1940.
10. Serizawa, T., Matsukuma, D., Nanameki, K., Uemura, M., Kurusu, F., and Akashi, M. (2004) *Macromolecules*, **37**, 6531.
11. Kharlampieva, E., Erel-Unal, I., and Sukhishvili, S.A. (2007) *Langmuir*, **23**, 175.
12. Wang, L., Wang, X., Xu, M.F., Chen, D.D., and Sun, J.Q. (2008) *Langmuir*, **24**, 1902.
13. Kong, W., Zhang, X., Gao, M.L., Zhou, H., Li, W., and Shen, J.C. (1994) *Macromol. Rapid Commun.*, **15**, 405.
14. Lvov, Y., Ariga, K., Ichinose, I., and Kunitake, T. (1995) *J. Am. Chem. Soc.*, **117**, 6117.
15. Lvov, Y., Lu, Z., Chenkman, J.B., Zu, X., and Rusling, J.F. (1998) *J. Am. Chem. Soc.*, **120**, 4073.
16. Serizawa, T., Yamaguchi, M., and Akashi, M. (2002) *Macromolecules*, **35**, 8656.
17. Johnston, A.P.R., Read, E.S., and Caruso, F. (2005) *Nano Lett.*, **5**, 953.
18. Haynie, D.T., Zhang, L., Rudra, J.S., Zhao, W., Zhong, Y., and Palath, N. (2005) *Biomacromolecules*, **6**, 2895.
19. Gao, M.Y., Gao, M.L., Zhang, X., Yang, Y., Yang, B., and Shen, J.C. (1994) *Chem. Commun.*, 2777.
20. Schmitt, J. and Decher, G. (1997) *Adv. Mater.*, **9**, 61.
21. Mamedov, A.A., Belov, A., Giersig, M., Mamedova, N.N., and Kotov, N.A. (2001) *J. Am. Chem. Soc.*, **123**, 7738.
22. He, J.-A., Valluzzi, R., Yang, K., Dolukhanyan, T., Sung, C., Kumar, J., Tripathy, S.K., Samuelson, L., Balogh, L., and Tomalia, D.A. (1999) *Chem. Mater.*, **11**, 3268.
23. Huo, F.W., Xu, H.P., Zhang, L., Fu, Y., Wang, Z.Q., and Zhang, X. (2003) *Chem. Commun.*, 874.
24. Zhang, X., Gao, M.L., Kong, X.X., Sun, Y.P., and Shen, J.C. (1994) *Chem. Commun.*, 1055.
25. Ariga, K., Lvov, Y., and Kunitake, T. (1997) *J. Am. Chem. Soc.*, **119**, 2224.
26. Saremi, F. and Tieke, B. (1998) *Adv. Mater.*, **10**, 388.
27. Tedeschi, C., Caruso, F., Möehwald, H., and Kirstein, S. (2000) *J. Am. Chem. Soc.*, **122**, 5841.
28. Advincula, R.C., Fells, E., and Park, M.K. (2001) *Chem. Mater.*, **13**, 2870.
29. Ma, N., Zhang, H., Song, B., Wang, Z., and Zhang, X. (2005) *Chem. Mater.*, **17**, 5065.
30. Cho, J., Hong, J., Char, K., and Caruso, F. (2006) *J. Am. Chem. Soc.*, **128**, 9935.

31. Qi, B., Tong, X., and Zhao, Y. (2006) *Macromolecules*, **39**, 5714.
32. Nguyen, P.M., Zacharia, N.S., Verploegen, E., and Hammond, P.T. (2007) *Chem. Mater.*, **19**, 5524.
33. Liu, X.K., Zhou, L., Geng, W., and Sun, J.Q. (2008) *Langmuir*, **24**, 12986.
34. Lvov, Y., Decher, G., and Sukhourukov, G. (1993) *Macromolecules*, **26**, 5396.
35. Lvov, Y., Haas, H., Decher, G., Möehwald, H., Mikhailov, A., Mtchedlishvily, B., Morgunova, E., and Vainshtein, B. (1994) *Langmuir*, **10**, 4232.
36. Stockton, W.B. and Rubner, M.F. (1997) *Macromolecules*, **30**, 2717.
37. Wang, L.Y., Wang, Z.Q., Zhang, X., Shen, J.C., Chi, L.F., and Fuchs, H. (1997) *Macromol. Rapid Commun.*, **18**, 509.
38. Laschewsky, A., Wischeerhoff, E., Denzinger, S., Ringsdorf, H., Delcorte, A., and Bertarand, P. (1997) *Chem. Eur. J.*, **3**, 34.
39. Clark, S.L. and Hammond, P. (2000) *Langmuir*, **16**, 10206.
40. Markarian, M.Z., Moussallem, M.D., Jomaa, H.W., and Schlenoff, J.B. (2007) *Biomacromolecules*, **8**, 59.
41. Li, Q., Quinn, J.F., and Caruso, F. (2005) *Adv. Mater.*, **17**, 2058.
42. Kim, H.N., Keller, S.W., Mallouk, T.E., Schmitt, J., and Decher, G. (1997) *Chem. Mater.*, **9**, 1414.
43. Wang, F., Ma, N., Chen, Q., Wang, W., and Wang, L. (2007) *Langmuir*, **23**, 9540.
44. Lee, H., Kepley, L.J., Hong, H.G., and Mallouk, T.E. (1988) *J. Am. Chem. Soc.*, **110**, 618.
45. Xiong, H.M., Cheng, M.H., Zhou, Z., Zhang, X., and Shen, J.C. (1998) *Adv. Mater.*, **10**, 529.
46. Shimazaki, Y., Mitsuishi, M., Ito, S., and Yamamoto, M. (1997) *Langmuir*, **13**, 1385.
47. Anzai, J., Kobayashi, Y., Nakamura, N., Nishimura, M., and Hoshi, T. (1999) *Langmuir*, **15**, 221.
48. Anzai, J. and Kobayashi, Y. (2000) *Langmuir*, **16**, 2851.
49. Suzuki, I., Egawa, Y., Mizukawa, Y., Hoshi, T., and Anzai, J. (2002) *Chem. Commun.*, 164.
50. Crespo-Biel, O., Dordi, B., Reinhoudt, D.N., and Huskens, J. (2005) *J. Am. Chem. Soc.*, **127**, 7594.
51. Ogawa, Y., Arikawa, Y., Kida, T., and Akashi, M. (2008) *Langmuir*, **24**, 8606.
52. Hiller, J.A., Mendelsohn, J.D., and Rubner, M.F. (2002) *Nat. Mater.*, **1**, 59.
53. Daubine, F., Cortial, D., Ladam, G., Atmani, H., Haikel, Y., Voegel, J.-C., Clezardin, P., and Bekirane-Jessel, N. (2009) *Biomaterials*, **30**, 6367.
54. Chung, A.J. and Rubner, M.F. (2002) *Langmuir*, **18**, 1176.
55. Wood, K.C., Boedicker, J.Q., Lynn, D.M., and Hammond, P.T. (2005) *Langmuir*, **21**, 1603.
56. Ren, K.F., Ji, J., and Shen, J.C. (2006) *Bioconjug. Chem.*, **17**, 77.
57. Sun, Y.P., Zhang, X., Sun, C.Q., Wang, B., and Shen, J.C. (1996) *Macromol. Chem. Phys.*, **197**, 147.
58. Lvov, Y. and Caruso, F. (2001) *Anal. Chem.*, **73**, 4212.
59. Wang, Y., Joshi, P.P., Hobbs, K.L., Johnson, M.B., and Schmidtke, D.W. (2006) *Langmuir*, **22**, 9776.
60. Calvo, E.J., Danilowicz, C., and Wolosiuk, A. (2002) *J. Am. Chem. Soc.*, **124**, 2452.
61. Lvov, Y., Yamada, S., and Kunitake, T. (1997) *Thin Solid Films*, **300**, 107.
62. Balasubramanian, S., Wang, X.G., Wang, H.C., Yang, K., Kumar, J., Tripathy, S.K., and Li, L. (1998) *Chem. Mater.*, **10**, 1554.
63. Van Cott, K.E., Guzy, M., Neyman, P., Brands, C., Heflin, J.R., Gibson, H.W., and Davis, R.M. (2002) *Angew. Chem. Int. Ed.*, **41**, 3236.
64. Kang, E.-H., Jin, P.C., Yang, Y.Q., Sun, J.Q., and Shen, J.C. (2006) *Chem. Commun.*, 4332.
65. Kang, E.-H., Bu, T.J., Jin, P.C., Sun, J.Q., Yang, Y.Q., and Shen, J.C. (2007) *Langmuir*, **23**, 7594.
66. Lowman, G.M., Tokuhisa, H., Lutkenhaus, J.L., and Hammond, P.T. (2004) *Langmuir*, **20**, 9791.
67. Guldi, D.M., Zilbermann, I., Anderson, G., Kotov, N.A.,

Tagmatarchise, N., and Prato, M. (2005) *J. Mater. Chem.*, **15**, 114.
68. Guldi, D.M., Rahman, G.M.A., Prato, M., Jux, N., Qin, S., and Ford, W. (2005) *Angew. Chem. Int. Ed.*, **44**, 2015.
69. Leväsalmi, J.-M. and McCarthy, T.J. (1997) *Macromolecules*, **30**, 1752.
70. Krasemann, L. and Tieke, B. (1998) *J. Membr. Sci.*, **150**, 23.
71. Bruening, M.L. and Sullivan, D.M. (2002) *Chem. Eur. J.*, **8**, 3833.
72. Park, M.-K., Deng, S., and Advincula, R.C. (2004) *J. Am. Chem. Soc.*, **126**, 13723.
73. Ball, V., Voegel, J.-C., and Schaaf, P. (2005) *Langmuir*, **21**, 4129.
74. Donath, E., Sukhorukov, G.B., Caruso, F., Davis, S.A., and Möehwald, H. (1998) *Angew. Chem. Int. Ed.*, **37**, 2202.
75. Caruso, F., Yang, W., Trau, D., and Renneberg, R. (2000) *Langmuir*, **16**, 8932.
76. Caruso, F., Caruso, R.A., and Möehwald, H. (1998) *Science*, **282**, 1111.
77. Sukhorukov, G.B., Antipov, A.A., Voigt, A., Donath, E., and Möehwald, H. (2001) *Macromol. Rapid Commun.*, **22**, 44.
78. Kang, E.-H., Liu, X.K., Sun, J.Q., and Shen, J.C. (2006) *Langmuir*, **22**, 7894.
79. Zhai, L., Cebeci, F.C., Cohen, R.E., and Rubner, M.F. (2004) *Nano Lett.*, **4**, 1349.
80. Yoo, D., Shiratori, S.S., and Rubner, M.F. (1998) *Macromolecules*, **31**, 4309.
81. Shiratori, S.S. and Rubner, M.F. (2000) *Macromolecules*, **33**, 4213.
82. Mendelsohn, J.D., Barrett, C.J., Chan, V.V., Pal, A.J., Mayes, A.M., and Rubner, M.F. (2000) *Langmuir*, **16**, 5017.
83. Lindor, M.R., Auch, M., and Möehwald, H. (1998) *J. Am. Chem. Soc.*, **120**, 178.
84. Mamedov, A.A., Kotov, N.A., Prato, M., Guldi, D.M., Wicksted, J.P., and Hirsch, A. (2002) *Nat. Mater.*, **1**, 190.
85. Lee, S.W., Kim, B.-S., Chem, S., Yang, S.-H., and Hammond, P.T. (2009) *J. Am. Chem. Soc.*, **131**, 671.
86. Joly, S., Kane, R., Radzilowski, L., Wang, T., Wu, A., Cohen, R.E., Thomas, E.L., and Rubner, M.F. (2000) *Langmuir*, **16**, 1354.
87. Lösche, M., Schmitt, J., Decher, G., Bouwman, W.G., and Kjaer, K. (1998) *Macromolecules*, **31**, 8893.
88. Schmitt, J., Grünewald, T., Decher, G., Pershan, P.S., Kjaer, K., and Lösche, M. (1993) *Macromolecules*, **26**, 7058.
89. Kellogg, G.J., Mayes, A.M., Stockton, W.B., Ferreira, M., and Rubner, M.F. (1996) *Langmuir*, **12**, 5109.
90. Yang, S.Y. and Rubner, M.F. (2002) *J. Am. Chem. Soc.*, **124**, 2100.
91. Sukhishvili, S.A. and Granick, S. (2000) *J. Am. Chem. Soc.*, **122**, 9550.
92. Sukhishvili, S.A. and Granick, S. (2002) *Macromolecules*, **35**, 301.
93. Serizawa, T., Nanameki, K., Yamamoto, K., and Akashi, M. (2002) *Macromolecules*, **35**, 2184.
94. Berbreiter, D.E. and Caraway, J.W. (1996) *J. Am. Chem. Soc.*, **118**, 6092.
95. Hu, T. and Wu, C. (2001) *Macromolecules*, **34**, 6802.
96. Jaber, J.A. and Schlenoff, J.B. (2005) *Macromolecules*, **38**, 1300.
97. Matsuda, T., Saito, Y., and Shoda, K. (2007) *Biomacromolecules*, **8**, 2345.
98. Keller, S.W., Johnson, S.A., Yonemoto, E.H., Brigham, E.S., and Mallouk, T.E. (1995) *J. Am. Chem. Soc.*, **117**, 12879.
99. Chan, E., Lee, D., Ng, M., Wu, G., Lee, K., and Yu, L. (2002) *J. Am. Chem. Soc.*, **124**, 12238.
100. Such, G.K., Quinn, J.F., Quinn, A., Tjipto, E., and Caruso, F. (2006) *J. Am. Chem. Soc.*, **128**, 9318.
101. Kohli, P., Taylor, K.K., Harris, J.J., and Blanchard, G.J. (1998) *J. Am.Chem. Soc.*, **120**, 11962.
102. Sukhishvili, S.A. (2005) *Curr. Opin. Colloid Interface Sci.*, **10**, 37.
103. Ibarz, G., Döhne, L., Donath, E., and Möehwald, H. (2001) *Adv. Mater.*, **13**, 1324.
104. Kovacevic, D., van der Burgh, S., de Keizer, A., and Stuart, M.A.C. (2002) *Langmuir*, **18**, 5607.
105. Fery, A., Schöler, B., Cassagneau, T., and Caruso, F. (2001) *Langmuir*, **17**, 3779.

106. Dubas, S.T. and Schlenoff, J.B. (2001) *Langmuir*, **17**, 7725.
107. Sukhorukov, G.B., Fery, A., Brumen, M., and Möehwald, H. (2004) *Phys. Chem. Chem. Phys.*, **6**, 4078.
108. Antipov, A.A., Sukhorukov, G.B., Leporatti, S., Radtchenko, I.L., Donath, E., and Möehwald, H. (2002) *Colloids Surf., A Physicochem. Eng. Asp.*, **535**, 198.
109. Antipov, A.A., Sukhorukov, G.B., and Möehwald, H. (2003) *Langmuir*, **19**, 2444.
110. Nolte, A.J., Takane, N., Hindman, E., Caynor, W., Rubner, M.F., and Cohen, R.E. (2007) *Macromolecules*, **40**, 5479.
111. Vázques, E., Dewitt, D.M., Hammond, P.T., and Lynn, D.M. (2002) *J. Am. Chem. Soc.*, **124**, 13992.
112. Fredin, N.J., Zhang, J., and Lynn, D.M. (2005) *Langmuir*, **21**, 5803.
113. Liu, X., Yang, J.W., Miller, A.D., Nack, E.A., and Lynn, D.M. (2005) *Macromolecules*, **38**, 7907.
114. Funhoff, A.M., van Nostrum, C.F., Janssen, A.P.C.A., Fens, M.H.A.M., Crommelin, D.J.A., and Hennink, W.E. (2004) *Pharm. Res.*, **21**, 170.
115. Veron, L., Ganee, A., Charreyre, M.T., Pichot, C., and Delair, T. (2004) *Macromol. Biosci.*, **4**, 431.
116. Luten, J., Akeroyd, N., Funhoff, A., Lok, M.C., Talsma, H., and Hennink, W.E. (2006) *Bioconjug. Chem.*, **17**, 1077.
117. Zhang, J. and Lynn, D.M. (2007) *Adv. Mater.*, **19**, 4218.
118. Choi, J. and Rubner, M.F. (2001) *J. Macromol. Sci., Pure Appl. Chem.*, **A38**, 1191.
119. Zhai, L., Nolte, A.J., Rubner, M.F., and Cohen, R.E. (2004) *Macromolecules*, **37**, 6113.
120. Glinel, K., Déjugnat, C., Prevot, M., Schöler, B., Schönhoff, M., and Klitzing, R. (2007) *Colloids Surf., A*, **303**, 3.
121. Petrov, A.I., Antipov, A.A., and Sukhorukov, G.B. (2003) *Macromolecules*, **36**, 10079.
122. Déjugnat, C. and Sukhorukov, G.B. (2004) *Langmuir*, **20**, 7265.
123. Déjugnat, C. and Sukhorukov, G.B. (2006) in *Responsive Polymer Materials: Design and Applications* (ed. S. Minko), Blackwell Publishing Professional, Ames, IA. p. 229.
124. Déjugnat, C., Halozan, D., and Sukhorukov, G.B. (2005) *Macromol. Rapid Commun.*, **26**, 961.
125. Mauser, T., Déjugnat, C., and Sukhorukov, G.B. (2004) *Macromol. Rapid Commun.*, **25**, 1781.
126. Mauser, T., Déjugnat, C., Möehwald, H., and Sukhorukov, G.B. (2006) *Langmuir*, **22**, 5888.
127. Mauser, T., Déjugnat, C., and Sukhorukov, G.B. (2006) *J. Phys. Chem. B*, **110**, 20246.
128. Kharlampieva, E. and Sukhishvili, S.A. (2003) *Langmuir*, **19**, 1235.
129. Kharlampieva, E. and Sukhishvili, S.A. (2004) *Langmuir*, **20**, 10712.
130. Cho, J. and Caruso, F. (2003) *Macromolecules*, **36**, 2845.
131. Ono, S.S. and Decher, G. (2006) *Nano Lett.*, **6**, 592.
132. Serpe, M.J., Jones, C.D., and Lyon, L.A. (2003) *Langmuir*, **19**, 8759.
133. Kharlampieva, E., Kozlovskaya, V., Tyutina, J., and Sukhishvili, S.A. (2005) *Macromolecules*, **38**, 10523.
134. Jaber, J.A. and Schlenoff, J.B. (2005) *Macromolecules*, **38**, 1300.
135. Corbitt, T.S., Sommer, J.R., Chemburu, S., Ogawa, K., Ista, L.K., Lopez, C.P., Whitten, D.G., and Schanze, K.S. (2009) *ACS Appl. Mater. Interfaces*, **1**, 48.
136. Kozlovskaya, V., Ok, S., Sousa, A., Libera, M., and Sukhishvili, S.A. (2003) *Macromolecules*, **36**, 8590.
137. Quinn, J.F. and Caruso, F. (2004) *Langmuir*, **20**, 20.
138. Chunder, A., Sarkar, S., Yu, Y., and Zhai, L. (2007) *Colloids Surf., B*, **58**, 172.
139. Swiston, A.J., Cheng, C., Um, S.H., Irvine, D.J., Cohen, R.E., and Rubner, M.F. (2008) *Nano Lett.*, **8**, 4446.
140. Skirtach, A.G., Antipov, A.A., Shchukin, D.G., and Sukhorukov, G.B. (2004) *Langmuir*, **20**, 6988.
141. Tao, X., Li, J., and Möehwald, H. (2004) *Chem. Eur. J.*, **10**, 3397.
142. Angelatos, A.S., Radt, B., and Caruso, F. (2005) *J. Phys. Chem. B*, **109**, 307.

143. Volodkin, D.V., Delcea, M., Möehwald, H., and Skirtach, A.G. (2009) *ACS Appl. Mater. Interfaces*, **1**, 1705.
144. Chemburu, S., Corbitt, T.S., Ista, L.K., Ji, E., Fulghum, J., Lopez, G.P., Ogawa, K., Schanze, K.S., and Whitten, D.G. (2008) *Langmuir*, **24**, 11053.
145. Jiang, C., McConney, M.E., Singamaneni, S., Merrick, E., Chen, Y., Zhao, J., Zhang, L., and Tsukruk, V.V. (2006) *Chem. Mater.*, **18**, 2632.
146. Jiang, C., Rybak, B.M., Marutsya, S., Kladitis, P., and Tsukruk, V.V. (2005) *Appl. Phys. Lett.*, **86**, 121912.
147. Barrett, C.J., Mamiya, J., Yager, K.G., and Ikeda, T. (2007) *Soft Matter*, **3**, 1249.
148. Van Cott, K.E., Guzy, M., Neyman, P., Brands, C., Heflin, J.R., Gibson, H.W., and Davis, R.M. (2002) *Angew. Chem. Int. Ed.*, **41**, 3236.
149. Rochon, P., Batalla, E., and Natansohn, A. (1995) *Appl. Phys. Lett.*, **66**, 136.
150. Kim, D.Y., Tripathy, S.K., Li, L., and Kumar, J. (1995) *Appl. Phys. Lett.*, **66**, 1166.
151. Lee, S.-H., Balasubramaniam, S., Kim, D.Y., Wiswanathan, N.K., Bian, S., Kumar, J., and Tripathy, S.K. (2000) *Macromolecules*, **33**, 6534.
152. Wang, H., He, Y., Tuo, X., and Wang, X. (2004) *Macromolecules*, **37**, 135.
153. Wu, L., Tuo, X., Cheng, H., Chen, Z., and Wang, X. (2001) *Macromolecules*, **34**, 8005.
154. Thompson, B.C. and Fréchet, J.M.J. (2008) *Angew. Chem. Int. Ed.*, **47**, 58.
155. Luo, C., Guldi, D.M., Maggini, M., Menna, E., Mondini, S., Kotov, N.A., and Prato, M. (2000) *Angew. Chem. Int. Ed.*, **39**, 3905.
156. Mattoussi, H., Rubner, M.F., Zhou, F., Kumar, J., Tripathy, S.K., and Chiang, L.Y. (2000) *Appl. Phys. Lett.*, **77**, 1540.
157. Li, H., Li, Y., Zhai, J., Cui, G., Liu, H., Xiao, S., Liu, Y., Lu, F., Jiang, L., and Zhu, D. (2003) *Chem. Eur. J.*, **9**, 6031.
158. Durstock, M.F., Spry, R.J., Baur, J.W., Taylor, B., and Chiang, L.Y. (2003) *J. Appl. Phys.*, **94**, 3253.
159. Mwaura, J.K., Pinto, M.R., Witker, D., Ananthakrishnan, N., Schanze, K.S., and Reynolds, J.R. (2005) *Langmuir*, **21**, 10119.
160. Chen, H.-Y., Hou, J., Zhang, S., Liang, Y., Yang, G., Yang, Y., Yu, L., Wu, Y., and Li, G. (2009) *Nat. Photonics*, **3**, 649.
161. Schmide, D.J., Cebeci, F.C., Kalcioglu, Z.I., Wyman, S.G., Ortiz, C., Van Vliet, K.J., and Hammond, P.T. (2009) *ACS Nano*, **3**, 2207.
162. Boulmedais, F., Tang, C.S., Keller, B., and Vörös, J. (2006) *Adv. Funct. Mater.*, **16**, 63.
163. Fou, A.C., Onitsuka, O., Ferriera, M., Hsieh, B., and Rubner, M.F. (1996) *J. Appl. Phys.*, **79**, 7501.
164. Onitsuka, O., Fou, A.C., Ferriera, M., Hsieh, B., and Rubner, M.F. (1996) *J. Appl. Phys.*, **80**, 4067.
165. Durstock, M. and Rubner, M.F. (1997) *Proc. SPIE*, **3148**, 126.
166. Ma, W., Iyer, P.K., Gong, X., Liu, B., Moses, D., Bazan, G.C., and Heeger, A.J. (2005) *Adv. Mater.*, **17**, 274.
167. Laurent, D. and Schlenoff, J.B. (1997) *Langmuir*, **13**, 1552.
168. DeLongchamp, D.M. and Hammond, P.T. (2001) *Adv. Mater.*, **13**, 1455.
169. DeLongchamp, D.M., Kastantin, M., and Hammond, P.T. (2003) *Chem. Mater.*, **15**, 1575.
170. Cutler, C.A., Bouguettaya, M., and Reynolds, J.R. (2002) *Adv. Mater.*, **14**, 684.
171. Lee, B., Seshadri, V., Palko, H., and Sotzing, G.A. (2005) *Adv. Mater.*, **17**, 1792.
172. DeLongchamp, D.M. and Hammond, P.T. (2004) *Chem. Mater.*, **16**, 4799.
173. DeLongchamp, D.M. and Hammond, P.T. (2004) *Adv. Funct. Mater.*, **14**, 224.
174. Moriguchi, I. and Fendler, J.H. (1998) *Chem. Mater.*, **10**, 2205.
175. Liu, S., Kurth, D.G., Möehwald, H., and Volkmer, D. (2002) *Adv. Mater.*, **14**, 225.
176. Xue, B., Peng, J., Xin, Z., Kong, Y., Li, L., and Li, B. (2005) *J. Mater. Chem.*, **15**, 4793.
177. Inoue, H., Sato, K., and Anzai, J. (2005) *Biomacromolecules*, **6**, 27.

178. Inoue, H. and Anzai, J. (2005) *Langmuir*, **21**, 8654.
179. Sato, K., Imoto, Y., Sugama, J., Seki, S., Inoue, H., Odagiri, T., Hoshi, T., and Anzai, J. (2005) *Langmuir*, **21**, 797.
180. Picart, C., Schneider, A., Etienne, O., Mutterer, J., Schaaf, P., Egles, C., Jessel, N., and Voegel, J.-C. (2005) *Adv. Funct. Mater.*, **15**, 1771.
181. Zhou, L., Yang, J., Estavillo, C., Stuart, J.D., Schenkman, J.B., and Rusling, J.F. (2003) *J. Am. Chem. Soc.*, **125**, 1431.
182. Kohli, N., Srivastava, D., Sun, J., Richardson, R.J., Lee, I., and Worden, R.M. (2007) *Anal. Chem.*, **79**, 5196.
183. Zhao, W., Xu, J.-J., Shi, C.-G., and Chen, H.-Y. (2005) *Langmuir*, **21**, 9630.
184. Calvo, E.J., Danilowicz, C.B., and Wolosiuk, A. (2005) *Phys. Chem. Chem. Phys.*, **7**, 1800.
185. Hodak, J., Etchenique, R., Calvo, E.J., Singhal, K., and Bartlett, P.N. (1997) *Langmuir*, **13**, 2708.
186. Forzani, E.S., Perez, M.A., Teijelo, M.L., and Calvo, E.J.R. (2002) *Langmuir*, **18**, 9867.
187. Ma, Y.J., Dong, W.F., Hempenius, M.A., Möehwald, H., and Vancso, G.J. (2006) *Nat. Mater.*, **5**, 724.
188. Grieshaber, D., Voros, J., Zambelli, T., Ball, V., Schaaf, P., Voegel, J.C., and Boulmedais, F. (2008) *Langmuir*, **24**, 13668.
189. Hoda, N. and Larson, R.G. (2009) *J. Phys. Chem. B*, **113**, 4232.
190. Sun, B., Jewell, C.M., Nathaniel, N.J., and Lynn, D.M. (2007) *Langmuir*, **23**, 8542.
191. Porcel, C., Lavalle, P., Ball, V., Decher, G., Senger, B., Voegel, J.-C., and Schaaf, P. (2006) *Langmuir*, **22**, 4376.
192. Schlenoff, J.B., Dubas, S.T., and Farhat, T. (2000) *Langmuir*, **16**, 9968.
193. Izquierdo, A., Ono, S.S., Voegel, J.-C., Schaaf, P., and Decher, G. (2005) *Langmuir*, **21**, 7558.
194. Kolasinska, M., Krastev, R., Gutberlet, T., and Warszynski, P. (2009) *Langmuir*, **25**, 1224.

Further Reading

Boudou, T., Crouzier, T., Ren, K., Blin, G., and Picart, C. (2010) *Adv. Mater.*, **22**, 441.

Hammond, P.T. (2004) *Adv. Mater.*, **16**, 1271.

Johnston, A.P.R., Cortez, C., Angelatos, A.S., and Caruso, F. (2006) *Curr. Opin. Colloid Interface Sci.*, **11**, 203.

Kerdjoudh, H., Berthelemy, N., Boulmedais, F., Stoltz, J.-F., Menu, P., and Voegel, J.C. (2010) *Soft Mater*, **6**, 3722.

Londe, G., Chunder, A., Wesser, A., Zhai, L., and Cho, H.J. (2008) *Sens. Actuators B*, **132**, 431.

Lutkenhaus, J.L. and Hammond, P.T. (2007) *Soft Mater*, **3**, 804.

Lynn, D.M. (2007) *Adv. Mater.*, **19**, 4118.

Serpe, M.J., Yarmey, K.A., Nolan, C.M., and Lyon, L.A. (2005) *Biomacromolecules*, **6**, 408.

Zhang, J., Chua, L.S., and Lynn, D.M. (2004) *Langmuir*, **20**, 8015.

8
Photorefractive Polymers

Kishore V. Chellapan, Rani Joseph, and Dhanya Ramachandran

8.1
Introduction

When perturbed with electromagnetic radiation of appropriate wavelength and intensity, photorefractive (PR) systems alter their local refractive index properties. To achieve PR properties, photoconductivity and electro-optical response are required. This chapter describes the physics and chemistry of PR systems to stimuli such as electric field and light. Various processes such as photogeneration of charge carriers, charge transport, electro-optic (EO) effect, and grating formation are described in the introductory sections. Material requirements, experimental techniques, and types of PR systems are discussed in the context of the recent progress in PR polymers and their applications.

The PR effect is observed in materials showing both photoconductivity and an electric-field-dependent refractive index. The first observation of PR effect in a polymer system was made in the EO polymer bisphenol-A-diglycidylether 4-nitro-1,2-phenylenediamine doped with the charge-transporting molecule diethylamino-benzaldehyde diphenylhydrazone [1]. PR media can be used to store a replica of the incident nonuniform intensity pattern as a refractive index modulation. Illumination with uniform intensity distribution of appropriate wavelength will erase any grating that has already been written inside the material, making them suitable for reversible and real-time holography. Many applications such as multiplexed data storage [2], holographic filters [3], neural networks [4], and updatable 3D display [5] have been demonstrated on a laboratory scale, but commercial products are not yet available mainly because of the strict requirements on material quality and the complexity of the optics required to implement these systems. Early searches for better materials had been concentrated on inorganic crystals, but after the invention of polymer PR materials, significant research efforts have directed to this class of materials as well due to the ease of processability, which is an advantage over inorganic crystals. Diffraction efficiency approaching 100% and two-beam coupling (TBC) gain coefficient of $>400\,\text{cm}^{-1}$ have been achieved in PR polymers. The gain coefficients in hybrid systems combining liquid crystals or low-molecular-weight glass-forming materials and polymers have

Handbook of Stimuli-Responsive Materials. Edited by Marek W. Urban
Copyright © 2011 WILEY-VCH Verlag GmbH & Co. KGaA, Weinheim
ISBN: 978-3-527-32700-3

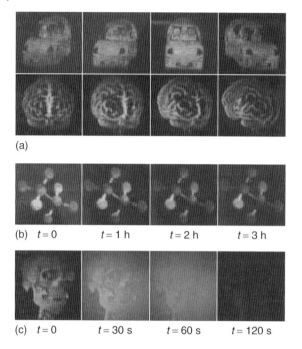

Figure 8.1 Images recorded from a distance of 75 cm from the display at different angles. (a) 3D hologram of a sports car and the hologram of a human brain recorded onto the same area after erasing the first. (b) Pictures at different times demonstrating the persistence of the hologram. (c) Erasure of the image using uniform exposure. Reprinted by permission from Macmillan Publishers Ltd: Nature, Ref. [5], Copyright 2008.

shown values reaching 3700 cm^{-1} [6, 7]. One of the demonstrated applications is an updatable holographic display which effectively utilized the persistence of recorded gratings and their erasure with uniform illumination. To achieve charge generation and transport, a copolymer of tetraphenyldiaminobiphenyl (TPD) and carbaldehyde aniline with an acrylic backbone was used in the composite. To lower the glass transition temperature, 9-ethyl carbazole was used and fluorinated dicyanostyrene was used to achieve nonlinear optical (NLO) properties. Figure 8.1 shows the 3D images recorded on this PR polymer-based display of size 4 × 4 in. [5].

Efforts are also being directed to PR polymers sensitive at fiber optic communication wavelengths. A potential application of PR polymers with infrared (IR) sensitivity is time-gated holographic imaging (TGHI), which can be used to extract information about subsurface features of biological samples. The complexity of polymeric systems arise from the difficulty in combining charge-transport and EO properties into an organized polymer matrix. The introductory sections of this chapter are intended to provide an understanding of the fundamental processes and the role of molecular units in the formation of PR gratings. The remaining sections illustrate successfully demonstrated PR polymer systems.

8.2
The Photorefractive Effect in Polymers

As the name implies, PR effect is light-induced modulation of the refractive index of a medium. Both photoconductivity, which refers to increase in electrical conductivity when the material is exposed to electromagnetic radiation of appropriate energy, and EO response, which refers to a change in therefractive index with electric field, are required for a medium to be PR.

8.2.1
Charge Generation and Transport in Polymers

Photoconducting materials are usually insulators in the absence of light but become more conductive upon irradiation with light, which means that excess movable charge carriers are generated in the material on exposure. The generation of mobile charge carriers in response to illumination requires light absorption by the polymer at that wavelength. At the same time, a successful application demands that the polymer be less absorbing at the operating wavelength. In order to meet these contradicting requirements, polymers with wider optical gaps are selected and charge-generating (CG) molecules (sensitizers) sensitive at the wavelength range of interest are doped into the polymer to extend the photosensitivity of the polymer to longer wavelengths in a controlled manner. If the polymer is hole (electron) transporting, the CG molecule is an electron acceptor (donor) which forms radical cations on photoexcitation. The charge transport is due to a series of electron-transfer processes between charged and neutral units or essentially a series of oxidation/reduction steps of molecular units involved. Excitation with light leads to the promotion of an electron from the highest occupied molecular orbital (HOMO) of the CG molecule to its lowest unoccupied molecular orbital (LUMO). The hole in the HOMO of the molecule can be filled by an electron from the HOMO of the charge-transporting polymer, thereby creating a mobile hole in the polymer backbone. Photoinduced charge transfer (CT) initiated by a sensitizer depends on many factors. The following condition must be satisfied for an efficient CT [8].

$$I_{D^*} - A_s - U_c \leq 0 \tag{8.1}$$

Here, I_{D^*} is the ionization potential of the excited state of the polymer (donor), A_s is the electron affinity of the sensitizer molecule, and U_c denotes the Coulomb attraction between the separated radicals. Polarizability of the surrounding matrix, morphology of the polymer which introduces a large separation between the donor and the acceptor, or a potential barrier preventing the separation of the electron–hole pair are factors that may inhibit an energetically allowed CT process [8, 9].

For electron-acceptor-type CG molecules, the electron density on the ring system is low due to the presence of strong electron-withdrawing groups such as $-NO_2$ and $-CN$. The chemical structures of selected electron acceptors viz,

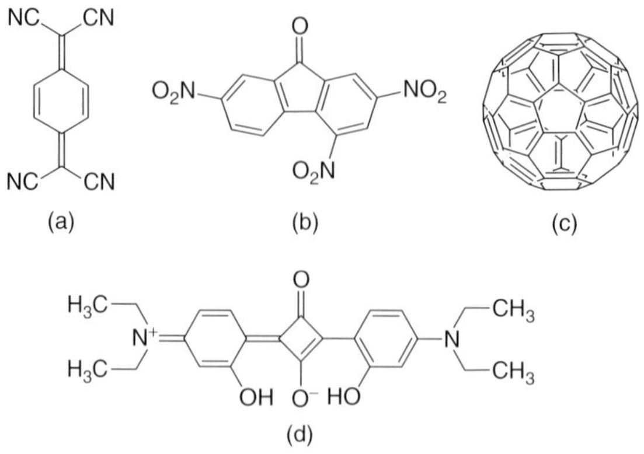

Figure 8.2 Chemical structures of charge-generating molecules: (a) 7,7,8,8-tetracyanoquinodimethane; (b) 2,4,7-trinitrofluorenone (TNF); (c) C_{60}; and (d) squarylium dye.

7,7,8,8-tetracyanoquinodimethane (a), 2,4,7-trinitrofluorenone (TNF) (b), C_{60} (c), and squarylium dye (d) are shown in Figure 8.2.

C_{60} molecule and its more soluble derivative [6,6]-phenyl C_{61}-butyric acid methyl ester (PCBM) are electron-acceptor molecules often utilized in photovoltaic devices. The ability of electron transfer from C_{60} to other molecules has been the subject of intense research and results indicate that photoinduced CT happens on a timescale of picoseconds [10]. Electron acceptors can form CT complexes with donor-type molecules and polymers that gives rise to new absorption bands in the visible and near-IR region of the spectrum. The influence of electron acceptor C_{60} on the performance of a PR polymer was demonstrated [11] by the large enhancement in steady state diffraction efficiency and initial rate of grating growth, while not creating undesired absorption effects due to ionized sensitizer molecules. Long-wavelength absorption, low reduction potential, and high triplet yield make C_{60} a better hole generator for polymer photoconductors with smaller oxidation potential than C_{60} [11]. Another class of sensitizers are phthalocyanines (Pcs) which exhibit high-absorption cross section in the visible and near-IR regions. A TBC gain coefficient of 350 cm^{-1} with response times in the millisecond range has been reported in a poly(N-vinylcarbazole) (PVK)-based PR composite sensitized with (2,9,16,23-tetra-*tert*-butyl-phthalocyaninato) zinc (ZnPc) and bis(*p*-methylbenzoate) (2,9,16,23-tetra-*tert*-butyl-phthalocyaninato) silicon (SiPc) derivatives [12]. A comparison of the photoconductivity of PVK-based PR composites sensitized with ZnPc, SiPc, and C_{60} is shown in Figure 8.3.

Other methods of sensitization include the use of semiconductor quantum dots (CdS or CdSe) or functionalizing the polymer with sensitizing moieties such as transition metal complexes [13]. For example, the studies performed by covalently attaching the ruthenium (Ru) complex Ru(bpy)$_2$(*m*-COOH-4-methylbpy)(PF$_6$)$_2$ to PVK using Friedel–Crafts acylation showed that the photogeneration efficiency of

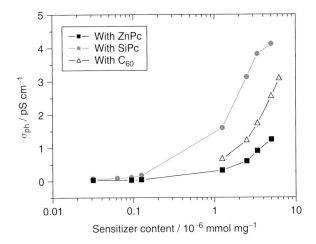

Figure 8.3 Photoconductivity of composites containing different concentrations of ZnPc (black squares), SiPc (gray circles), or C_{60} (open triangles), measured at 22 V μm^{-1}. Reprinted with permission from Ref. [12]. Copyright 2009 American Chemical Society.

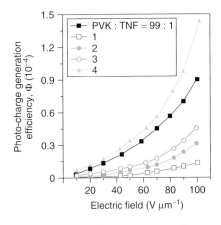

Figure 8.4 Photocharge generation efficiency of the PVK–Ru complexes 1 (open squares), 2 (closed circles), 3 (open circles), 4 (closed triangles), and PVK/TNF composite (closed squares) as a function of external electric-field strength. Reproduced with permission from RSC Publishing.

PVK–Ru complexes increased with increasing w/w% of the Ru-complex [14]. The electric-field dependence of the photogeneration efficiency of complexes containing 0.028, 0.036, 0.070, and 0.891 w/w% of the Ru-complex, labeled 1, 2, 3, and 4 respectively, are shown in Figure 8.4 along with that of a PVK–TNF system.

The role of the sensitizer is vital in the formation of the space charge field, but the concentration of the sensitizer cannot be increased with an intention to increase the PR performance. Addition of up to 1 w/w% sensitizer was found to increase the PR properties [15], but higher concentrations were found to reduce the refractive index modulation and increase the response time due to the higher density of traps produced by photoreduction of the sensitizer [16]. Thus, the optimal concentration of sensitizer in most of the reported works is around 1 w/w%, and this has the added advantage of reduced absorption as well.

Electronic interactions between the neighboring macromolecules are weak. Consequently, the valence, conduction, and exciton bands are narrow. As the dielectric constant is low, the coulombic attractions between the electrons and holes is high, giving high recombination rates and lower quantum yield of photogeneration, and external electric field is necessary for efficient charge generation and transport [17]. Photogeneration of carriers in polymers is a multistep process involving the creation of intermediate states rather than direct generation of carriers following the absorption of light [18, 19]. Also, in contrast to inorganic crystalline materials, carrier transport in polymers proceeds through hopping within a spatially random (positional disorder) and energetically disordered system of localized states [20]. The terms *positional* and *energetic* disorder imply that the distance between hopping sites and the energy required to hop from one localized state to the other vary significantly. Carrier mobility in polymers is field and temperature dependent. In the Bässler model, the mobility is temperature dependent with $\log \mu \propto 1/T^2$, and field dependent according to $\log \mu \propto E$ [21]. The photogenerated charge carriers move within the bulk of the polymer under the influence of an electric field until they are trapped. PVK, typically obtained by polymerization of N-vinylcarbazole (NVK), is the most widely studied polymeric photoconductor, and the first practical application in xerographic machines (IBM;1970) utilized PVK sensitized with TNF. A variety of polymers containing aromatic and heteroaromatic side groups or backbones were designed and synthesized as photoconductors [22].

Photoconducting polymers can be broadly classified into two types of systems. One is the so-called molecularly doped polymers (MDPs) in which the molecules giving rise to carrier generation and transport are dispersed in an inert polymer matrix such as poly(methyl methacrylate), polyvinylalcohol, polyvinylchloride, polystyrene, bisphenol-A-polycarbonate, and so on. If electron-acceptor-type CG molecules are used, charge-transporting capability is achieved by the addition of molecules with high electron density in the ring system. Examples of some charge-transport molecules, N-ethylcarbazole (ECZ) (a), 4-(N,N'-diethylamino)benzaldehydediphenyl hydrazone (b), and N,N'-diphenyl-N,N'-bis(3-methylphenyl)-[1,1'-biphenyl]-4,4'-diamine (c), are shown in Figure 8.5.

The term *molecular doping* comes from the fact that the dopants retain their identity in the host polymer. Usually, the concentration of active molecules in these systems range from 1 to 50% and thus the term *doping* is different from that used in the context of inorganic semiconductors [23]. MDPs have the advantage of design flexibility as each component is separately added to the host but the host polymer does not participate in charge transport, reducing the mobility of the carriers [24]. The upper limit of doping is usually determined by the onset of phase separation of the components, which is a disadvantage of MDPs. There are reports

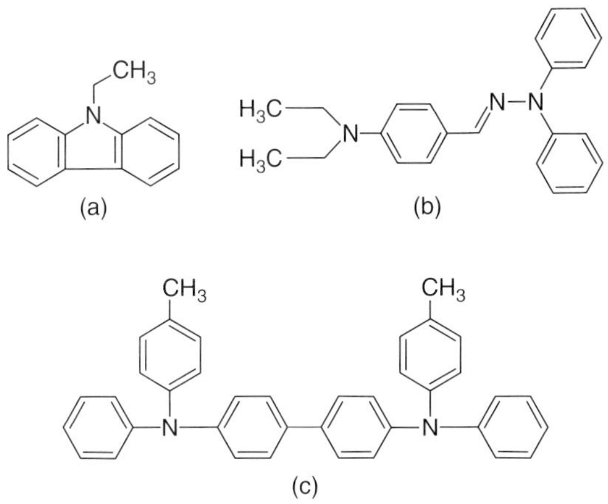

Figure 8.5 Chemical structures of some charge-transporting molecules: (a) N-ethylcarbazole; (b) 4-(N,N'-diethylamino)benzaldehydediphenyl hydrazone; and (c) N,N'-diphenyl-N,N'-bis(3-methylphenyl)-[1,1'-biphenyl]-4,4'-diamine.

of using this design approach for the fabrication of electroluminescent devices and electrophotographic photoreceptors as well [25].

Another type of charge-transporting polymers has the transport unit covalently attached to the polymer main chain. This reduces the passive volume of the host as the polymer itself takes part in the charge-transport process. PVK, first synthesized in 1934 by the polymerization NVK, is the first-known polymer photoconductor with relatively stable radical cations and high charge-carrier mobility [26]. Polymerization of NVK can be carried out by cationic, free-radical, CT, coordination, or electrochemical process [27, 28].

Photoconductivity in carbazole-containing polymers can be enhanced by introducing pendant dimeric carbazole units, in the side chain and main chain by cationic and step-growth polymerization of corresponding monomers. The ease of introducing different substituents into the carbazole ring, low cost, high thermal, and photochemical stability make carbazole-based compounds attractive as charge-transporting polymers. Some examples of carbazole-based photoconducting polymers PVK (a), poly(N-vinylcarbazole-co-styrene) (b), poly(trans-1-(3-vinyl)carbazolyl)-2-(9-carbazolyl)cyclobutane) (c), poly-(N-(2-carbazolyl) ethyl acrylate) (d), and a fully functionalized carbazole-containing polymer (e), are shown in Figure 8.6. A higher charge-carrier mobility observed in some of these polymers has been attributed to the lack of excimer-forming sites in the matrix [29, 30].

PVK is a photoconducting host for many high-performing PR composites with diffraction efficiencies approaching 100%, TBC gain of $\approx 400 \, \text{cm}^{-1}$ and refractive index contrasts of 0.01. But carrier mobility in PVK is only on the order of $10^{-7} \, \text{cm}^2 \, (\text{V s})^{-1}$ and the glass transition temperature (T_g) is very high,

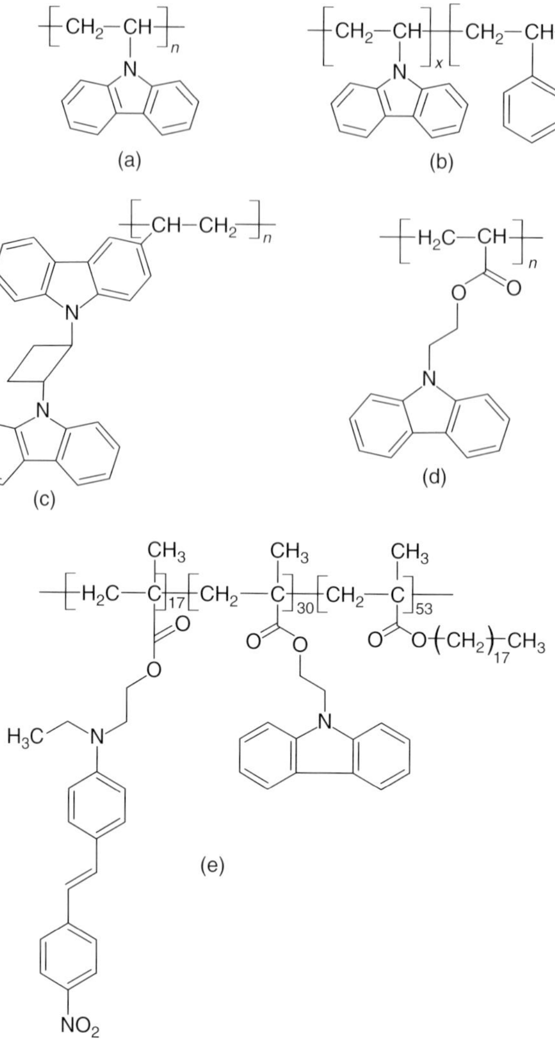

Figure 8.6 Chemical structures of pendant or in-chain electronically isolated photoconducting polymers: (a) PVK; (b) poly(N-vinylcarbazole-co-styrene); (c) poly(trans-1-(3-vinyl)carbazolyl)-2-(9-carbazolyl)cyclobutane); (d) poly-(N-(2-carbazolyl)ethyl acrylate); and (e) fully functionalized carbazole-containing polymer.

requiring the use of plasticizers. Siloxane polymer with pendant carbazole groups, which has a lower intrinsic T_g and hole-transporting properties similar to that of PVK has been utilized to prepare PR composites with refractive index modulation of $\sim 10^{-2}$ [31]. Wright et al. reported a PR system based on poly (methyl-bis-(3-methoxyphenyl)-(4-propylphenyl)amine)siloxane (MM-PSX-TAA),

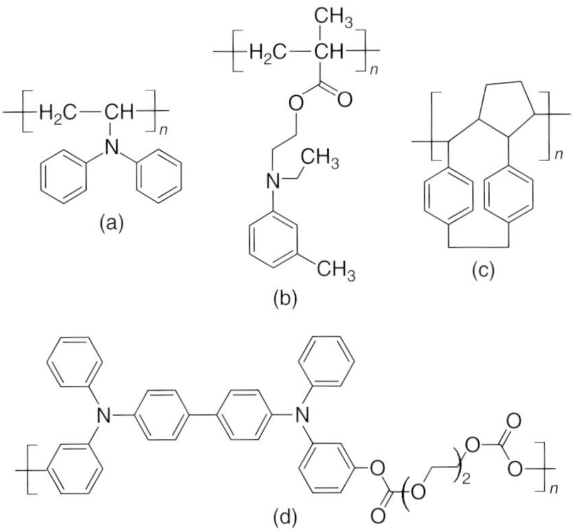

Figure 8.7 Chemical structures of photoconducting polymers: (a) poly(N-vinyldiphenylamine); (b) poly(E,E-[6,2]-paracyclophane-1,5-diene); (c) poly(2-(N-ethyl-N-3-tolylamino)ethyl methacrylate); and (d) a condensation polymer containing triphenyldiamine unit.

a charge-transporting polymer system with a siloxane backbone with pendant triarylamine groups, doped with the NLO chromophore 4-di(2-methoxyethyl) aminobenzylidene malononitrile (AODCST) and the plasticizer butyl benzyl phthalate (BBP) with response times in the millisecond range [32]. Other examples of photoconducting polymers include poly(N-vinyldiphenylamine), poly(vinylarylamines), and polymers containing triphenyldiamine moieties [33–35]. These polymers have been found to exhibit charge-carrier mobilities that exceed the values of PVK. The chemical structures of poly(N-vinyldiphenylamine) (a), poly(E,E-[6,2]-paracyclophane-1,5-diene) (b), poly(2-(N-ethyl-N-3-tolylamino)ethyl methacrylate) (c), and triphenyldiamine containing condensation polymer (d), are shown in Figure 8.7.

If a photoconducting polymer is illuminated with nonuniform light, such as an interference pattern, charge carriers are generated at the brighter regions. Owing to the concentration gradients, or owing to an applied electric field, the charges surviving geminate recombination move toward darker regions of the pattern leaving behind their countercharge. The mobility of one type of charge carrier is higher than for its countercharge. These mobile charges are eventually immobilized at electrically active defects or sites called *traps*, capable of trapping a mobile carrier. For example, addition of molecules with ionization potential lower than that of the polymer host will give rise to hole trapping. The hole remains trapped at this location until an electron from a neighboring electron-rich unit gains sufficient energy and moves to the trap. Trapping and detrapping are respectively oxidation and reduction of the spatially distributed redox sites. It has been reported that at higher number densities, such trap states can form alternate transport levels through which carrier transport occurs [36]. The distance over which the carriers move is estimated to be of the order of a few micrometers. Over time, a periodic space charge field is generated, which resembles the interference pattern. It must be noted that this space charge field is phase-shifted with respect to the initial intensity pattern due to the above-discussed migration of charges.

8.2.2
Electro-Optic Response in Polymers

In PR crystals, the refractive index is altered mainly by the linear EO effect, which results in a refractive index modulation of the form given in Equation 8.2 in response to the space charge field.

$$\Delta n(x) = -\frac{1}{2} n^3 r_{\text{eff}} E_{\text{sc}}(x) \tag{8.2}$$

where n is the average refractive index of the material and r_{eff} is the effective EO coefficient. In polymers, a field-dependent refractive index is achieved by the addition of highly polarizable molecules, usually referred to as *nonlinear optical chromophores*, to the host polymer. These molecules, with the general structure depicted in Figure 8.8 have an electron-donating-type group linked to an electron-acceptor-type group using a π-conjugated system such as benzene, azobenzene, stillbene, biphenyl, heterocycle, polyenes, or tolans. The mismatch in electron affinities lead to a delocalization of the π-electrons resulting in a molecular dipole with the acceptor side denser in electrons.

Common electron-acceptor groups attached to the π-conjugated system are $-NO_2$, $-NO$, $-CN$, $-COOH$, $-CONH_2$, $-CONHR$, $-CONR_2$, $-CHO$, $-SO_2R$, $-COR$, $-CF_3$, $-COCH_3$, $-CH=C(CN)_2$, $-SO_2$ $-NH_2$, $-N_2^+$, and $-NH_2^+$. Electron-donor groups can be $-NH_2$, $-NHCH_3$, $-N(CH_3)_2$, $-NHR$, $-N_2H_3$, $-F$, $-Cl$, $-Br$, $-I$, $-SH$, $-SR$, OR, $-CH_3$, $-OH$, $-NHCOCH_3$, $-OCH_3$, $-SCH_3$, $-OC_6H_5$, and $-COOCH_3$. A typical example of an NLO molecule is *para*-nitroaniline (pNA), in which the electron-donor NH_2 group is connected to electron-acceptor NO_2 group by an aromatic benzene ring.

The parameters that can be changed at the molecular level are the relative electron affinities of the donor and acceptor groups and the length as well as the nature of the conjugated segment connecting the donor to the acceptor. The chemical structures of some of the NLO chromophores such as 4-(dimethylamino)-4'-nitrostilbene (DANS) (a), Disperse Red 1 (b), 1-(2'-ethylhexyloxy)-2, 5-dimethyl-4-(4'-nitrophenylazo)benzene (EHDNPB) (c), N,N-diethyl-substituted

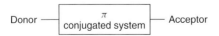

Figure 8.8 The general structure of a nonlinear optical chromophore.

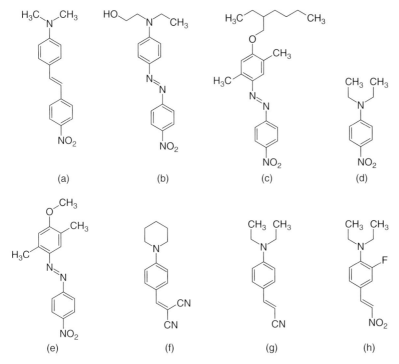

Figure 8.9 Chemical structure of selected nonlinear optical molecules: (a) 4-(dimethylamino)-4′-nitrostilbene (DANS); (b) Disperse Red 1; (c) 1-(2′-ethylhexyloxy)-2,5-dimethyl-4-(4′-nitrophenylazo)benzene; (d) N,N-diethyl-substituted *para*-nitroaniline; (e) 2,5-dimethyl-4-(*p*-nitrophenylazo)anisole; (f) 4-piperidin-4-ylbenzylidenemalononitrile; (g) 4-N,N-diethylaminocinnamonitrile; and (h) 3-fluoro-4-N,N-diethylamino-β-nitro-styrene.

pNA (d), 2,5-dimethyl-4-(*p*-nitrophenylazo)anisole (DMNPAA) (e), 4-piperidin-4-ylbenzylidenemalononitrile (f), 4-N,N-diethylaminocinnamonitrile (g), and 3-fluoro-4-N,N-diethylamino-β-nitro-styrene (h) are shown in Figure 8.9.

At the microscopic level, the dependence of the dipole moment of such a molecule on an electric field is given by Equation 8.3

$$\mu_i = \mu_i(0) + \alpha_{ij} E_j + \beta_{ijk} E_j E_k + \gamma_{ijkl} E_j E_k E_l + \ldots \tag{8.3}$$

where α, β, and γ are the polarizability, first hyperpolarizability, and second hyperpolarizability, respectively. Polymers doped with such chromophores have been shown to possess high EO coefficients [37]. Most of the molecules with large first hyperpolarizability also have a strong permanent dipole moment and an ensemble of such molecules with random distribution will have no macroscopic second-order NLO effects. Therefore, poling techniques have to be employed to break the centrosymmetry of the systems under study. Polymers with high glass transition temperatures (T_g) are poled at temperatures near the T_g by orienting the molecular dipoles along an applied electric field. In order to preserve the alignment, the temperature is lowered with the electric field applied. Burland et al. [38] provide a detailed description of poling techniques and other applications. Once poled, in high T_g polymers, the dipoles are considered frozen and the refractive index modulation is mainly due to the EO effect. Polymers with the T_g near the room temperature, can be poled *in situ* during the PR grating formation as the dipoles have orientational freedom due to the flexibility of the host. This leads to the modulation of the birefringence of the material in addition to the EO modulation, thereby resulting in a stronger PR effect. This phenomenon, called the *orientational enhancement*, was first reported by Moerner et al. [39]. A calculation of the changes in the bulk linear and nonlinear polarizabilities of such a system can be performed assuming a Maxwell–Boltzmann distribution for the dipoles in the host [40].

In order to facilitate orientational enhancement, plasticizer molecules are used to lower the T_g of the polymer manifested by higher flexibility. These molecules function through a varying degree of solvating action on the polymer. They are inserted between the polymer chains pushing them apart, which causes a reduction in the intermolecular cohesive forces. The addition of a plasticizer leads to an increase in the orientational mobility of the NLO chromophores due to the lowered T_g, which, in turn, gives rise to higher orientational birefringence (OB). The response time has been found to decrease as the T_g is reduced in a PR composite based on PVK sensitized with TNF as the photoconductor, DMNPAA as the NLO molecule, and ECZ as the plasticizer [41]. The chemical structure of selected common plasticizer molecules used in PR systems are shown in Figure 8.10.

Instead of using a passive molecule such as dioctylphthalate, active plasticizers that may contribute to either the charge-transport or EO effect can be utilized. The molecule ECZ, widely used in PR polymers, is a plasticizer that contributes to charge transport as well. It has been reported that an inert plasticizer reduces the T_g more efficiently than a charge-transporting plasticizer, but reduces the charge-carrier mobility and the PR phase shift. The molecule 4-4′-*n*-pentylcyanobiphenyl (5CB) has been reported to lower the T_g of a PVK-based PR system from 230 to 25 °C

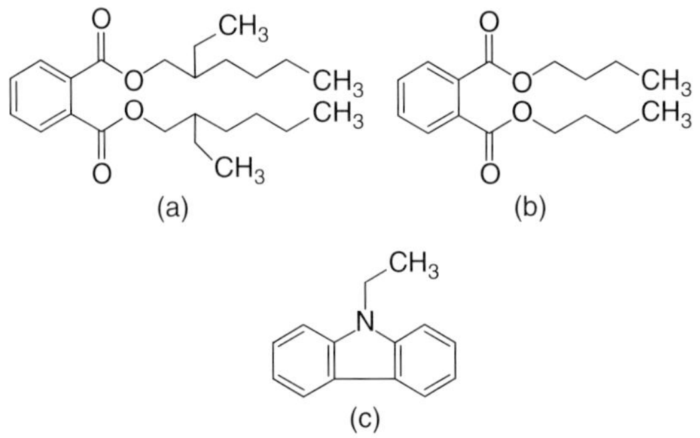

Figure 8.10 Chemical structure of inert and charge-transporting plasticizers: (a) dioctylphthalate; (b) dibutylphthalate; and (c) *N*-ethylcarbazole.

while contributing to the refractive index modulation [42]. The need for a plasticizer was eliminated by using a liquid NLO molecule achieving diffraction efficiencies of more than 60% with tens of millisecond of response times [43].

The role of push–pull molecules in PR systems is to provide refractive index modulation in response to an electric field. In polymers with a low glass transition temperature (T_g), where the molecules are relatively free to rotate due to a higher free volume, the chromophores can reorient in the applied/generated electric field. Because of the anisotropy in polarizability of the chromophores, such a reorientation will induce high refractive index anisotropy termed as *orientational birefringence*. The ability of a chromophore molecule to induce refractive index change is usually expressed as the figure of merit (FOM) defined by Equation 8.4 [44],

$$\text{FOM} = \frac{1}{M}\left[9\mu_g\beta + \frac{2\mu_g^2 \Delta\alpha}{k_B T}\right] \tag{8.4}$$

where M is the molecular mass, μ_g is the ground state dipole moment, β is the first-order hyperpolarizability, $\Delta\alpha$ is the anisotropy in linear polarizability, k_B is the Boltzmann constant, and T is the absolute temperature. The first term in Equation 8.4 represents the contribution from the linear EO effect and the second term represents the contribution of OB.

If a molecule is embedded in a polymer host with high T_g and poled, the dominant component to the refractive index modulation results from the Pockels effect. In this case, the FOM is given by Vannikov and Girishina [45],

$$\text{FOM}_{\text{PEO}} = 9\mu_g\beta_0, \text{ where } \beta_0 = \frac{6\mu_{ge}^2 \Delta\mu}{E_{ge}^2} \tag{8.5}$$

Here, μ_{ge} is the transition dipole moment between the ground and excited states, $\Delta\mu$ is the difference between the dipole moments in the excited and ground states, and E_{ge} is the transition energy. It should be noted that the calculated values of FOM of different NLO molecules can be used to predict the refractive index modulation only if they affect the space charge field formation equally and they all have equal orientational properties under the internal space charge field [46]. Apparently, this means that enhanced performance can be achieved from NLO molecules with lower FOM by tuning the properties of the host. As mentioned previously, the molecules must be properly oriented so that a centrosymmetry does not exist. The degree of alignment or the polar order in the sample is usually expressed as an order parameter [38]. The change in absorbance induced by poling can be used to estimate the polar order (φ) in the system. The order parameter can be estimated as $1 - (A_\perp/A_0)$, where A_0 and $A_\perp$ are the absorbance of unpoled and poled samples respectively, measured with the electric field of the probing light perpendicular to the poling direction [38, 47].

Concentrations of NLO molecules can be optimized on the basis of the intended application but there are trade-offs between PR performance and the allowed NLO content. As NLO molecules have high dipole moment, high concentration levels will adversely affect the carrier mobility. Also, phase separation, leading to dimers and higher order aggregates may occur at high concentrations, which affects the EO response. The energy of the HOMO levels of NLO molecules relative to the charge-transporting host influences the PR performance. For example, increasing the concentration of the NLO molecule with HOMO closest to that of PVK increases the PR speed, whereas another NLO molecule with higher HOMO level showed little improvement [48]. Doping with NLO molecules with ionization potential (I_p) lower than the transport polymer introduces deep traps in the system. Although such traps are required for longer storage times, higher density of traps might be detrimental as it directly affects the mobility of the carriers. Also, a random distribution of dipoles increases the energetic disorder in the system by an amount proportional to the square root of the dipole concentration and to the strength of the dipole moment [49]. Mobility measurements conducted on PR composites consisting of PVK doped with the NLO chromophore EHDNPB and sensitizer TNF showed that the charge transport was through the NLO molecules rather than the PVK backbone. The mobility of the carriers remained unchanged when the PVK host was replaced by the dielectric polycarbonate [50].

8.2.3
Formation of Photorefractive Gratings

As discussed above, the space charge field is phase-shifted with respect to the incident intensity pattern due to the migration of photogenerated charges. Thus, the final refractive index modulation, due to this space charge field, is phase-shifted with respect to the initial intensity distribution. Figure 8.11 shows the processes

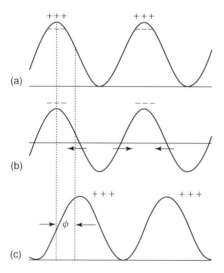

Figure 8.11 Mechanism of formation of photorefractive grating; (a) illumination pattern creating charge carriers; (b) charge redistribution and direction of internal space charge field; and (c) final refractive index grating.

involved in establishing a PR grating inside a PR medium. The first step (a) is the illumination of the medium with an interference pattern that creates charge carriers at brighter regions. During step (b), charges are redistributed and an internal space charge field is established. Step (c) shows the refractive index grating amplitude that has been phase-shifted by Φ with respect to the initial intensity pattern.

The phase shift is considered to be a unique property of PR gratings, which results in asymmetric energy exchange between two mutually coherent light beams interfering inside a PR material. This so-called TBC effect is described in the following section. The refractive index of the material can be also altered by other light-triggered phenomena such as thermal effects, photopolymerization, photochromic effects, or transient gratings resulting from nonlinear processes. However, the term *photorefractive effect* is reserved for a refractive index modulation produced only by the above-described mechanism involving light-induced charge generation, transport, and a field-induced modulation of the refractive index.

8.3
The Two-Beam Coupling Effect

As described above, the PR effect has the unique signature of a phase shift between the incident light intensity pattern and the resulting refractive index grating. Thus, if two light beams interfere inside a PR medium, one of the beams gains intensity at the expense of the other. This effect is called the *photorefractive two-beam coupling*. An experimental geometry for studying the TBC is shown in Figure 8.12.

The writing beams (I_1 and I_2) self diffract from the grating that they created inside the material. This is true for all types of holographic grating but for the non-PR case, the outgoing beams $I_{1,\text{Trans}}$, $I_{2,\text{Diffr}}$ and $I_{2,\text{Trans}}$, $I_{1,\text{Diffr}}$, interfere equally in both directions. In the case of phase-shifted gratings, such as in a PR experiment, this interference is constructive in one beam direction and destructive in the other due to the phase differences involved. This asymmetric process leads to the amplification of one beam at the expense of the other. The energy exchange in a PR medium is quantified by the gain coefficient Γ which can be estimated from the

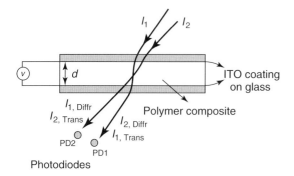

Figure 8.12 Light beam paths in a two-beam coupling experiment.

intensities $I_{1,\text{out}}$ and $I_{2,\text{out}}$ measured by the photodiodes PD1 and PD2 respectively, using Equation 8.6 [44].

$$\Gamma = \frac{\cos\theta_1}{d} \ln\left[\frac{\beta_p I_{1,\text{out}}}{I_{2,\text{out}}}\right] \tag{8.6}$$

where θ_1 is the angle of incidence of I_1 and $\beta_p = I_2/I_1$. The gain coefficient is related to the refractive index modulation Δn and the phase shift Φ by the following equation.

$$\Gamma = \frac{4\pi}{\lambda} \frac{\Delta n}{m} \sin(\Phi) \tag{8.7}$$

where λ is the wavelength and m is the modulation depth of the interference pattern given by $m = 2\beta_p^{0.5}/(1+\beta_p)$. From the above equation, it is clear that the phase difference Φ between the intensity pattern and the resulting grating plays a major role in the energy transfer process. If the TBC effect is strong enough such that the gain in intensity is greater than the losses suffered by the beam due to absorption and reflection on passing through the PR device, beam amplification can occur. Such an amplifying effect cannot be seen with other types of local gratings unless the light pattern is translated or the medium has large refractive index contrast.

8.4
High-Performance Photorefractive Polymers

The PR effect, which was once considered an undesirable phenomenon, is now being looked at as a promising candidate for reversible holographic data storage. Commercial applications of PR polymers require faster response times, low applied fields, and high stability. As no single material could provide all these properties, a variety of materials are being examined. One of the advantages of PR polymer systems over other grating recording materials such as photopolymers and photothermoplastics is the lack of postprocessing to fix the image in the medium. PR gratings are reversible enabling erasure of the stored information by uniform illumination. The main requirements for PR effect, namely, photoconductivity and EO response, can be achieved in a single material system in a variety of methods ranging from simple guest–host systems where the active molecules dispersed in an inert host to fully functional polymers where all these functionalities are provided by suitable molecular segments covalently attached to the main chain of the polymer host. Fully functionalized systems are stable against phase separation but these are not the best-performing systems. The following sections describe high-performance PR polymer systems reported so far.

8.4.1
Guest–Host Polymer Systems

The guest–host method is the simplest method to prepare a PR polymer system as it offers a variety of material combinations. Guest–host systems with glass

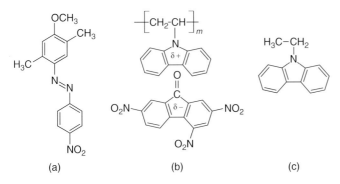

Figure 8.13 The chemical structure of the molecules used in the first high-efficiency photorefractive composite. (a) DMNPAA; (b) CT complex of PVK with TNF; and (c) ECZ.

transition (T_g) temperature near room temperature have shown high TBC gain coefficients due to the field-induced reorientation of the dipolar chromophores [45]. First-time observation of the highest PR diffraction efficiency (near 100%) was in a guest–host system reported by Meerholz et al. [51]. Figure 8.13 shows the molecules used in this system.

DMNPAA can be synthesized from p-nitrobenzenediazonium tetrafluoroborate and 2,5-dimethyl aniline by diazotization reaction. Similar to other azo dye molecules, this molecule also undergoes a trans–cis–trans photoisomerization process, which has been utilized for multilayer polarization-encoded optical data storage [52].

The EO response functions of the composite are shown in Figure 8.14. The molecular parameters estimated are the dipole moment $\mu_0 = 5.5 \pm 0.5 \times 10^{-18}$ esu, anisotropy in polarizability $\Delta \alpha = 3.9 \pm 1 \times 10^{-23}$ cm^3, and the hyperpolarizability $\beta = 48 \pm 10 \times 10^{-18}$ esu [53]. A detailed characterization of this chromophore and some other NLO molecules can be found in Ref. [46]. The composition of the PR system was DMNPAA : PVK : ECZ : TNF = 50 : 33 : 16 : 1 w/w%. The glass transition temperature of the composite was near the room temperature, which facilitated the rotational orientation of the DMNPAA molecules in the system. The TBC gain of 220 cm^{-1} at an electric field of 90 V μm^{-1} exceeded the absorption loss in the material giving a net optical gain of 207 cm^{-1}, for p-polarized light. The large diffraction efficiency observed in the system was limited by reflection and absorption losses in the PR device. The high performance of this system was ascribed to the low T_g and the higher loading of the NLO molecule with large $\Delta\alpha$ [51].

Another important PR polymer system based on PVK is the composite PVK : PDCST: BBP : C$_{60}$ in the ratio 49.5 : 35 : 15 : 0.5 wt%. At an electric field of 120 V μm^{-1}; this PR system exhibited a TBC gain coefficient of 200 cm^{-1} and a response time of 50 ms at 1 W cm^{-2} writing intensity [54]. Later, the NLO molecules in this system were replaced by the dicyanostyrene derivatives AODCST and 4-(azepan-1-yl)benzylidenemalononitrile (7-DCST) resulting in a very fast response time of ∼5 ms at 1 W cm^{-2} and at 100 V μ m^{-1} while maintaining the

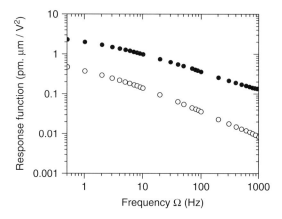

Figure 8.14 Effective electro-optic response functions of the DMNPAA : PVK : ECZ : TNF system estimated with frequency-dependent electro-optic measurements. Response functions measured at the fundamental frequency (solid circles) and its second harmonic (open circles) of different frequencies of the applied voltage. Reprinted with permission from Ref. [53]. Copyright 1996, American Institute of Physics.

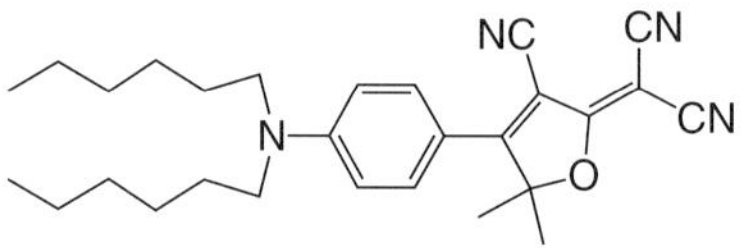

Figure 8.15 The chemical structure of 2-dicyanomethylen-3-cyano-5,5-dimethyl-4-(4′-dihexylaminophenyl)-2,5-dihydrofuran (DCDHF-6).

optical gain [55]. The effect of the NLO molecule on the performance of the PR system PVK : NLO chromophore : BBP : C_{60} was subjected to a detailed analysis by varying the structure of the NLO molecules. A series of DCST derivatives and a series of cyano ester styrene (CEST) derivatives, both with varying amine donor were used for this purpose. These systems showed very fast response times making them suitable for video-rate applications. The results indicated that, for DCST derivatives, photoconductivity was the rate-determining step rather than the orientational dynamics of the NLO molecules. In the case of CEST derivatives, no correlation between photoconductivity and speed was observed [56]. In another modification, the NLO molecule was changed to 2-dicyanomethylen-3-cyano-5,5-dimethyl-4-(4′-dihexylaminophenyl)-2,5-dihydrofuran (DCDHF-6) with the structure shown in Figure 8.15 and gain coefficients reaching 400 cm^{-1} were obtained at an electric field of 100 V μm^{-1} [57].

Sensitizer molecules in the PVK : PDCST : BBP : C_{60} system were replaced by the molecules ZnPc and SiPc to utilize the light-harvesting properties of Pcs [12]. The chemical structures of ZnPc and SiPc are shown in Figure 8.16. As the oxidation

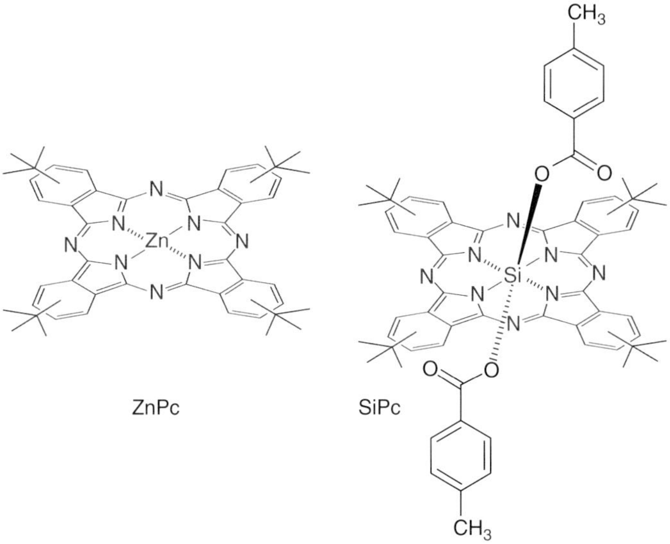

8.4 High-Performance Photorefractive Polymers

Figure 8.16 The chemical structures of ZnPc and SiPc molecules.

and reduction potentials of Pcs can be tuned by changing the central atom and the peripheral substituents, a sensitizer meeting the requirements of a particular photoconducting host may be designed.

The likelihood of aggregation of the molecules in the composite was reduced by the increased solubility due to the peripheral tertbutyl substituents. The measured photoconductivity values were clearly higher for composites with SiPc than that of composites with ZnPc or C_{60}. Similar results followed in PR experiments as well, yielding better performance for composite with SiPc as sensitizer. A TBC gain coefficient of 350 cm^{-1} and response time of 100 ms was obtained. The diffraction efficiency was 91% [12].

Despite the recognized drawbacks such as increased response time under continuous illumination and crystallization of the components, PVK has been the choice of photoconductive host. It has been suggested that chromophores with ionization potential higher than that of PVK could stabilize the response time [58]. In order to avoid these problems, new PR systems with other photoconducting hosts were investigated. The system based on poly (acrylic tetraphenyldiaminobiphenyl) (PATPD), a polymer with a hole-transporting TPD-type group as pendant to a polyacrylate backbone, has been found to be stable against performance reduction under long-term exposure [59]. This system with PATPD/7-DCST/ECZ/C_{60} (49.5 : 35 : 15 : 0.5 w/w%) showed millisecond response times with a TBC gain of 150 cm^{-1}. The structure of PATPD and 7-DCST are shown in Figure 8.17.

By changing the composition of the above-described system to 54.5 : 25 : 20 : 0.5 w/w%, the transparency of the system was adjusted so that PR experiments could be performed at 532 and 633 nm. This enabled the composite to be suitable for

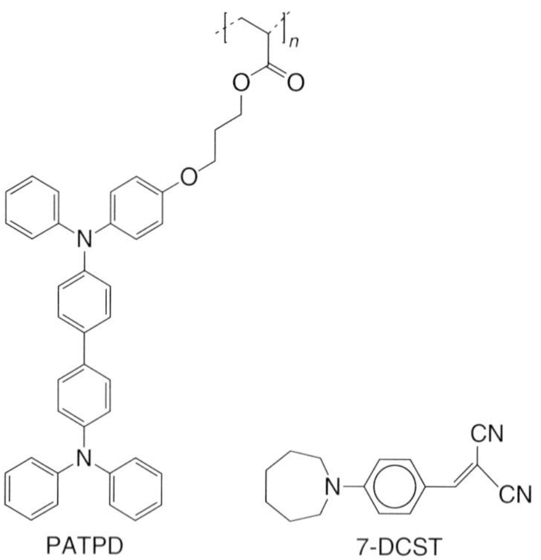

8 Photorefractive Polymers

Figure 8.17 Charge-transporting polymer and the NLO molecule used in a stable PR system with a non-PVK backbone.

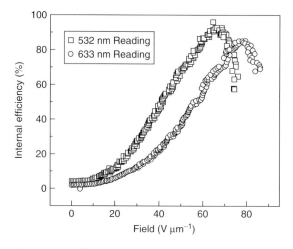

Figure 8.18 Diffraction efficiency as a function of electric field for the PR system PATPD/7-DCST/ECZ/C_{60}, (54.5 : 25 : 20 : 0.5 w/w%).

recording color information in the green and red regions [60]. Figure 8.18 shows the dependence of diffraction efficiency of this system on the applied electric field.

By employing single-pulse writing, a response time of 0.3 ms was obtained. The high voltage requirement of the PATPD/7-DCST/ECZ/C_{60} was reduced by using thinner devices made from a modified composite with another molecule (*N,N*-di-*n*-butylaniline-4-yl) l-dicyanomethylidene-2cyclohexene (DBDC)

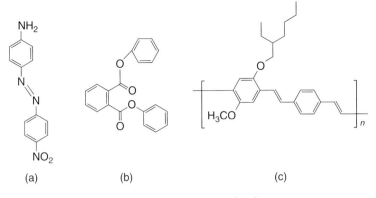

Figure 8.19 Chemical structures of (a) 4-(4-nitrophenylazo) aniline (DO3), (b) diphenyl phthalate (DPP), and (c) p-PMEH-PPV.

for stability. Devices of 20-μm thickness made from the new composite PATPD/DBDC/7-DCST/ECZ/C_{60} (39.3 : 40 : 10 : 10 : 0.7 w/w%) showed an internal diffraction efficiency of 80% at a lower applied voltage [61]. In addition to the reduced voltage requirement, the dynamics of the index modulation showed a fast time constant of 10 ms, indicating possible use in applications requiring video-rate speeds.

Attempts have been also made to prepare PR polymer systems based on conjugated polymers with delocalized π-electron system which is known to help in obtaining higher carrier mobilities. For example, PR polymer composites based on the conjugated polymer poly[o(p)-phenylenevinylene-alt-2-methoxy-5-(2-ethylhexyloxy)-p-phenylenevinylene] (p-PMEH-PPV) have been reported. Pre-poled samples of the composition p-PMEH-PPV : DPP : DO3 : C_{60} (74 : 20 : 5 : 1 w/w%) showed a high optical gain of 403 cm^{-1} with no electric field applied [62]. Chemical structures of the plasticizer molecule DPP (diphenyl phthalate), NLO molecule DO3 (4-(4-nitrophenylazo) aniline), and the polymer constituting this system are shown in Figure 8.19.

It should be noted that the highest reported gain of 3700 cm^{-1} in a polymer-based system was observed in a hybrid system with nematic liquid crystal molecules sandwiched between PVK : TNF layers [6].

8.4.2
Functionalized Polymer Systems

Fully functionalized systems are expected to be more stable against phase separation and crystallization as the molecular units responsible for the PR requirements are covalently attached to the main chain. Usually, these type of systems are designed to possess low T_g and an NLO moiety linked to the main chain via a flexible link. Low T_g can be achieved by the introduction of different alkyl side chains. One of the

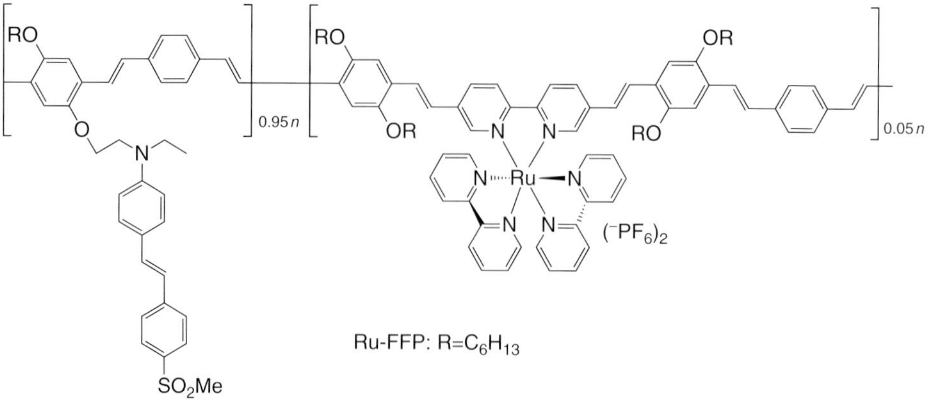

Figure 8.20 Fully functionalized PR polymer containing ruthenium complex (Ru-FFP).

best-performing fully functionalized systems is shown in Figure 8.20. This system contained Ru(II)-tri(bispyridyl) complex as the photogenerator moiety [63].

The metal-to-ligand charge transfer (MLCT) processes taking place in the Ru-complex, a process that has been extensively studied, was the primary reason for selecting this moiety. The conjugated polymer backbone was selected owing to the higher carrier mobilities observed in these types of polymers. The polymer was synthesized by following the Heck coupling method. This system, after poling, exhibited a net optical gain of more than 200 cm^{-1} under zero electric field but the response time was on the order of several hundreds of seconds. This long response time was ascribed to the undesired structural defects created in the conjugated backbone by the Heck coupling reaction [63].

Another fully functionalized system is shown in Figure 8.21. It is synthesized following the Stille coupling reaction and is composed of an NLO chromophore

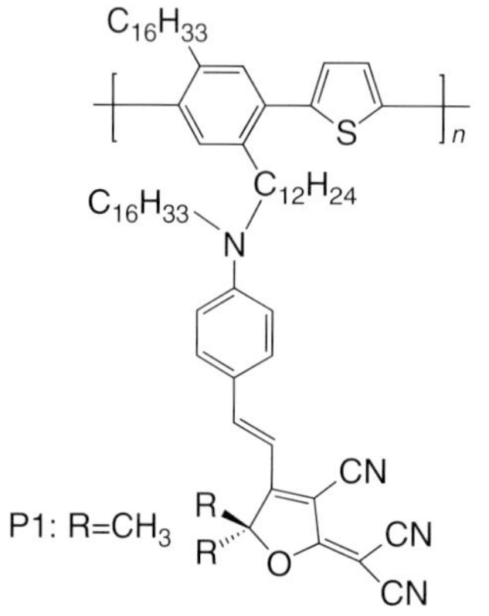

Figure 8.21 Fully functionalized PR polymer containing tricyanodihydrofuran derivative chromophore, P1.

attached onto conjugated poly(p-phenylene-thiophene) backbone [64]. This system, which operates at a wavelength of 780 nm, showed a gain coefficient of $\sim$180 cm^{-1}, with external diffraction efficiency of 68% at an electric field of 50 V μm^{-1} [64].

Although there are many fully functionalized polymer systems, the net gain from most of these systems is not as high as that of the guest–host systems and these systems have longer response times due to their more rigid structures. A large number of high-performance PR polymers have been reported in the literature. The current focus is on improving the stability of materials while maintaining faster responses and high diffraction efficiencies at lower electric fields.

8.5
Experimental Techniques

Experiments conducted on polymer sandwiched between two glass plates coated with transparent conducting electrodes such as the indium tin oxide (ITO) is a typical method to evaluate PR polymers. Usually, the ITO plate is etched with a mixture of hydrochloric acid and nitric acid at a ratio of 3 : 1, respectively, to define conducting regions, and spacers of desired thickness (30–100 μm) are used to assemble a cell. The PR material is filled into the space between the conducting plates of the cell-utilizing capillary forces. Etching helps to define the conducting areas to be covered with the polymer so that device breakdown due to leakage currents can be avoided. Device preparation depends on the specific PR composite, the main requirements are the application of an electric field and transparency to light.

8.5.1
Photoconductivity and Electro-Optic Responses

Photoconductivity can be estimated by a DC photocurrent method that measures the light-induced change in the current through the sample under illumination. If a photoconductor is irradiated with appropriate radiation, carrier generation, and recombination will take place. If J_{ph} is the photocurrent density (A cm^{-2}) measured under an electric field E (V cm^{-1}), the photoconductivity (Ω^{-1} cm^{-1}) is given by

$$\sigma_{ph} = \frac{J_{ph}}{E} \qquad (8.8)$$

The number of generated charges per absorbed photon, called the *internal quantum efficiency*, is given by Equation 8.9 [65].

$$\Phi_{int} = \frac{hc}{e\lambda \ln(\alpha) L} \frac{J_{ph}}{I} \qquad (8.9)$$

Here I is the intensity of incident light, L is the thickness of the photoconductor, α is the absorption coefficient, and λ is the wavelength of light. If I_{ph} is the photocurrent, P_0 is the light power density, A is the illuminated area, and V is the applied voltage, the change in conductivity per incident light intensity, called the *photoconductive sensitivity* (S cm W^{-1}) is given by

$$S = \frac{I_{ph}L}{P_0 AV} \tag{8.10}$$

The mobility of the carriers is measured with the time-of-flight (TOF) technique [66]. In this technique, the molecule under study is sandwiched between two electrodes, of which one is transparent to allow illumination of the sample. If needed, a carrier generation layer can be used to photogenerate and inject carriers into the polymer being studied. Illumination of the sample using a highly absorbed pulse of light through one transparent electrode creates a sheet of charge carriers, which drift under the electric field. The duration of the pulse should be lower than the time taken by the carriers to reach the other electrode. Current through the sample is monitored as the potential drop across a load resistor connected in series with the sample and power supply. The time taken for the arrival of carriers at the other electrode is determined from the knee observed in the current transient. If t_r is the time taken by the carriers under an electric field of magnitude $E = v/d$, where v is the applied voltage and d is the thickness, the *mobility*, defined as the velocity of the carriers under unit electric field, is calculated using Equation 8.11

$$\mu = \frac{d}{t_r E} \tag{8.11}$$

Other methods for the determination of carrier mobility include holographic time of flight (HTOF) and extraction current transients [67, 68]. The density of traps may be estimated with near-IR optical absorption in the case of C_{60}-sensitized PVK systems [69] or can be estimated from thermally stimulated discharge current measurements [70].

The relevant EO coefficients can be extracted from a measurement of the electric-field-induced phase shift between the two orthogonal components of the electric field vector of polarized light after the light passes through the sample. In low T_g PR systems, the refractive index modulation is due to birefringence, Pockel's effect, and Kerr effect [53]. The light-induced or field-induced birefringence $n_p - n_s$ can be calculated by keeping the sample in between two crossed polarizers and measuring the transmitted intensity as a function of time or applied field as desired [71]. If the contribution from birefringence is dominant, for characterizing the EO response, a quantity called the *electro-optic response* function is used [72].

8.5.2
Two-Beam Coupling

As explained in Section 8.3, the TBC gain coefficient is an important parameter that quantifies the strength of energy transfer between the interfering beams. Translating the interference pattern inside the sample at a rate much faster than the response time of the material can give valuable information about the PR phase shift Φ and the maximum refractive index modulation, which, in turn, can be used extract the trap density in the composite [73, 74]. Figure 8.22 shows the steady-state

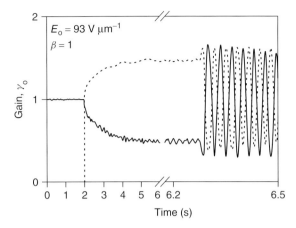

Figure 8.22 Two-beam coupling in PVK/PDCST/BBP/C$_{60}$ showing energy transfer from beam 1 (solid curve) to beam 2 (dashed curve) as beam 2 is turned on at $t = 2$ s. At $t = 6.3$ s, the interference pattern is translated quickly. Reprinted with permission from Ref. [74]. Copyright 1998 Optical Society of America.

TBC in a PVK-based system with 4-piperidinobenzylidene-malononitrile (PDCST) as NLO chromophore; an out-of-phase oscillation of detected intensities is observed when the interference pattern is translated quickly. On the basis of the dependence of Φ on the grating spacing and the applied electric field, a trap density of 1.5×10^{17} cm^{-3} was estimated in this system [74].

As was shown in Figure 8.12, the experiment is usually performed in a tilted geometry so that the applied electric field has a nonzero component along the grating vector. An ideal geometry is one in which the interfering beams create a grating along the thickness of the sample, in which case the applied electric field is along the grating vector. Such methods have been reported in the literature [75]. Transient energy transfers may be observed due to other absorption-related effects but the PR TBC effect persists in steady state. In low T_g PR systems, the direction energy transfer can be changed by switching the direction of applied field or by changing the polarization of the interfering beams. The intensity of the interfering beams is measured and must be verified for the dependence on the magnitude and direction of applied electric field. Figure 8.23 shows the dependence of the gain coefficient in a fast PR composite MM-PSX-TAA/AODCST/C$_{60}$ [32].

On the basis of the dependence of Γ on various parameters such as magnitude of electric field, intensity of the light beams, and the interbeam angles, one can deduce information about the processes and governing physical parameters at the microscopic level. A TBC effect has been observed without an applied electric field in high T_g-poled polymers and sol–gels [63, 76].

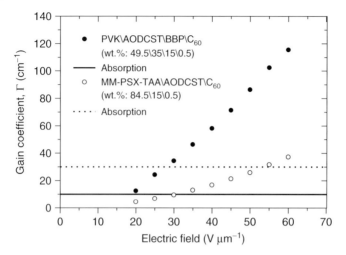

Figure 8.23 Electric-field dependence of the gain coefficient of composites of PVK/AODCST/BBP/C$_{60}$ (filled circles) and MM-PSX-TAA/AODCST/C$_{60}$ (15 w/w% AODCST, open circles). $I =$ 100 mW cm^{-2}, and $\lambda = 647$ nm. The solid (dotted) line represents the absorption coefficient of the PVK/AODCST/BBP/C$_{60}$ (MM-PSX-TAA/AODCST/C$_{60}$) sample. Reprinted with permission from Ref. [32]. Copyright 2003 American Chemical Society.

8.5.3
Diffraction Efficiency and Response Time

The diffraction efficiency of a PR polymer system can be measured in the four-wave mixing (FWM) geometry. A grating is written inside the material with two mutually coherent beams and a probe beam is made to counterpropagate one of the writing beams. The probe beam is usually weaker or has a polarization direction orthogonal to that of the writing beams so that it does not interact with the writing beams or affect the grating. The fourth wave, generated by the diffraction of the probing beam, is detected by a photodiode and the diffracted intensity is recorded. The external diffraction efficiency η^{ext} is the ratio of the intensity of the diffracted beam to that of the probing beam. The internal diffraction efficiency is given by $h^{\text{int}} = I_4/(I_4+I_{3'})$. The experimental geometry of an FWM experiment is shown in Figure 8.24.

The diffraction efficiency depends on the thickness of the device as well as the applied voltage. A simplified form of the dependence is given by

$$\eta \propto \sin^2\left[\frac{\pi \Delta n d}{\lambda(\cos\theta_1 \cos\theta_2)^{\frac{1}{2}}}\right] \tag{8.12}$$

where θ_1 and θ_2 are the writing beam angles in the polymer. The writing beams in an FWM experiment are usually s-polarized due to the fact that beam-coupling effects are minimal for this polarization state. The diffraction efficiency is one of the important parameters often quoted with performance figures of all holographic

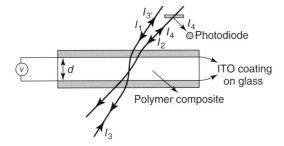

Figure 8.24 Beam paths in the four-wave mixing experiment. The diffracted light beam (J_4) is made to fall on the photodiode using a semi silvered mirror.

recording materials as it denotes the available intensity for image formation. The PR response speed is usually estimated from the grating buildup dynamics by fitting the experimental data with a stretched exponential or a biexponential function. It has been reported that if the T_g is high, the grating buildup time is limited by the rotational alignment of the NLO chromophores under the local electric field. Studies done on the PR polymer system PVK : TNF : DMNPAA : ECZ with varying chromophore content showed that, irrespective of the chromophore content, the response times of the composites were similar when the T_g of the systems were near room temperature (Figure 8.25) [77].

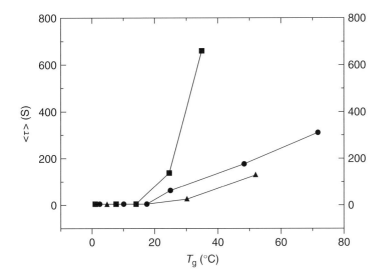

Figure 8.25 Dependence of the average response times τ on T_g for different chromophore concentrations in a high-performance PR polymer-based on PVK – 30 w/w% (triangles), 40 w/w% (circles), and 50 w/w% (squares). Reproduced with permission from Ref. [77]. Copyright 1998 Optical Society of America.

It is recognized that the dynamics of grating formation in a multicomponent PR system depends on many factors including the properties of components such as the charge-transporting polymer, sensitizer, plasticizer as well as experimental conditions such as geometry of the experiment, intensity, and wavelength of writing beams.

8.6
Conclusions

PR polymer systems have emerged as promising candidates for holographic data storage and other applications. After the first observation in 1991, the understanding of the effect has evolved considerably over the years yielding more and more high-performance PR polymer systems. Response times of the order of milliseconds and diffraction efficiencies approaching 100% have been demonstrated. Such systems are suitable for real-time image processing and other video-rate applications. Stability issues with the these systems are also being addressed; this is expected to lead to material systems capable of offering stable performance figures in the near future.

References

1. Ducharme, S., Scott, J.C., Twieg, R.J., and Moerner, W.E. (1991) Observation of the photorefractive effect in a polymer. *Phys. Rev. Lett.*, **66**, 1846–1849.
2. Steckman, G.J., Bittner, R., Meerholz, K., and Psaltis, D. (2000) Holographic multiplexing in photorefractive polymers. *Opt. Commun.*, **185**, 13–17.
3. Petrov, V.M., Lichtenberg, S., Petter, J., Tschudi, T., Chamrai, A.V., Bryksin, V.V., and Petrov, M.P. (2003) Optical on-line controllable filters based on photorefractive crystals. *J. Opt. A: Pure Appl. Opt.*, **5**, S471–S476.
4. Frauel, Y., Pauliat, G., eVilling, A., and Roosen, G. (2001) High-capacity photorefractive neural network implementing a kohonen topological map. *Appl. Opt.*, **40**, 5162–5169.
5. Tay, S., Blanche, P.A., Voorakaranam, R., Tunç, A.V., Lin, W., Roku-tanda, S., Gu, T., Flores, D., Wang, P., Li, G., Hilaire, P.S., Thomas, J., Norwood, R.A., Yamamoto, M., and Peyghambarian, N. (2008) An updatable holographic three-dimensional display. *Nature*, **451**, 694–698.
6. Bartkiewicz, S., Matczyszyn, K., Miniewicz, A., and Kajzar, F. (2001) High gain of light in photoconducting polymer-nematic liquid crystal hybrid structures. *Opt. Commun.*, **187**, 257–261.
7. Angelone, R., Ciardelli, F., Colligiani, A., Greco, F., Masi, P., Romano, A., Ruggeri, G., and Stehlé, J.L. (2008) Unconditionally stable indole-derived glass blends having very high photorefractive gain: the role of intermolecular interactions. *Appl. Opt.*, **47**, 6680–6691.
8. Brabec, C., Johannson, H., Padinger, F., Neugebauer, H., Humme-len, J., and Sariciftci, N. (2000) Photoinduced FT-IR spectroscopy and CW-photocurrent measurements of conjugated polymers and fullerenes blended into a conventional polymer matrix. *Sol. Energy Mater. Sol. Cells*, **61**, 19–33.
9. Lof, R.W., van Veenendaal, M.A., Koopmans, B., Jonkman, H.T., and Sawatzky, G.A. (1992) Band gap, excitons, and Coulomb interaction in solid C_{60}. *Phys. Rev. Lett.*, **68**, 3924–3927.
10. Smilowitz, L., Sariciftci, N.S., Wu, R., Gettinger, C., Heeger, A.J., and Wudl, F.

(1993) Photoexcitation spectroscopy of conducting-polymer-C_{60} composites: photoinduced electron transfer. *Phys. Rev. B*, **47**, 13835–13842.

11. Silence, S.M., Walsh, C.A., Scott, J.C., and Moerner, W.E. (1992) C_{60} sensitization of a photorefractive polymer. *Appl. Phys. Lett.*, **61**, 2967–2969.

12. Gallego-Gomez, F., Quintana, J.A., Villalvilla, J.M., Diaz-Garcia, M.A., Martin-Gomis, L., Fernandez-Lazaro, F., and Sastre-Santos, A. (2009) Phthalocyanines as efficient sensitizers in low-T_g hole-conducting photorefractive polymer composites. *Chem. Mater.*, **21**, 2714–2720.

13. Li, X., Chon, J.W.M., and Gu, M. (2008) Nanoparticle-based photorefractive polymers. *Aust. J. Chem.*, **61**, 317–323.

14. Oh, J.W., Choi, C.S., Jung, Y., Lee, C., and Kim, N. (2009) Photoconducting polymers containing covalently attached ruthenium complexes as a photosensitizer. *J. Mater. Chem.*, **19**, 5765–5771.

15. Ostroverkhova, O., Moerner, W.E., He, M., and Twieg, R.J. (2003) High-performance photorefractive organic glass with near-infrared sensitivity. *Appl. Phys. Lett.*, **82** (21), 3602–3604.

16. David Van Steenwinckel, E.H. and Persoons, A. (2001) Dynamics and steady-state properties of photorefractive poly(n-vinylcarbazole)-based composites sensitized with (2,4,7-trinitro-9-fluorenylidene)malononitrile in a 0–3 wt% range. *J. Chem. Phys.*, **114**, 9557.

17. Mozumder, A. (1974) Effect of an external electric field on the yield of free ions. I. General results from the onsager theory. *J. Chem. Phys.*, **60** (11), 4300–4304.

18. Arkhipov, V.I., Bässler, H., Deussen, M., Göbel, E.O., Kersting, R., Kurz, H., Lemmer, U., and Mahrt, R.F. (1995) Field-induced exciton breaking in conjugated polymers. *Phys. Rev. B*, **52**, 4932–4940.

19. Arkhipov, V.I. and Bässler, H. (2004) Exciton dissociation and charge photogeneration in pristine and doped conjugated polymers. *Phys. Status Solidi (a)*, **201**, 1152–1187.

20. Arkhipov, V.I., Heremans, P., Emelianova, E.V., and Bässler, H. (2005) Effect of doping on the density-of-states distribution and carrier hopping in disordered organic semiconductors. *Phys. Rev. B*, **71**, 045214–045221.

21. Bässler, H. (2000) Charge transport in random organic semiconductor, in *Semiconducting Polymers – Chemistry, Physics and Engineering* (eds G. Hadziioannou and P.F. van Hutten), Wiley-VCH Verlag GmbH, Germany, p. 365.

22. Mylnikov, V.S. (1994) *Photoconducting Polymers*, Advances in Polymer Science, vol. 115, Springer, Berlin/Heidelberg.

23. Sworakowski, J. and Ulański, J. (2003) Electrical properties of organic materials. *Annu. Rep. Prog. Chem. Sect. C*, **99**, 87–125.

24. Stolka, M., Yanus, J.F., and Pai, D.M. (1984) Hole transport in solid solutions of a diamine in polycarbonate. *J. Phys. Chem.*, **88**, 4707–4714.

25. Dongge Ma, D.W., Hong, Z., Zhao, X., Jing, X., and Wang, F. (1997) Bright blue electroluminescent devices utilizing poly(N-vinyl carbazole) doped with fluorescent dye. *Synth. Met.*, **91**, 331–332.

26. Hoegl, H. (1965) On photoelectric effects in polymers and their sensitization by dopants. *J. Phys. Chem.*, **69**, 755–766.

27. Reppe, W., Keyssner, E., and Dorrer, E. (1934) *DRP*, **664**, 231.

28. Grazulevicius, J., Strohriegl, P., Pielichowski, J., and Pielichowski, K. (2003) Carbazole-containing polymers: synthesis, properties and applications. *Prog. Polym. Sci.*, **28**, 1297.

29. Oshima, R., Uryu, T., and Seno, M. (1985) Improved hole mobility of polyacrylate having a carbazole chromophore. *Macromolecules*, **18**, 1043.

30. Uryu, T., Okhawa, H., and Oshima, R. (1987) Synthesis and high hole mobility of isotactic poly(2-N-carbazolylethyl acrylate). *Macromolecules*, **20**, 712.

31. Schloter, S., Hofmann, U., Strohriegl, P., Schmidt, H.W., and Haarer, D. (1998) High-performance polysiloxane-based photorefractive polymers with nonlinear optical azo,

stilbene, and tolane chromophores. *J. Opt. Soc. Am. B*, **15**, 2473–2475.
32. Wright, D., Gubler, U., and Moerner, W.E. (2003) Photorefractive properties of poly(siloxane)-triarylamine-based composites for high-speed applications. *J. Phys. Chem. B*, **107**, 4732–4737.
33. Stolka, M., Pai, D.M., Refner, D.S., and Yanus, J.C. (1983) Photoconductivity and hole transport in polymers of aromatic amine-containing methacrylates. *J. Polym. Sci. Polym. Chem. Ed.*, **21**, 969.
34. Bellmann, E., Shaheen, S.E., Grubbs, R.H., Marder, S.R., Kippelen, B., and Peyghambarian, N. (1999) Organic two-layer light-emitting diodes based on high-T_g hole-transporting polymers with different redox potentials. *Chem. Mater.*, **11**, 399.
35. Kido, J., Harada, G., and Nagai, K. (1996) Organic electroluminescent device with aromatic amine-containing polymer as a hole transport layer (II): Poly(arylene ether sulfone)-containing tetraphenylbenzidine.. *Polym. Adv. Technol.*, **7**, 31.
36. Malliaras, G.G., Krasnikov, V.V., Bolink, H.J., and Hadziioannou, G. (1995) Control of charge trapping in a photorefractive polymer. *Appl. Phys. Lett.*, **66**, 1038–1040.
37. Hayden, L.M., Sauter, G.F., Ore, F.R., and Pasillas, P.L. (1990) Second-order nonlinear optical measurements in guest-host and side-chain polymers. *J. Appl. Phys.*, **68**, 456–465.
38. Burland, D.M., Miller, R.D., and Walsh, C.A. (1994) Second-order nonlinearity in poled-polymer systems. *Chem. Rev.*, **94**, 31–75.
39. Moerner, W.E., Silence, S.M., Hache, F., and Bjorklund, G.C. (1994) Orientationally enhanced photorefractive effect in polymers. *J. Opt. Soc. Am. B*, **11** (2), 320–330.
40. Wu, J.W. (1991) Birefringent and electro-optic effects in poled polymer films: steady-state and transient properties. *J. Opt. Soc. Am. B*, **8** (1), 142–152.
41. Bittner, R., Däubler, T.K., Neher, D., and Meerholz, K. (1999) Influence of glass-transition temperature and chromophore content on the steady-state performance of poly(N-vinylcarbazole)-based photorefractive polymers. *Adv. Mater.*, **11**, 123–127.
42. Zhang, J. and Singer, K.D. (1998) Homogeneous photorefractive polymer/nematogen composite. *Appl. Phys. Lett.*, **72** (23), 2948–2950.
43. Choi, C.S., Moon, I.K., and Kim, N. (2009) New photorefractive polymer composites doped with liquid nonlinear optical chromophores. *Macromol. Res.*, **17**, 874–878.
44. Ostroverkhova, O. and Moerner, W.E. (2004) Organic photorefractives: mechanisms, materials, and applications. *Chem. Rev.*, **104**, 3267–3314.
45. Vannikov, A.V. and Girishina, A.D. (2003) The photorefractive effect in polymeric systems. *Russ. Chem. Rev.*, **72**, 471–488.
46. Moylan, C.R., Wortmann, R., Twieg, R.J., and McComb, I.H. (1998) Improved characterization of chromophores for photorefractive applications. *J. Opt. Soc. Am. B*, **15**, 929–932.
47. Sugihara, O., Kunioka, S., Nonaka, Y., Aizawa, R., Koike, Y., Kinoshita, T., and Sasaki, K. (1991) Second-harmonic generation by Cerenkov-type phase matching in a poled polymer waveguide. *J. Appl. Phys.*, **70**, 7249–7252.
48. Ostroverkhova, O. and Singer, K.D. (2002) Space-charge dynamics in photorefractive polymers. *J. Appl. Phys.*, **92** (4), 1727–1743.
49. Goonesekera, A. and Ducharme, S. (1999) Effect of dipolar molecules on carrier mobilities in photorefractive polymers. *J. Appl. Phys.*, **85**, 6506–6514.
50. West, D., Rahn, M., Im, C., and Bassler, H. (2000) Hole transport through chromophores in a photorefractive polymer composite based on poly(N-vinylcarbazole). *Chem. Phys. Lett.*, **326**, 407–412.
51. Meerholz, K., Volodin, B.L., Kippelen, B., and Peyghambarian, N. (1994) A photorefractive polymer with high optical gain and diffraction efficiency near 100%. *Nature*, **371**, 497–499.
52. Li, X., Chon, J.W.M., Wu, S., Evans, R.A., and Gu, M. (2007) Rewritable polarization-encoded multilayer data storage in

2,5-dimethyl-4-(p-nitrophenylazo)anisole doped polymer. *Opt. Lett.*, **32**, 277–279.

53. Kippelen, B., Sandalphon, Meerholz, K., and Peyghambarian, N. (1996) Birefringence, Pockels, and Kerr effects in photorefractive polymers. *Appl. Phys. Lett.*, **68**, 1748–1750, doi: 10.1063/1.116653.

54. Jepsen, A.G., Thompson, C.L., Twieg, R.J., and Moerner, W.E. (1997) High performance photorefractive polymer with improved stability. *Appl. Phys. Lett.*, **70** (12), 1515–1517.

55. Wright, D., Diaz-Garcia, M.A., Casperson, J.D., DeClue, M., Moerner, W.E., and Twieg, R.J. (1998) High-speed photorefractive polymer composites. *Appl. Phys. Lett.*, **73**, 1490–1492.

56. Diaz-Garcia, M.A., Wright, D., Casperson, J.D., Smith, B., Glazer, E., and Moerner, W.E. (1999) Photorefractive properties of poly(N-vinyl carbazole) based composites for high speed applications. *Chem. Mater.*, **11**, 1784–1791.

57. Wright, D., Gubler, U., Roh, Y., Moerner, W.E., He, M., and Twieg, R.J. (2001) High-performance photorefractive polymer composite with 2-dicyanomethylen-3-cyano-2,5-dihydrofuran chromophore. *Appl. Phys. Lett.*, **79** (26), 4274–4276.

58. Herlocker, J.A., Fuentes-Hernandez, C., Ferrio, K.B., Hendrickx, E., Blanche, P.A., Peyghambarian, N., Kippelen, B., Zhang, Y., Wang, J.F., and Marder, S.R. (2000) Stabilization of the response time in photorefractive polymers. *Appl. Phys. Lett.*, **77**, 2292–2294.

59. Thomas, J., Fuentes-Hernandez, C., Yamamoto, M., Cammack, K., Matsumoto, K., Walker, G., Barlow, S., Kippelen, B., Meredith, G., Marder, S., and Peyghambarian, N. (2004) Bistriarylamine polymer-based composites for photorefractive applications. *Adv. Mater.*, **16** (22), 2032–2036.

60. Thomas, J., Eralp, M., Tay, S., Li, G., Wang, P., Yamamoto, M., Schülzgen, A., Norwood, R., and Peyghambarian, N. (2006) *Photorefractive Polymers with Sub-millisecond Response Time*, vol. 6335, SPIE, p. 633503.

61. Eralp, M., Thomas, J., Li, G., Tay, S., Schülzgen, A., Norwood, R.A., Peyghambarian, N., and Yamamoto, M. (2006) Photorefractive polymer device with video-rate response time operating at low voltages. *Opt. Lett.*, **31** (10), 1408–1410.

62. Suh, D., Park, O., Ahn, T., and Shim, H. (2002) Large two-beam coupling in the p-PMEH-PPV/DPP/DO3/C-60. *Jpn. J. Appl. Phys.*, **41** (4A), L428–L430.

63. Peng, Z., Gharavi, A.R., and Yu, L. (1997) Synthesis and characterization of photorefractive polymers containing transition metal complexes as photosensitizer. *J. Am. Chem. Soc.*, **119**, 4622–4632.

64. You, W., Cao, S., Hou, Z., and Yu, L. (2003) Fully functionalized photorefractive polymer with infrared sensitivity based on novel chromophores. *Macromolecules*, **36**, 7014–7019.

65. Däubler, T.K., Kulikovsky, L., Neher, D., Cimrová, V., Hummelend, J., Mecher, E., Bittner, R., and Meerholz, K. (2002) in *Nonlinear Optical Transmission Processes and Organic Photorefractive Materials*, Proceedings of the SPIE, vol. 4462 (eds C.M. Lawson and K. Meerholz), SPIE, pp. 206–216.

66. West, D., Rahn, M., Im, C., and Bässler, H. (2000) Hole transport through chromophores in a photorefractive polymer composite based on poly(N-vinyl carbazole). *Chem. Phys. Lett.*, **326**, 407–412.

67. Malliaras, G.G., Krasnikov, V.V., Bolink, H.J., and Hadziioannou, G. (1995) Holographic time-of-flight measurements of the hole-drift mobility in a photorefractive polymer. *Phys. Rev. B*, **52**, R14324–R14327.

68. Juška, G., Genevičius, K., Arlauskas, K., Österbacka, R., and Stubb, H. (2002) Charge transport at low electric fields in π-conjugated polymers. *Phys. Rev. B*, **65**, 233 208.

69. Grunnet-Jepsen, A., Wright, D., Smith, B., Bratcher, M.S., DeClue, M.S., Siegel, J.S., and Moerner, W.E. (1998) Spectroscopic determination of trap density in C_{60}-sensitized photorefractive polymers. *Chem. Phys. Lett.*, **291**, 553–561.

70. Chang, C.J. and Whang, W.T. (1997) Trap characteristics study of photorefractive polymer materials by thermal

71. Termine, R., Aiello, I., Godbert, N., Ghedini1, M., and Golemme1, A. (2008) Light-induced reorientation and birefringence in polymeric dispersions of nano-sized crystals. *Opt. Express*, **16**, 6910–6920.
72. Sandalphon, Kippelen, B., Meerholz, K., and Peyghambarian, N. (1996) Ellipsometric measurements of poling birefringence, the Pockels effect, and the Kerr effect in high-performance photorefractive polymer composites. *Appl. Opt.*, **35**, 2346–2354.
73. Sutter, K. and Günter, P. (1990) Photorefractive gratings in the organic crystal 2-cyclooctylamino-5-nitropyridine doped with 7,7,8,8-tetracyanoquinodimethane. *J. Opt. Soc. Am. B*, **7** (12), 2274–2278.
74. Grunnet-Jepsen, A., Thompson, C.L., and Moerner, W.E. (1998) Systematics of two-wave mixing in a photorefractive polymer. *J. Opt. Soc. Am. B*, **15**, 905–913.
75. Kwon, O.P., Montemezzani, G., Günter, P., and Lee, S.H. (2004) High-gain photorefractive reflection gratings in layered photoconductive polymers. *Appl. Phys. Lett.*, **84**, 43–45.
76. Darracq, B., Canva, M., Chaput, F., Boilot, J.P., Riehl, D., Lévy, Y., and Brun, A. (1997) Stable photorefractive memory effect in sol-gel materials. *Appl. Phys. Lett.*, **70**, 292–294.
77. Bittner, R., Bräuchle, C., and Meerholz, K. (1998) Influence of the glass-transition temperature and the chromophore content on the grating buildup dynamics of poly(N-vinylcarbazole)-based photorefractive polymers. *Appl. Opt.*, **37**, 2843–2851.

9
Photochromic Responses in Polymer Matrices
Dhanya Ramachandran and Marek W. Urban

9.1
Introduction

Many concepts of photochromic systems have been inspired by nature. An illustrative example is the reversible shape changes of the photoreceptor molecule rhodopsin in eyes that produces a cascade of molecular rearrangements responding to light illumination [1]. Upon photoexcitation, all-trans retinal undergoes isomerization to 11-cis across the double bond, thus causing shape changes. Although this event occurs at the angstrom (Å) level, there are many molecular interconnected events and responses that are not fully understood. Similar processes have been observed in certain molecules in materials chemistry, although on a much smaller and at a less-orchestrated scale. These are known as *photochromic molecules*, which were discovered over 150 years ago in an organic crystal of tetracene, and later on, in inorganic and organometallic materials [2–4]. In spite of a long research history, a quest for new photochromic materials continues. Color changes induced by ultraviolet (UV) radiation caused by trans $\rightleftharpoons$ cis isomerization, ring opening–closing reactions, inter- or intramolecular charge transfer, or bimolecular cycloaddition reactions are of particular interest. To make use of their photochromic behavior in various applications, photochromic entities are often incorporated into polymer matrices.

While creating individual molecules and understanding their electronic properties facilitated further advances in understanding molecular mechanisms governing photochromism, useful applications require fabrication into films, sheets, fibers, or beads. This is typically achieved by incorporating photochromic molecules into macromolecular matrices by doping or dispersing, or covalently attaching them onto a polymer backbone. Since photoisomerization of chromophores alter equilibrium conformations, the dimensions of the surrounding polymer matrix play an essential role in spatial rearrangements. The dimensional classifications of the matrix effect have been extensively examined [5], with notable applications in ophthalmic lens, optical storage media, photosensors, and biomedical devices. Furthermore, the kinetics of conformational changes of polymers with photochromic moieties depends upon their physical or chemical incorporation into the matrix [6]. As one

Handbook of Stimuli-Responsive Materials. Edited by Marek W. Urban
Copyright © 2011 WILEY-VCH Verlag GmbH & Co. KGaA, Weinheim
ISBN: 978-3-527-32700-3

might anticipate, photochromic transitions are slower in polymers with a lower free volume content due to the limited space-reducing segmental motions, but may be altered by local polarity, intermolecular interactions, or steric restrictions [7–9]. Another limiting factor is molecular aggregations of a photochromic entity that may also influence the kinetics of responses [10].

9.2
Photochromic Polymeric Systems

As pointed out earlier, photochromic chromophores can be incorporated into polymer matrices either by dispersing them or covalently attaching them to a polymer backbone, and many studies focused on the fundamental understanding of mechanisms governing photochromic polymers as well as on applications ranging from photoswitching to optical data-storage devices, sensors or light-driven reactors, and artificial muscles. Although we realize that there are many outstanding photochromic materials, and scientific and patent literature provides many detailed and excellent overviews, this chapter focuses on selected examples that influence molecular structures, conformational changes, environmental effects, and kinetics of photochromic responses in polymeric matrices.

9.2.1
Influence of Molecular Structures

The effect of molecular properties such as the length and mobility of polymer chains as well as the nature of the substituents play a significant role on the responses of photochromic systems to electromagnetic radiation. The quantum yield (Φ) of photoinduced isomerization quantifies the efficiency with which the conformational change (or ring cyclization) takes place. The main factors are environments of polymer matrices and polarity [11]. The Φ-values appear to be enhanced for systems with less bulky substituents [12, 13]. As trans isomers absorb light, photoinduced trans→cis conversion at the photostationary state may be as high as 70%, depending upon associated functional groups [14]. For cis→trans isomerization to occur effectively, the molecules require enough space around them, and for bulkier molecules, to be able to exhibit rotational mobility, higher free volumes are required. Optical storage studies conducted on poly[(4-nitrophenyl)[4-[[2-(methacryloyoxy)ethyl]ethylamino]phenyl]diazene] (pDR1M) and poly[(4-nitronaphthyl)[4[-[2-(methacryloyloxy)ethyl]ethylamino]phenyl]diazene] (pNDR1M) showed that the dynamics of writing depends on the size of photochromic moiety, while the maximum achievable modulation and the rate of relaxation are size independent [15]. The decrease in the rate of writing, which chemically represents trans–cis–trans conformation changes resulted from the reduced quantum yields of photoisomerization attributed to the presence of bulkier naphthyl groups. In contrast, phenyl rings enhanced quantum yields because of their smaller critical volume. Similar trends were observed in other photochromic systems as well [16]. Most of the applications of photochromic materials make use of photoinduced reactions in the solid state, which take place in the nanorange free volume of the matrix, where the free volume occupied by the van der Waals envelope of molecular groups is sufficient. Interestingly enough, photochromic probes have been also utilized to estimate microstructure and free volume distribution inside the polymer networks [17, 18].

For fulgides, where reversible photoinduced ring cyclization is responsible for photochromic reactions, an important requirement is that the carbon atoms that form C-C single bonds come in close contact upon photoirradiation. Thus, conformational freedom of the neighboring molecular groups affects photoisomerization quantum yields, such as in diarylethene derivatives with heterocyclic aryl groups, where photochemical ring-cyclization and ring-opening reactions occur.

The effect of polymer matrices on photochromic response of diarylethene-type species has been investigated by incorporating cis-1,2-bis(2-methylbenzothiophene-3-yl)-1,2-dicyanoethene (BTCN), 1,2-bis(2-methylbenzothiophene-3-yl)maleic anhydride (BTMA), and 1,2-bis(2-methylbenzothiophene-3-yl)perfluorocyclopropene (BTF6) into polyethylene (PE), poly(methylmethacrylate) (PMMA), and polyethylene glycol grafted polysiloxane (PEG-g-PS) [19]. The quantum yield Φ of photochromic ring cyclization was highest for PEG-g-PS, which is the matrix with the highest polarity. The observed behavior was contrasted with that of a solution phase, where a higher polarity of the solvent resulted in lower quantum yields. Also, higher Φ-values in PEG-g-PS for lowest glass transition temperature (T_g) networks was observed, indicating again that a higher free volume is significant in photochromic responses and appear to be a prerequisite for effective photochromic responses over the polarity of the matrix.

Along the same lines, increased flexibility of side polymer chains also reduces the movement restrictions resulting in a greater conformational freedom. Addition of flexible oligomer plasticizers will lower the T_g, thus increasing their switching speeds [20]. Although the use of plasticizing species may be beneficial, they are usually not chemically attached to the matrix, which may result in phase separation or other adverse effects leading to diffusion or undesirable stratification. One of the permanent structural modifications to increase the flexibility of the polymer chains is the attachment of alkyl spacers between the chromophore and the polymer backbone [21]. Covalent attachment of photochromic molecules is advantageous, such as shown in polyurethane elastomers that exhibit mechanochromic properties, achieved by chemically incorporating azobenzene oligomer into a polyurethane copolymer [22].

Among other factors that may influence the photochromic effect, covalent insertion of inert spacers may minimize the steric hindrance of pendant chromophores, thus providing enhanced mobility to move away responsive species from a bulky polymer backbone [14]. Again, enhanced free volume by the incorporation of alkyl spacers in PMMA was shown to be beneficial to enhance the responses [23]. More complex behavior is shown in liquid crystalline polymers, where in addition to the length of the spacer, light-induced orientation behavior depends on the enthalpic stability of the mesophase [24].

The molecular weight of photochromic polymers is another factor that influences the rate of photoisomerization and thermal relaxations [21], but there are systems with marginal molecular-weight dependence [14]. Just as photochromic responses result from light-induced conformational changes of molecular segments that may be preceded by photochemical ring-opening or ring-closing reactions, azobenzene-functionalized poly(methyl phenylsilane) (PMPS) is an example in which light-induced conformational changes occur, but reversible reactions are induced by heating [25]. The efficiency of light-induced conformational changes in PMPS matrix containing pendant 4-nitroazobenzene group increases with the decreasing (-Si-O-) units with respect to azobenzene moiety. Also, conformational changes of polymer chains affect the rate of isomerization reaction of photochromic molecules, as evidenced by kinetic analysis of the isomerization of spiropyran-modified poly(L-glutamic acid) (PSLG) synthesized by a dicyclohexylcarbodiimide (DCC) coupling reaction [26]. Compared to that of the random coil structure, isomerization of the spiropyran side chains in PSLG in α-helix structure was significantly restrained.

9.2.2
Environmental Effects

Photochromic reactions in polymers also depend on the polarity of the polymer matrix. For example, when dispersed in or covalently attached to polymer matrices, the stability of polar merocyanine (MC) isomers in a polar matrix will be enhanced. Consequently, the reverse reaction rates in polar matrices will be reduced. Interactions between polar polymer matrices and photochromic 1,3,3-trimethylindoline-6′-nitrobenzopyrylospiran (SP) molecules resulted in a blueshift of the absorption spectrum of SP. As shown in Figure 9.1, for the poly(n-butyl methacrylate) (PnBMA) matrix with a lower T_g, larger blueshifts are observed compared to that of PMMA and styrene–butadiene–styrene (SBS) matrices. This behavior was attributed to polar and ionic interactions between PnBMA matrices and SP [27], the magnitude of these interactions though will vary depending upon the T_g of the matrix.

As one would anticipate, the dependence of electronic properties of photochromic chromophores upon environment is more pronounced in solvents, where hypsochromic or bathochromic shifts of the absorption and fluorescence spectra were observed and are strongly influenced by the polarity of solvent molecules [28]. Chromophore properties are also influenced by the physical properties of the matrix, such as viscosity, conductivity, pH, solubility, or phase-transition temperatures [29, 30]. For example, the reversible light-induced insolubility of polystyrene with pendant azobenzene in cyclohexane induced by irradiation was attributed to the reduction in solvent–polymer interactions, as the azobenzene chromophore transforms from the less polar trans form to the more polar cis form [30].

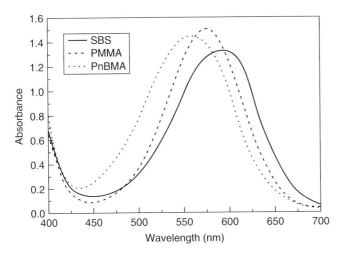

Figure 9.1 Effect of polymer matrix on the spectra of spiropyran. The absorption peak shifts toward lower wavelength in polar matrix. Larger blueshifts are observed for the poly(n-butyl methacrylate) (PnBMA) compared to poly(methyl methacrylate) (PMMA), and styrene–butadiene–styrene (SBS) matrices.

9.2.3
Kinetics of Photoreactions

Understanding mechanistic aspects of photoinduced reactions is necessary for the optimization of photochromic polymers. The rate of light-induced reactions and thermal reverse reactions, the magnitude of activation energy barriers, and the wavelength dependence are of particular importance, as they govern the response of the system.

The activation energies involved in thermal reversible reactions is typically estimated by monitoring the rate of the reverse reactions as a function of temperature. For solutions, this follows the Arrhenius equation [31]:

$$k = \nu \exp\left(-\frac{E}{RT}\right) \quad (9.1)$$

where k is the thermal decay rate, ν is a weak temperature-dependent frequency factor, E is the activation energy, R is the universal gas constant, and T is the absolute temperature [31]. The thermal reverse reaction usually follows a first-order kinetics, although there are exceptions from this behavior [32]. In many polymer matrices, thermal relaxation reactions follow a nonexponential kinetics of the form given by the equation

$$k(t) = k_0 \exp\left[-(t/\tau)^\beta\right] \quad (9.2)$$

where, τ is the relaxation time, and β ($0 < \beta \leq 1$) quantifies the extent of deviation from an exponential behavior. It has been proposed that the difference in decay kinetics between polymer matrices and solutions lies in the fact that, in polymer matrices, isomerization processes must thermally overcome a distribution of potential barriers owing to the spatially distributed polymer segments with variable dynamics [33]. Although the kinetics of relaxation reactions has been investigated, no satisfactory model describing the observed behaviors is available [34].

The kinetics of light-induced forward reactions in photochromic spiropyran derivative, l-β-(4-trifluoromethyl benzoyloxy)ethyl]-3,3-dimethyl-spiro[indoline-2,3'-[3H]-naphtho[2,1-b]-1,4-oxazine] (SO) showed that the forward reaction rates are faster in poly(ethylene terephthalate) (PET) compared to polycaprolactam (Nylon-6). As the polarities of both matrices are similar, those differences were attributed to the increased crystallinity of the latter, which retarded the rotation of the active molecules after the light-induced cleavage of the C_{spiro}–O bond [35]. The forward reaction rates also exhibit a temperature dependence, as observed for SP dispersed in PMMA and SBS matrices [27]. The rate of photoisomerization decreases at higher temperatures and, at 100 °C, the reverse and forward reaction rates are equal, resulting in no change of color upon light irradiation. The reverse reactions, however, do not follow the first-order kinetics, and instead, a mechanism involving two parallel first-order reactions was proposed [27].

The presence of undesirable side reactions associated with photochromic transitions may also reduce the number of cycles of forward and reverse reactions. This concept known as a *fatigue problem* is associated with photochromic molecules and is represented by the following reaction:

$$B' \xleftarrow{\Phi_s} A \underset{\Phi_r}{\overset{\Phi_f}{\rightleftharpoons}} B \tag{9.3}$$

where, a photochromic molecule A converts to isomer B upon irradiation, with a quantum yield Φ_f, whereby the associated side reactions exhibit quantum yield Φ_s, and B reverts back to A, with a quantum yield Φ_r.

In order to achieve an acceptable usage cycle greater than 10^3 times, the value of Φ_s must be less than 0.0001 (assuming that conversion to B is almost 100%) [36]. To increase the photochemical stability of photochromic molecules, which will also increase the fatigue factor, substituents with electron-donating and electron-withdrawing groups can be utilized. For example, as shown in Figure 9.2, the addition of methyl groups at 4 and 4' positions of thiophene ring in 1,2-di(2-dimethyl-5-phenylthiophen-3-yl) perfluorocyclopentene (b) increases the number of repeatable cycles in hexane from 80 to 200 in the absence of oxygen [37]. Synthetic modifications of photochromes have yielded enhanced fatigue factors [36]. When 2-(1-(2,5-dimethyl-3-furyl)ethylidene)-3-(2-adamantylidene) succinic anhydride was incorporated in polymer matrices, the increased fatigue resistance was due to the increased rigidity of the matrix that prevented the conversion of the molecules to the inactive isomeric form [38]. It appears that the presence of sufficient free volume to accommodate ring-opening reactions is essential, but the question of what the threshold of cross-linking density is for a given polymer system to maintain photochromic switching remains open.

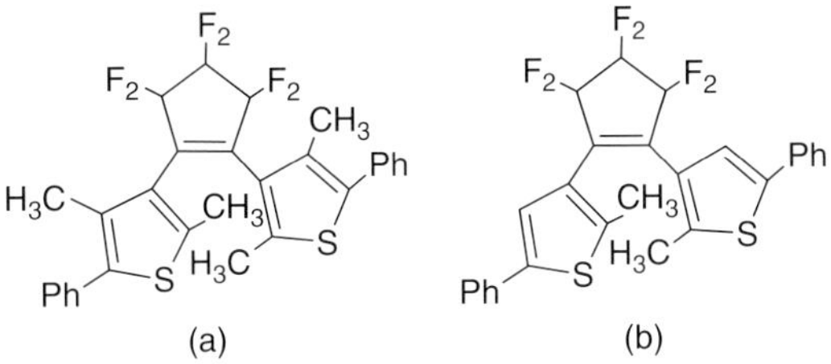

Figure 9.2 Molecular structures of (a) 4,4′-methyl-1,2-di-(2-dimethyl-5-phenylthiophen-3-yl) perfluorocyclopentene and (b) 1,2-di-(2-dimethyl-5-phenylthiophen-3-yl)perfluorocyclopentene.

9.3
Photochromic Systems

The performance of physically dispersed and covalently attached photochromic molecules in polymer matrices depends on the molecular structure, conformational changes, environment, and kinetic processes. Since there is no single molecule matrix pair that possesses all these attributes, the search for newer materials with higher response speeds and fatigue factor continues [39] and is primarily driven by potential applications. While the specificity and complexity of the matrices are the main road blocks, individual photochromic entities and their molecular and electronic properties typically are grouped on the basis of their scientific and technological importance into azobenzenes, spiropyrans, diarylethenes, and fulgides. The following sections provide details regarding their structural and electronic properties, which are responsible for photochromicity.

9.3.1
Azobenzenes

Azobenzenes are aromatic compounds containing two phenyl groups separated by an azo (N=N) group which is responsible for reversible trans ⇌ cis isomerization induced by light, and irreversible cis→trans induced by heat. The absorption spectrum of azobenzene exhibits a high intensity π-π^* transition in the UV region and a low intensity n-π^* transition in the visible region. The n-π^* transitions are responsible for colors that can be manipulated from the UV to the visible regions via modifying the ring substituents. On the basis of the substitution pattern and the relative energies between n-π^* and π-π^* states, the azo aromatics are generally categorized into three classes: azobenzene-, amino azobenzene-, and pseudo stillbene-type molecules. While unsubstituted azobenzene represents the simple form, the second class consists of azobenzenes substituted with electron-donating groups, particularly the amino groups at ortho and para positions. The third class represents azobenzenes substituted with an electron donor and electron acceptor groups at 4- and 4′-positions generally known as *pseudostillbenes*. The n-π^* and π-π^* transitions in amino-substituted azobenzenes are almost completely overlapped, whereas pseudostillbene-type molecules, because of a strong asymmetric electron distribution between the groups of opposite charges, exhibit π-π^* absorption shift to the lower energy and n-π^* to higher. It should be noted that the factors influencing the relative energies of n-π^* and π-π^* states result from π-conjugation and the nature of substitution pattern on the aromatic ring. The greater the extent of conjugation and the presence of strong electron-donating and electron-withdrawing substituents, the lower the energy gap, hence shifting absorption bands toward higher wavelength. Reversible isomerization between trans and cis conformations as well as molecular and structural changes resulting from dispersing and/or covalently attaching azobenzenes are probably the most challenging and, at the same time, difficult to control processes.

The mechanism of trans–cis photoisomerization in azobenzenes proceed either via rotation about the azo bond with the rupture of the π-character under π-π^* excitation, or via inversion mechanism under n-π^* excitation, where the π-bond remains intact. Both mechanisms are shown in Figure 9.3. On the other hand, thermally induced reversible cis–trans back reaction proceeds through a rotational mechanism. The trans–cis photoconversion is almost completely reversible in picoseconds without any side reactions, whereas thermal cis → trans conversion usually has timescales ranging from nano- to milliseconds, or even hours, depending upon the substituent and the local environment. A variety of azobenzene systems have been designed and synthesized, including cyclodextrins [40–42], admantanes [43], polycyclics [44], bacteriorodophsin [45], and crown ethers [46].

One of the drawbacks of doping polymer matrices with azobenzenes that exhibit photoinduced motions is that at higher concentration levels, phase separation

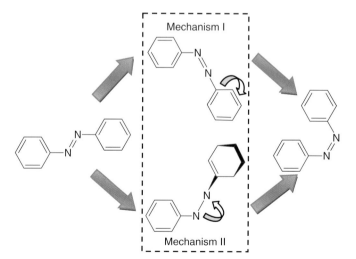

Figure 9.3 Mechanism of trans → cis isomerization in azobenzene.

and microcrystallization may occur. One approach to eliminate this adverse effect was to covalently attach azobenzene chromophores to a polymer matrix. Both side- and main-chain azobenzene-substituted polymers with good stability, rigidity, and ease of processability involve synthetic routes either using direct polymerization of azobenzene-functionalized monomers or postfunctionalization, where the chromophores are introduced to reactive precursor polymers without harsh polymerization conditions. For side- and main-chain azobenzene polymers, imides [47], esters [48], isocyanates [49], acrylates [50], methacrylates [51], urethanes [52], ethers [53], organometallic ferrocene polymers [54], dendrimers [55], and conjugated polydiacetylenes [56] have been utilized. Selected examples of azobenzene-containing polymer repeating units of (i) side chain, (ii) epoxy, (iii) organometallic ferrocene, (iv) main chain, and (v) conjugated polydiacetylene are illustrated in Figure 9.4.

9.3.2
Spiropyrans

Spiropyrans are photochromic materials that exist in two forms: SP (colorless) and MC (colored), and these structures are shown in Figure 9.5. The MC form exhibits a characteristic absorption band due to extended π-electron conjugation in the visible region. In polar solvents, it exists in any or all of the following four complicated zwitterionic forms by changing its conformation about the conjugated bond: trans–transoid–trans (TTT), trans–transoid–cis (TTC), cis–transoid–cis (CTC), and cis–transoid–trans (CTT). Furthermore, MC exhibits a strong tendency to aggregate in stacks, giving rise to parallel (head-to-head, J-aggregates) and antiparallel (head-to-tail, H-aggregates) molecular dipoles [57]. The open MC form converts back to the closed SP form either by irradiation in the visible region or by thermal exposure. Contrary to azobenzenes, where a number of reversible trans–cis isomerization cycles are possible without any degradation, in spiropyans, the number of cycles is limited, thereby limiting the applicability. Knowledge of photochemical and thermal processes in the ring opening and closure pathways lead to systems with better photochemical and thermal stability. The techniques applied include protonation of the MC-phenoxide moiety [58], synthetic modification of the SP with crown ethers [59], or a 7-trifluoromethylquinoline group [60], and by intramolecular bidentate metal ion chelation [61].

Time-resolved spectroscopic studies provide useful information on the dynamics and nature of intermediates involved in reversible photochemical and thermal ring openings as well as closure reactions in spiropyran compounds. Figure 9.5 depicts these reactions, where the first step involves cleavage of the C–O bonds between the spiro carbon and the oxygen, followed by orthogonal SP to planar MC conversions. The C–O bond cleavage occurs on a timescale of picoseconds or even faster. It has been proposed that the mechanism of photochemical ring-opening reaction involves the immediate formation of a singlet excited state, intermediate species, or a "metastable" species in less than 100 fs. A part of the metastable species restores the initial C–O bond back-formation in a few picoseconds, and the remaining portion forms a mixture of transient MC conformers with a decay time constant of

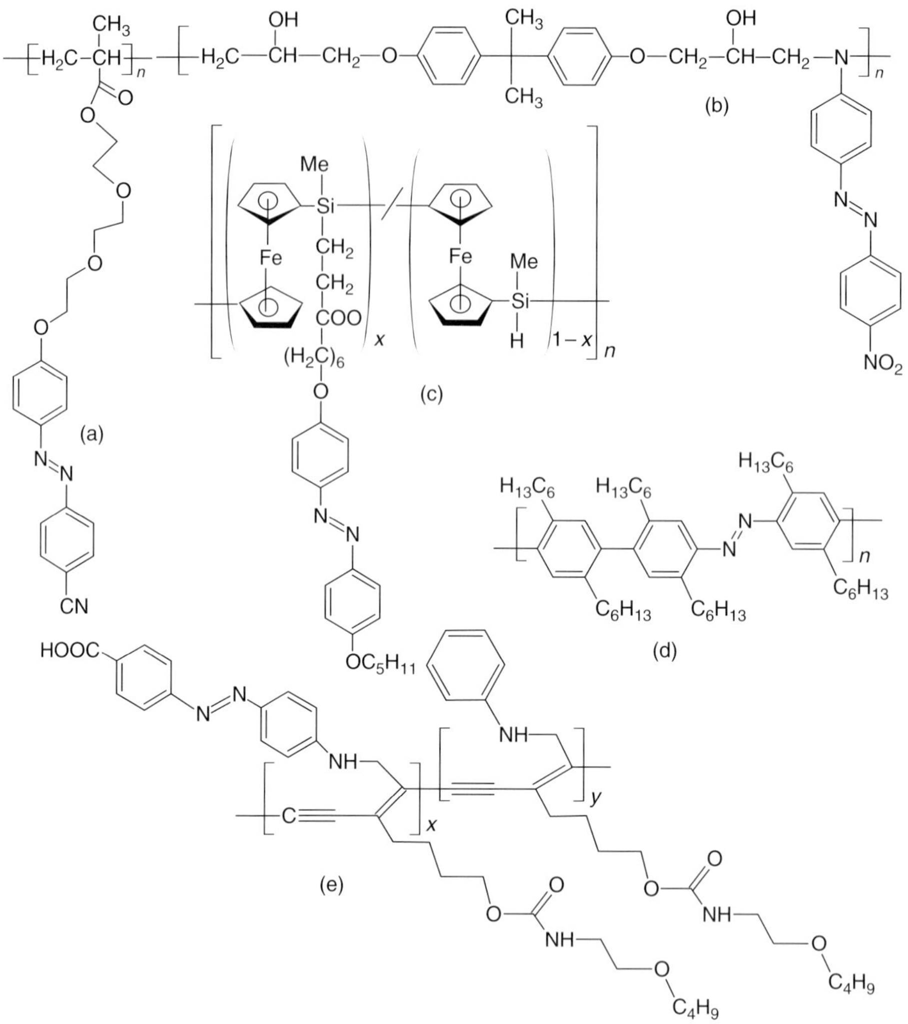

Figure 9.4 Selected examples of azobenzene-based polymers.

100 ps. The mechanism of photochromic transformation of spiropyrans is shown in Figure 9.5.

The most common synthetic methods for designing spiropyrans involve condensation of a heterocyclic quaternary salt with an alkyl group at the vicinal position to the heteroatom or condensation of 2-hydroxyarenealdehyde with the corresponding heterocyclic methylene base. As a majority of bases have the tendency to dimerize, their use is undesirable during the synthesis. Benzothiazoline, benzodithiole, and benzopyrans can be synthesized by using alkylimmonium, oxonium, and thionium salts as precursors, respectively. One of the characteristic features of spiropyrans is a strong absorption, maximum around the 290–380 nm regions due to the

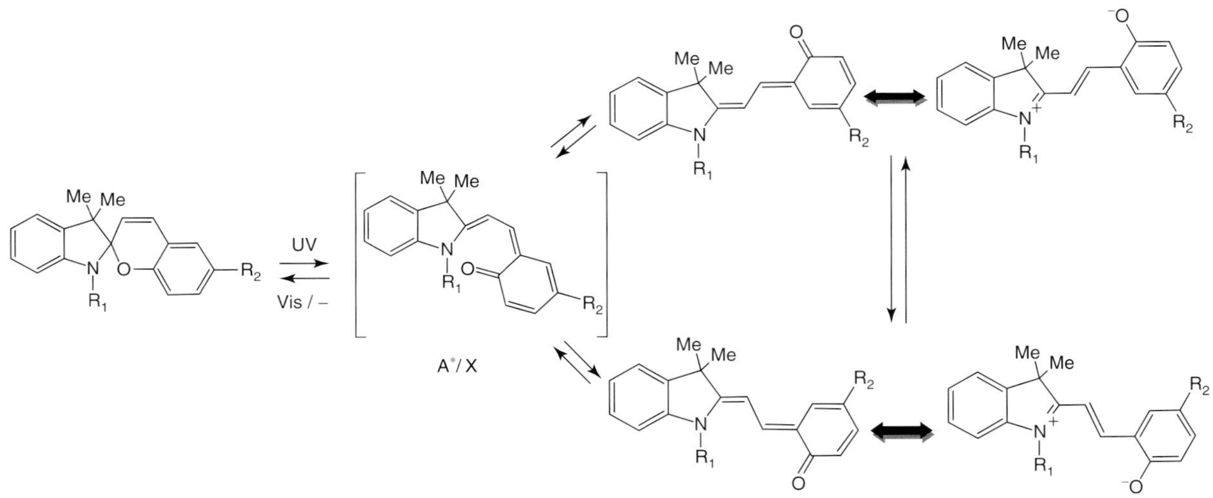

Figure 9.5 Mechanism of ring-opening and ring-closing in spiropyrans.

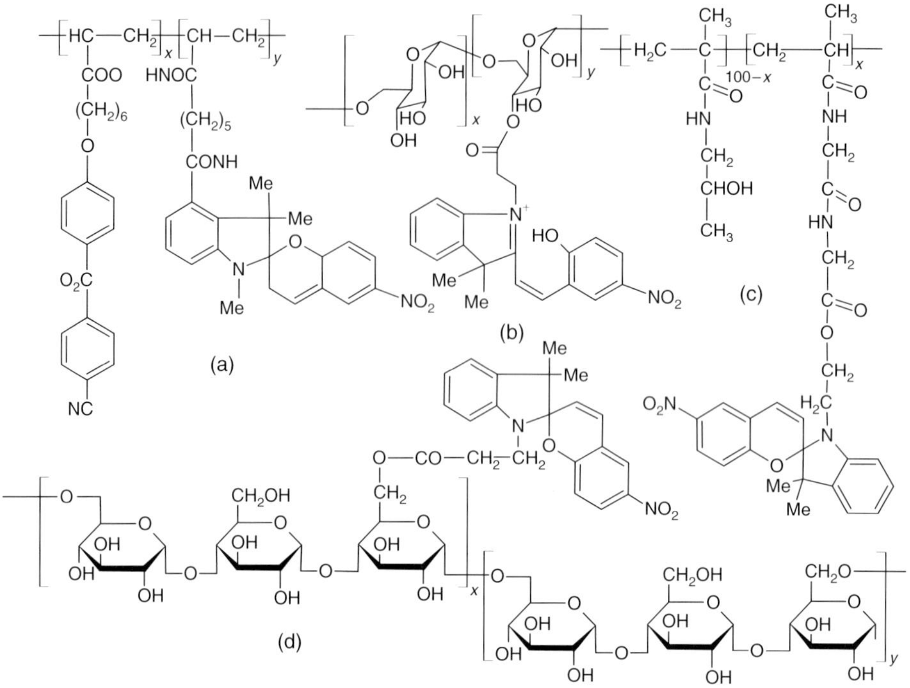

Figure 9.6 Selected examples of spiropyran-based polymers.

presence of benzopyran moiety. A number of attempts to incorporate spiropyran molecules into polymer systems have been reported, which include photochromic studies of spiropyrans in polymer matrices [62], liquid crystal polymers containing spiropyran as mesophases [63], and spiropyran-grafted dextran [64], pullulan [65], polydimethylsiloxane [66], and other synthetic polymers [67]. Selected examples are illustrated in Figure 9.6.

9.3.3
Diarylethenes

Diarylethenes with heterocyclic aryl groups undergo photochemical electrocyclization cycloreversion reactions between 1,3,5-hexatriene (open form) and 1,3-cyclohexadiene (closed form) entities. This is illustrated in Figure 9.7. These entities are the thermally irreversible organic photochromic compounds that are resistant to fatigue. Their thermal reversibility depends upon the aromatic stabilization energies of the aryl groups. For low aromatic stabilization energy aryl groups, such as thiophene, thiazole, and furans, colored closed-ring structures are thermally stable and do not return to the corresponding open-ring structures. In contrast to the aryl groups having high aromatic stabilization energy, such as indoles, phenyls, and pyrroles, the colored ring structures are thermally unstable [36]. One example is 1,2-bis(2-cyano-1,5-dimethyl-4-pyrrolyl)perfluorocyclopentene,

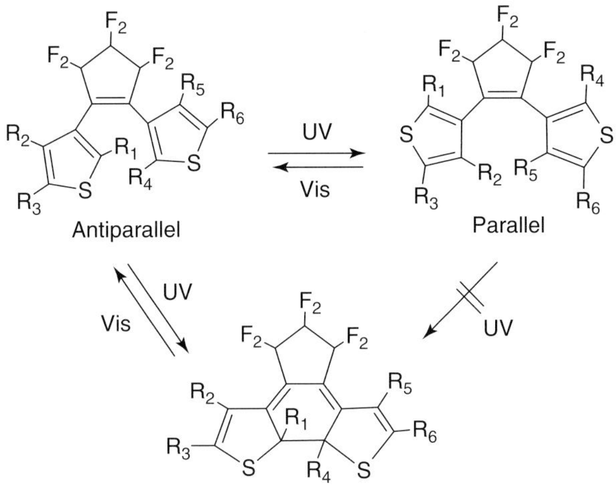

Figure 9.7 UV/vis-induced parallel and antiparallel conformations in diarylethenes.

which turns blue upon UV irradiation. This unstable blue-colored closed ring returns to the colorless open-ring form with a half life time of 37 s, which requires a thermal fading activation energy of 32.5 kJ mol^{-1} [68]. Furthermore, the introduction of bulky substituents at the reacting positions of dithenylethenes decreases the thermal stability of the closed-ring isomers at higher temperatures [69]. An extraordinarily high thermal stability of $t_{1/2} = 4.7 \times 10^5$ years was shown for a closed-ring isomer of 1,2-bis(5-methyl-2-phenylthiazol-4-yl) perfluorocyclopentene at 30 °C [70]. The ortho isomers of dithia-(dithienylethena)phane derivatives with dithienylethene moieties and benzene bridges also exhibit good thermal stability [71]. Thermal stability can be enhanced by the oxidation of sulfide bonds in terarylene derivatives [72]. Also, the presence of electron-withdrawing and bulky substituents decreases the thermal stability of the closed-ring structures by weakening the central carbon-carbon bonds [73].

The colorless open form of diarylethenes is a twisted π-system in conjugation with two aromatic rings, whereas the colored closed form is a planar π-system. The absorption spectra of both forms exhibit band shifts resulting from the ring substituents in the closed-ring isomer, whereas for the open-ring isomer the shift depends on cycloalkene structure. In order to achieve sensitivity in the visible and near infrared regions, diarylethenes can be manipulated by the substitution of electron-donating or electron-withdrawing groups. Closed isomers of diarylethenes with oligothiophenes and polyene groups showed enhancement of the absorption coefficient and a shift toward longer wavelength regions. As shown in Figure 9.7, the two rings of diarylethene system exhibit parallel (mirror symmetry) and antiparallel (C_2 symmetry) conformations. However, photocyclization reactions proceed only via antiparallel conformation. In addition, the quantum yield of diarylethene derivatives depends upon the ratio of the two conformers and is expected to increase with the increase of the antiparallel conformation. However, incorporation of bulky substituents and forceful confinement of the molecule in the parallel conformation with a cubic or rodlike antiparallel conformation will enhance the antiparallel content. The switching dynamics of ring opening and closure investigated by time-resolved spectroscopy showed that the switching occurs in the picosecond timescale [74].

Photochromic diarylethenes can be incorporated into polymers by following free-radical polymerization [75], polycondensation [76], Fridel–Crafts alkylation [77], or oxidative polymerization [78]. The side-chain and main-chain homopolymers, copolymers, block copolymers, and π-conjugated polymers synthesized by the above-mentioned methods have shown good photochromic properties in films and solution phase owing to the higher content of diarylethene chromophores in the polymer system compared to that in simple guest–host systems [79–81]. Selected examples are illustrated in Figure 9.8.

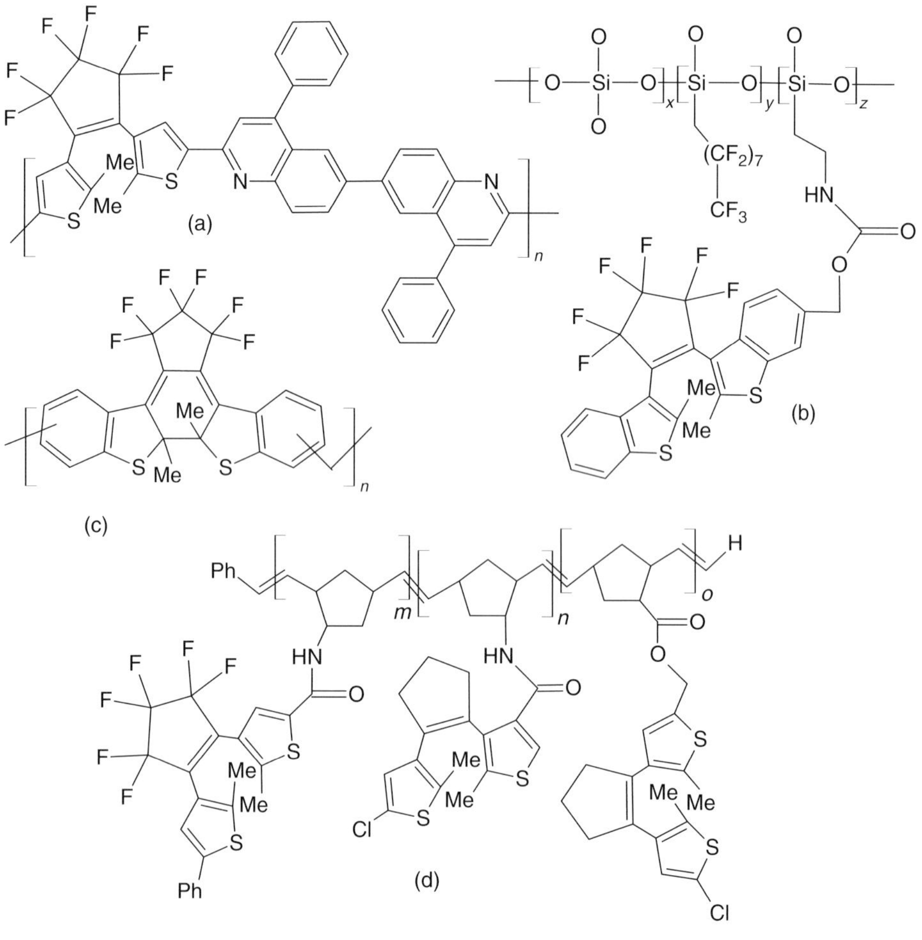

Figure 9.8 Selected examples of diarylethene-based polymers.

9.3.4
Fulgides

Fulgides are photochromic compounds that exist in colorless open E-form and photocyclized colored C-form. As shown in Figure 9.9, they also exist in the Z-form which is a geometrical isomer of the E-form that undergoes photochemical E–Z isomerization. The C-form undergoes photoisomerization to the E-form upon irradiation in the visible region known as *Woodward–Hoffmann pericyclic photoreaction*, whereas the E-form undergoes photoisomerization to the Z-form by UV irradiation. Various coloration mechanisms, such as E–Z-photoisomerization of the double bond [82], formation of colored radical intermediates [83], photochemical change between the mesomeric form [84], and photochemical 6π-electrocyclization of the hexatriene moiety [85] have been proposed.

With a few exceptions, fulgides are mostly thermally irreversible photochromic compounds and the thermal ring opening is usually accompanied with hydrogen rearrangement and dehydrogenative aromatization, which are the side reactions [86, 87]. In order to prevent thermal side reactions, benzylidene and isopropylidene groups can be replaced with mesitylmethylene groups [88] and the absence of hydrogen as well as the presence of vicinal methyl groups on the ring-closing carbon atom inhibit these reactions [89]. 2,5-Dimethyl-3-furyl moiety was the first thermally irreversible photochromic fulgide applied in reversible optical recording data storage media [90]. The presence of a heteroaromatic ring as well as substituents other than hydrogen on the ring resulted in thermally irreversible photochromic systems, also causing steric hindrance that plays an important role in the quantum-yield enhancement. The E-to-Z isomerization quantum yields are smaller for fulgides with larger alkyl groups (*n*-propyl < ethyl < methyl) [91] but the process does not occur in systems with isopropyl [92] and tert-butyl ring substituents [93]. Their good photochromic performance is attributed to thermal stability of both isomers at room temperature, stability against photochemical fatigue, high reaction quantum yields, and high reaction efficiencies. Absorption of fulgides can be shifted toward the visible region by enhancing the electron-donating ability, and by introducing heteroaromatic groups. Electron-donating substituents such as methoxy, dimethylamino, and methylthio groups in the 5-position of the

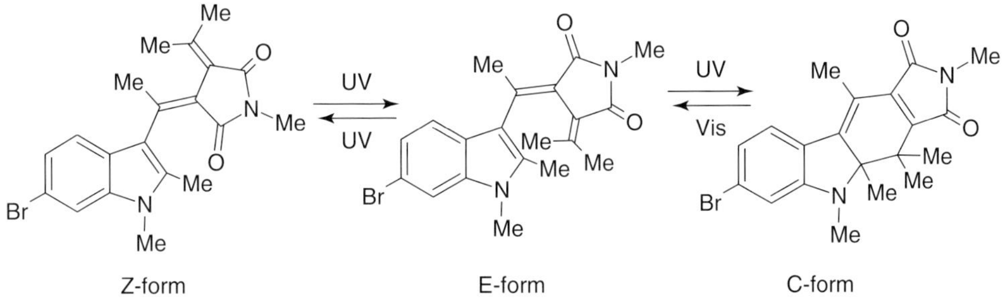

Figure 9.9 UV/vis induced fulgides in Z-, E-, and C-forms.

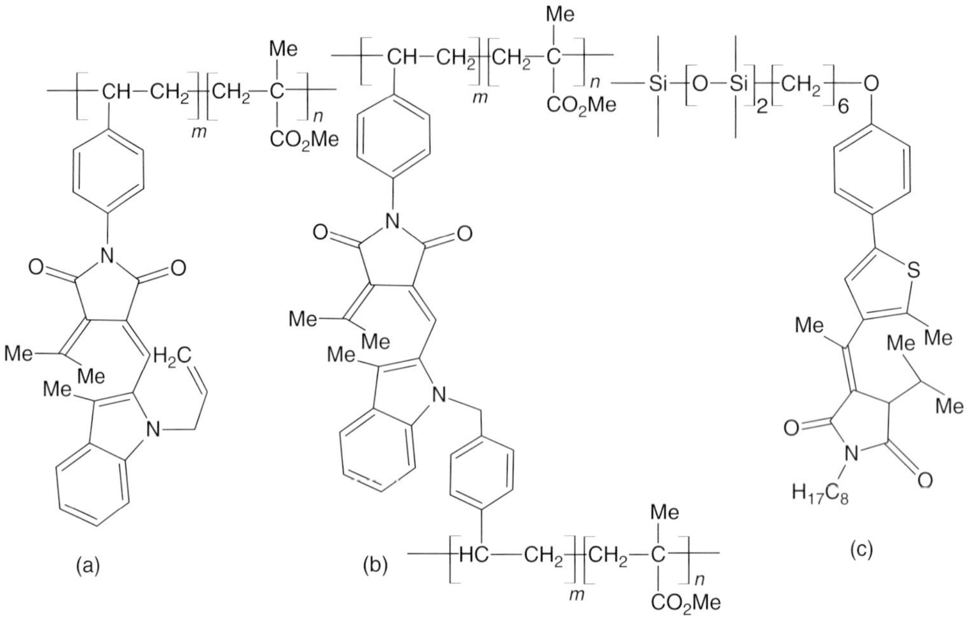

Figure 9.10 Selected examples of fulgide-based polymers.

indole ring facilitate a remarkable shift toward visible region, whereas the 6-position results in higher molar absorptivity [94]. In addition, the increased strength of the electron-donor substituents lowers the UV and visible quantum efficiencies, but the presence of electron-withdrawing substituents shifts the absorption toward lower wavelength regions [95].

Structural features and mechanisms of photoreactions in fulgides have been examined [96, 97], and the reported photochemical conversion timescales are in the nanosecond range [98]. In polymethyl methacrylate, polystyrene, and nitrocellulose polymer matrices, photocyclization processes occur in less than 10 ps [99]. Various synthetic routes such as Stobbe condensation [100], oxidative dimerization of 2-lithiocinnamic acid catalyzed by $FeCl_3$ [101, 102], substitution of 1,4-butynediol with carbon monoxide in the presence of Pd catalyst under elevated temperature and high pressure [103] have also been investigated for synthesizing fulgides. A number of homopolymers and copolymers composed of photochromic 2-indolyl fulgides or fulgimides were designed and synthesized [104]; selected examples are shown in Figure 9.10.

9.4
Outlook of Photochromic Materials

Although photochromic polymers are an important class of materials with many potential applications, and many excellent studies were devoted to this topic, recent

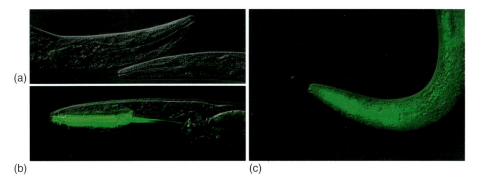

Figure 9.11 Fluorescence microscopy images of *Caenorhabditis elegans* incubated with (a) 10% dimethyl sulfoxide (DMSO), (b) photoswitch within the first 10 min, and (c) after 60 min.

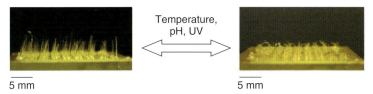

Figure 9.12 Optical images of poly(2-(N,N-dimethylamino)ethyl methacrylate-*co*-*n*-butyl acrylate-*co*-N,N-(dimethylamino) azobenzene acrylamide) copolymer that exhibit cilia-like morphology capable of reversible responses to temperature, pH, and electromagnetic radiation.

studies have shown that photochromic materials can be utilized in light-activated biological functions. One example is bis(pyridinium)(DTE), a water-soluble diarylethene derivative, which shows a higher degree of fatigue resistance. This molecule was used as a light-activated biological switch to induce paralysis in *Caeonorhabditis elegans*, a multicellular eukaryotic organism shown in Figure 9.11 [105]. Similar functionalized systems can be used in microbial separation or waste water treatments if the functional unit of the photochromic polymer system can be suitably activated by light-triggering ring-opening reactions. Of particular interest and importance are copolymerized photochromic molecules that exhibit not only multistimuli responsiveness but also unique shapes. Recent examples are cilia-like materials created from colloidal particles [106]. Figure 9.12 illustrates the response of cilia-like films to such stimuli as temperature, pH, or UV light. Particles of poly(2-(*N,N*-dimethylamino)ethyl methacrylate-*co*-*n*-butyl acrylate-*co*-*N,N*-(dimethylamino) azobenzene acrylamide) form films capable of generating surfaces with cilia-like features. While film morphological attributes allow the formation of wavy whiskers, chemical composition of the copolymer facilitates chemical, thermal, and electromagnetic responses manifested by simultaneous shape and color changes as well as excitation wavelength-dependent

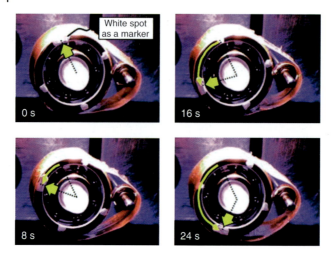

Figure 9.13 Rotation of the light-driven polymer motor with the liquid crystal elastomer-laminated film induced by simultaneous irradiation with UV (366 nm) and visible light (>500 nm) at room temperature.

fluorescence. These studies demonstrate that synthetically produced polymeric films can exhibit combined thermal, chemical, and electromagnetic sensing leading to locomotive and color responses that may find numerous applications in sensing devices, intelligent actuators, defensive mechanisms, and others. Another example is the creation of light-driven plastic motors that enable direct conversion of light to mechanical energy [107]. As illustrated in Figure 9.13, light-induced mechanical motion of azobenzene-functionalized liquid crystalline elastomer film is used to run a miniature pulley system. This concept of the battery-free motions and noncontact operation mechanical devices is another example of how incorporating functional and stimuli-responsive entities with photochromic properties may serve unique functions.

Acknowledgments

The authors are grateful to numerous industrial sponsors for funding this research. Also, partial support from ONR and NSF is acknowledged.

References

1. Birge, R. (1990) Photophysics and molecular electronic applications of the rhodopsins. *Annu. Rev. Phys. Chem.*, **41**, 683–733.
2. Fritsche, M. (1867) *C. R. Acad. Sci.*, **64**, 1035.
3. ter Meer, E. (1876) Uber dinitro-verbindungen der Fettreihe. *Ann. Chem.*, **181**, 1.
4. Hirshberg, Y. (1950) Photochromic dens la serie de la bianthrone. *C. R. Acad. Sci.*, **231**, 903.

5. Ichimura, K. (1989) Photoreactive polymeric materials. *Denshi Zairyo*, 71–76.
6. Krongauz, V.A. (1990) in *Photochromism: Molecules and Systems* (eds H. Dürr and H. Bouas-Laurent), Elsevier, Amsterdam, pp. 793–821.
7. Trancong, Q., Chikaki, S., and Kanato, H. (1994) Relationship between reorientational motions of a photochromic dopant and local relaxation processes of a glassy polymer matrix. *Polymer*, **35**, 4465–4469.
8. Liu, J., Deng, Q., and Jean, Y. (1993) Free-volume distributions of polystyrene probed by positron annihilation comparison with free volume theories. *Macromolecules*, **26**, 7149–7155.
9. Liu, F. and Urban, M.W. (2010) Recent advances and challenges in designing stimuli-responsive polymers. *Prog. Polym. Sci.*, **35**, 3–23.
10. Ichimura, K. (1990) photochromic materials and photoresists in *Photochromism: Molecules and Systems* (eds H. Dürr and H. Bouas-Laurent), Elsevier, Amsterdam, pp. 903–918.
11. Pietro, B. and Sandra, M. (1979) Cis-trans photoisomerization of azobenzene. Solvent and triplet donors effects. *J. Phys. Chem.*, **83**, 648–652.
12. Rau, H. and Yu-Quan, S. (1988) Photoisomerization of sterically hindered azobenzenes. *J. Photochem. Photobiol. A*, **42**, 321–327.
13. Morishima, Y., Tsuji, M., Kamachi, M., and Hatada, K. (1992) Photochromic isomerization of azobenzene moieties compartmentalized in hydrophobic microdomains in a microphase structure of amphiphilic polyelectrolytes. *Macromolecules*, **25**, 4406–4410.
14. Sudesh Kumar, G., Savariar, C., Safran, M., and Neckers, D.C. (1985) Chelating copolymers containing photosensitive functionalities. 3. Photochromism of cross-linked polymers. *Macromolecules*, **18**, 1525–1530.
15. Ho, M., Natansohn, A., and Rochon, P. (1995) Azo polymers for reversible optical storage 7. The effect of the size of the photochromic groups. *Macromolecules*, **28**, 6124–6127.
16. Fernandez, R., Mondragon, I., Galante, M., and Oyanguren, P. (2009) Influence of chromophore concentration and network structure on the photo-orientation properties of crosslinked epoxy-based azopolymers. *J. Polym. Sci. Polym. Phys.*, **47**, 1004–1014.
17. Anseth, K., Rothenberg, M., and Bowman, C. (1994) A photochromic technique to study polymer network volume distributions and microstructure during photopolymerizations. *Macromolecules*, **27**, 2890–2892.
18. Sato, M. and Yamashita, T. (2002) Quantitative estimation of free volume distribution of polymers with photochromic reactions. *J. Photopolym. Sci. Technol.*, **15**, 115–119.
19. Kwon, D., Shin, H., Kim, E., Boo, D., and Kim, Y. (2000) Photochromism of diarylethene derivatives in rigid polymer matrix: structural dependence, matrix effect, and kinetics. *Chem. Phys. Lett.*, **328**, 234–243.
20. Evans, R., Hanley, T., Skidmore, M., Davis, T., Such, G., Yee, L., Ball, G., and Lewis, D. (2005) The generic enhancement of photochromic dye switching speeds in a rigid polymer matrix. *Nat. Mater.*, **4**, 249–253.
21. Altomare, A., Andruzzi, L., Ciardelli, F., Solaro, R., and Tirelli, N. (1999) Methacrylic polymers containing permanent dipole azobenzene chromophores spaced from the main chain. ^{13}C NMR spectra and photochromic properties. *Macromol. Chem. Phys.*, **200**, 601–608.
22. Seog-Jun, K. and Reneker, D.H. (1993) A mechanochromic smart material. *Polym. Bull.*, **31**, 367–374.
23. Yamashita, T., Yoshida, N., and Wada, N. (2005) Free volume distribution of poly(alkyl methacrylate)s and the effect of inclusion. *J. Photopolym. Sci. Technol.*, **18**, 103–108.
24. Fischer, T., Laesker, L., Stumpe, J., and Kostromin, S.G. (1994) Photoinduced optical anisotropy in films of photochromic liquid crystalline polymers. *J. Photochem. Photobiol. A*, **80**, 453–459.

25. Horiuchi, H., Fukushima, T., Zhao, C., Okutsu, T., Takigami, S., and Hiratsuka, H. (2008) Conformational change of poly(methylphenylsilane) induced by the photoisomerization of pendant azobenzene moiety in the film state. *J. Photochem. Photobiol. A*, **198**, 135–143.
26. Katayama, I., Tezuka, Y., Yajima, H., and Ishii, T. (1995) Kinetic study of conformational transition accompanied by isomerization of spiropyrans bound to poly(l-glutamic acid) side chains. *J. Photopolym. Sci. Technol.*, **8**, 65–74.
27. Lin, J.S. (2003) Interaction between dispersed photochromic compound and polymer matrix. *Eur. Polym. J.*, **39**, 1693–1700.
28. Samanta, S. and Locklin, J. (2008) Formation of Photochromic spiropyran polymer brushes via surface-initiated, ring-opening metathesis polymerization: reversible photocontrol of wetting behavior and solvent dependent morphology changes. *Langmuir*, **24**, 9558–9565.
29. Irie, M. and Schnabel, W. (1981) Photoresponsive polymers – on the dynamics of conformational-changes of polyamides with backbone azobenzene groups. *Macromolecules*, **14**, 1246–1249.
30. Masahiro, I. and Hisami, T. (1983) Photoresponsive polymers. 5. Reversible solubility change of polystyrene having azobenzene pendant groups. *Macromolecules*, **16**, 210–214.
31. Sworakowski, J., Janus, K., and Nespurek, S. (2005) Kinetics of photochromic reactions in condensed phases. *Adv. Colloid Interface Sci.*, **116**, 97–110.
32. Vandewijer, P. and Smets, G. (1968) Photochromic polymers. *J. Polym. Sci.: Part C*, **22**, 231–245.
33. Richert, R. and Heuer, A. (1997) Rate-memory and dynamic heterogeneity of first-order reactions in a polymer matrix. *Macromolecules*, **30**, 4038–4041.
34. Such, G., Evans, R., Yee, L., and Davis, T. (2003) Factors influencing photochromism of spiro-compounds within polymeric matrices. *J. Macromol. Sci., Polym. Rev.*, **C43**, 547–579.
35. Wang, M., Yeh, C., and Hu, A. (1995) Photochromism of novel spirooxazine. 1. Investigation of the photocoloration in polymer films and fibers. *Polym. Int.*, **38**, 101–104.
36. Irie, M. (2000) Diarylethenes for memories and switches. *Chem. Rev.*, **100**, 1685–1716.
37. Irie, M., Lifka, T., Uchida, K., Kobatake, S., and Shindo, Y. (1999) Fatigue resistant properties of photochromic dithienylethenes: by-product formation. *Chem. Commun.*, **8**, 747–748.
38. Tork, A., Lafond, C., Pouraghajani, O., Bolte, M., Ritcey, A.M., and Lessard, R.A. (2002) Matrix effects on the photochemical fatigue resistance of fulgide-doped polymers. *Opt. Eng.*, **41**, 2310–2314.
39. Atsushi, K., Atsuhiro, T., Takeru, H., Toyoji, O., and Jiro, A. (2010) Fast photochromic polymers carrying [2.2] paracyclophane-bridged imidazole dimer. *Macromolecules*, **43**, 3764–3769.
40. Tomatsu, I., Hashidzume, A., and Harada, A. (2005) Photoresponsive hydrogel system using molecular recognition of alpha-cyclodextrin. *Macromolecules*, **38**, 5223–5227.
41. Takashima, Y., Nakayama, T., Miyauchi, M., Kawaguchi, Y., Yamaguchi, H., and Harada, A. (2004) Complex formation and gelation between copolymers containing pendant azobenzene groups and cyclodextrin polymers. *Chem. Lett.*, **33**, 890–891.
42. Yagai, S. and Kitamura, A. (2008) Recent advances in photoresponsive supramolecular self-assemblies. *Chem. Soc. Rev.*, **37**, 1520–1529.
43. Zarwell, S. and Ruck-Braun, K. (2008) Synthesis of an azobenzene-linker-conjugate with tetrahedrical shape. *Tetrahedron Lett.*, **49**, 4020–4025.
44. Ai, M., Groeper, S., Zhuang, W., Dou, X., Feng, X.L., Mullen, K., and Rabe, J.P. (2008) Optical switching studies of an azobenzene rigidly linked to a

45. Schmidt, B., Sobotta, C., Heinz, B., Laimgruber, S., Braun, M., and Gilch, P. (2005) Excited-state dynamics of bacteriorhodopsin probed by broadband femtosecond fluorescence spectroscopy. *Biochim. Biophys. Acta., Bioenerg.*, **1706**, 165–173.
46. Thuery, P., Nierlich, M., Lamare, E., Dozol, J.F., Asfari, Z., and Vicens, J. (2000) Bis(crown ether) and azobenzocrown derivatives of calix[4]arene. A review of structural information from crystallographic and modelling studies. *J. Inclusion Phenomena and Macrocyclic Chemistry*, **36**, 375–408.
47. Aleksandrova, E.L., Goikhman, M.Y., Subbotina, L.I., Romashkova, K.A., Gofman, I.F., Kudryavtsev, V.V., and Yakimanskii, A.V. (2003) Optical and photosensitive properties of comb-shaped polyamide-imides. *Semiconductors*, **37**, 821–824.
48. Wu, C.C., Gu, Q.C., Huang, Y., and Chen, S.X. (2003) The synthesis and thermotropic behaviour of an ethyl cellulose derivative containing azobenzene-based mesogenic moieties. *Liq. Cryst.*, **30**, 733–737.
49. Mayer, S. and Zentel, R. (1998) A new chiral polyisocyanate: an optical switch triggered by a small amount of photochromic side groups. *Macromol. Chem. Phys.*, **199**, 1675–1682.
50. Tian, Y.Q., Watanabe, K., Kong, X.X., Abe, J., and Iyoda, T. (2002) Synthesis, nanostructures, and functionality of amphiphilic liquid crystalline block copolymers with azobenzene moieties. *Macromolecules*, **35**, 3739–3747.
51. Andruzzi, L., Altomare, A., Ciardelli, F., Solaro, R., Hvilsted, S., and Ramanujam, P.S. (1999) Holographic gratings in azobenzene side-chain polymethacrylates. *Macromolecules*, **32**, 448–454.
52. Park, C.K., Zieba, J., Zhao, C.F., Swedek, B., Wijekoon, W.M.K.P., and Prasad, P.N. (1995) Highly cross-linked polyurethane with enhanced stability of 2nd-order nonlinear-optical properties. *Macromolecules*, **28**, 3713–3717.
53. He, X.H., Zhang, H.L., Yan, D.Y., and Wang, X.Y. (2003) Synthesis of side-chain liquid-crystalline homopolymers and triblock copolymers with p-methoxyazobenzene moieties and poly(ethylene glycol) as coil segments by atom transfer radical polymerization and their thermotropic phase behavior. *J. Polym. Sci., Part A: Polym. Chem.*, **41**, 2854–2864.
54. Kondo, T., Kanai, T., and Uosaki, K. (2001) Control of the charge-transfer rate at a gold electrode modified with a self-assembled mono-layer containing ferrocene and azobenzene by electro- and photochemical structural conversion of cis and trans forms of the azobenzene moiety. *Langmuir*, **20**, 6317–6324.
55. Junge, D.M. and McGrath, D.V. (1997) Photoresponsive dendrimers. *Chem. Commun.*, **9**, 857–858.
56. Sukwattanasinitt, M., Lee, D.C., Kim, M., Wang, X.G., Li, L., Yang, K., Kumar, J., Tripathy, S.K., and Sandman, D.J. (1999) New processable, functionalizable polydiacetylenes. *Macromolecules*, **32**, 7361–7369.
57. Minkin, V.I. (2004) Photo-, thermo-, solvato-, and electrochromic spiroheterocyclic compounds. *Chem. Rev.*, **104**, 2751–2776.
58. Fissi, A., Pieroni, O., Angelini, N., and Lenci, F. (1999) Photoresponsive polypeptides. photochromic and conformational behavior of spiropyran-containing poly(l-glutamate)s under acid conditions. *Macromolecules*, **32**, 7116–7121.
59. Kellmann, A., Tfibel, F., Pottier, E., Guglielmetti, R., Samat, A., and Rajzmann, M. (1993) Effect of nitro substituents on the photochromism of some spiro[indoline-naphthopyrans] under laser excitation. *J. Photochem. Photobiol. A*, **76**, 77–82.
60. Guo, X., Zhou, Y., Zhang, D., Yin, B., Liu, Z., Liu, C., Lu, Z., Huang, Y., and Zhu, D. (2004) 7-trifluoromethylquinoline-functionalized luminescent photochromic spiropyran with the stable merocyanine species both in solution and in the solid state. *J. Org. Chem.*, **69**, 8924–8931.

61. Wojtyk, J.T.C., Buncel, E., and Kazmaier, P.M. (1998) Effects of metal ion complexation on the spiropyran-merocyanine interconversion: development of a thermally stable photo-switch. *Chem. Commun.*, 1703–1704.
62. Tork, A., Boudreault, F., Roberge, M., Ritcey, A.M., Lessard, R.A., and Galstian, T.V. (2001) Photochromic behavior of spiropyran in polymer matrices. *Appl. Opt.*, **40**, 1180–1186.
63. Berkovic, G., Krongauz, V., and Weiss, V. (2000) Spiropyrans and spirooxazines for memories and switches. *Chem. Rev.*, **100**, 1741–1754.
64. Edahiro, J., Sumaru, K., Takagi, T., Shinbo, T., and Kanamori, T. (2006) Photoresponse of an aqueous two-phase system composed of photochromic dextran. *Langmuir*, **22**, 5224–5226.
65. Hirakura, T., Nomura, Y., Aoyama, Y., and Akiyoshi, K. (2004) Photoresponsive nanogels formed by the self-assembly of spiropyrane-bearing pullulan that act as artificial molecular chaperones. *Biomacromolecules*, **5**, 1804–1809.
66. Evans, R.A., Hanley, T.L., Skidmore, M.A., Davis, T.P., Such, G.K., Yee, L.H., Ball, G.E., and Lewis, D.A. (2005) The generic enhancement of photochromic dye switching speeds in a rigid polymer matrix. *Nat. Mater.*, **4**, 249–253.
67. Menju, A., Hayashi, K., and Irie, M. (1981) Photoresponsive polymers. 3. reversible solution viscosity change of poly(methacrylic acid) having spirobenzopyran pendant groups in methanol. *Macromolecules*, **14**, 755–758.
68. Uchida, K., Matsuoka, T., Sayo, K., Iwamoto, M., Hayashi, S., and Irie, M. (1999) Thermally reversible photochromic systems. Photochromism of a dipyrrolylperfluorocyclopentene. *Chem. Lett.*, **8**, 835–836.
69. Morimitsu, K., Shibata, K., Kobatake, S., and Irie, M. (2002) Dithienylethenes with a novel photochromic performance. *J. Org. Chem.*, **67**, 4574–4578.
70. Takami, S., Kobatake, S., Kawai, T., and Irie, M. (2003) Extraordinarily high thermal stability of the closed-ring isomer of 1,2-bis(5-methyl-2-phenylthiazol-4-yl)perfluorocyclopentene. *Chem. Lett.*, **32**, 892–893.
71. Hossain, M.K., Takeshita, M., and Yamato, T. (2005) Synthesis, structure, and photochromic properties of dithia(dithienylethena)phane derivatives. *Eur. J. Org. Chem.*, **13**, 2771–2776.
72. Jeong, Y.C., Gao, C., Lee, I.S., Yang, S.I., and Ahn, K.H. (2009) The considerable photostability improvement of photochromic terarylene by sulfone group. *Tetrahedron Lett.*, **50**, 5288–5290.
73. Gilat, S.L., Kawai, S.H., and Lehn, J.M. (2006) Light-triggered molecular devices: photochemical switching of optical and electrochemical properties in molecular wire type diarylethene species. *Chem. Eur. J.*, **1**, 275–284.
74. Hania, P., Telesca, R., Lucas, L., Pugzlys, A., van Esch, J., Feringa, B., Snijders, J., and Duppen, K. (2002) An optical and theoretical investigation of the ultrafast dynamics of a bisthienylethene based photochromic switch. *J. Phys. Chem. A*, **106**(37), 8498–8507.
75. Kim, E., Choi, Y., and Lee, M. (1999) Photoinduced refractive index change of a photochromic diarylethene polymer. *Macromolecules*, **32**(15), 4855–4860.
76. Holland, N., Hugel, T., Neuert, G., Cattani-Scholz, A., Renner, C., Oesterhelt, D., Moroder, L., Seitz, M., and Gaub, H. (2003) Single molecule force spectroscopy of azobenzene polymers: switching elasticity of single photochromic macromolecules. *Macromolecules*, **36**(6), 2015–2023.
77. Jeong, Y., Park, D., Kim, E., Yang, S., and Ahn, K. (2006) Polymerization of a photochromic diarylethene by Friedel-Crafts alkylation. *Macromolecules*, **39**(9), 3106–3109.
78. Uchida, K., Takata, A., Saito, M., Murakami, A., Nakamura, S., and Irie, M. (2003) Synthesis of novel

photochromic films by oxidation polymerization of diarylethenes containing phenol groups. *Adv. Funct. Mater.*, **13**(10), 755–762.

79. Shen, L., Ma, C., Pu, S.Z., Cheng, C.J., Xu, J.K., Li, L., and Fu, C.Q. (2009) Synthesis and properties of novel photochromic poly(methyl methacrylate-co-diarylethene)s. *New J. Chem.*, **33**, 825–830.

80. Nishi, H. and Kobatake, S. (2008) Photochromism and optical property of gold nanoparticles covered with low-polydispersity diarylethene polymers. *Macromolecules*, **41**, 3995–4002.

81. Nakashima, H. and Irie, M. (1997) Synthesis of silsesquioxanes having photochromic diarylethene pendant groups. *Macromol. Rapid Commun.*, **18**, 625–633.

82. Chakraborty, D.P., Sleigh, T., Stevenson, R., Swoboda, G.A., and Weinstein, B. (1996) Preparation and geometric isomerism of dipiperonylidenesuccinic acid and anhydride. *J. Org. Chem.*, **31**, 3342–3345.

83. Schäonberg, A. (1936) The photochemical formation of organic diradicals. Part III. Investigations on anthracene, the fulgides, thiophosgene and their derivatives. *Trans. Faraday Soc.*, **32**, 514–521.

84. Gheorghiu, C.V. (1947) *Bull. Ec. Polytech. Jassy*, **2**, 141–155.

85. Santiago, A. and Becker, R.S. (1968) Photochromic fulgides. Spectroscopy and mechanism of photoreactions. *J. Am. Chem. Soc.*, **90**, 3654–3658.

86. Crescente, O., Heller, H.G., and Oliver, S. (1979) Overcrowded molecules. Part 16. Thermal and photochemical reactions of (E,E)-bis(benzylidene)succinic anhydride. *J. Chem. Soc., Perkin Trans. 1*, **1**, 150–153.

87. Heller, H.G. and Oliver, S. (1981) Photochromic heterocyclic fulgides. Part 1. Rearrangement reactions of (E)-3-furylethylidene(isopropylidene)succinic anhydride. *J. Chem. Soc., Perkin Trans.*, **1**, 197–201.

88. Heller, H.G. and Megit, R.M. (1974) Overcrowded molecules. Part IX. Fatigue-free photochromic systems involving (E)-2-isopropylidene-3-(mesitylmethylene)succinic anhydride and n-phenylimide. *J. Chem. Soc., Perkin Trans. 1*, **1**, 923–927.

89. Darcy, P.J., Heller, H.G., Strydom, P.J., and Whittall, J. (1981) Photochromic heterocyclic fulgides. Part 2. Electrocyclic reactions of (e)-2, 5-dimethyl-3-furylethylidene(alkyl-substituted methylene)succinic anhydrides. *J. Chem. Soc., Perkin Trans.*, **1**, 202–205.

90. Feringa, B.L., Jager, W.F., and Lange, B. (1993) Organic materials for reversible optical data storage. *Tetrahedron*, **49**, 8267–8310.

91. Yokoyama, Y., Goto, T., Inoue, T., Yokoyama, M., and Kurita, Y. (1988) Fulgides as efficient photochromic compounds. Role of the substituent on furylalkylidene moiety of furylfulgides in the photoreaction. *Chem. Lett.*, **17**, 1049–1052.

92. Yokoyama, Y., Inoue, T., Yokoyama, M., Goto, T., Iwai, T., Kera, N., Hitomi, I., and Kurita, Y. (1994) Effects of steric bulkiness of substituents on quantum yields of photochromic reactions of furylfulgides. *Bull. Chem. Soc. Jpn.*, **67**, 3297–3303.

93. Kiji, J., Okano, T., Kitamura, H., Yokoyama, Y., Kubota, S., and Kurita, Y. (1995) Synthesis and photochromic properties of fulgides with a t-butyl substituent on the furyl- or thienyl-methylidene moiety. *Bull. Chem. Soc. Jpn.*, **68**, 616–619.

94. Yokoyama, Y., Tanaka, T., Yamane, T., and Kurita, Y. (1991) Synthesis and photochromic behavior of 5-substituted indolylfulgides. *Chem. Lett.*, **20**, 1125–1128.

95. Tomoda, A., Kaneko, A., Tsuboi, H., and Matsushima, R. (1993) Photochromism of heterocyclic fulgides. IV. Relationship between chemical structure and photochromic performance. *Bull. Chem. Soc. Jpn.*, **66**, 330–333.

96. Yoshioka, Y., Tanaka, T., Sawada, M., and Irie, M. (1989) Molecular and crystal structures of E- and Z-isomers of 2,5-dimethyl-3-furylethylidene (isopropylidene) succinic anhydride. *Chem. Lett.*, **18**, 19–22.

97. Kumar, V.A. and Venkatesan, K. (1993) Studies on photochromism of a thermally stable fulgide in the crystalline state: X-ray crystallographic investigation of (E)-2-isopropylidene-3-(1-naphthylmethylene)succinic anhydride. *Acta Crystallogr.*, **B49**, 896–900.
98. Lenoble, C. and Becker, R.S. (1986) Photophysics, photochemistry, and kinetics of photochromic fulgides. *J. Phys. Chem.*, **90**, 2651–2654.
99. Kurita, S., Kashiwagi, A., Kurita, Y., Miyasaka, H., and Mataga, N. (1990) Picosecond laser photolysis studies on the photochromism of a furylfulgide. *Chem. Phys. Lett.*, **171**, 553–557.
100. Fan, M. and Yu, L., and Zhao, W. (1999) in *Fulgide Family Compounds: Synthesis, Photochromism, and Applications in Organic Photochromic and Thermochromic Compounds*, vol. 1 (eds J.C. Crano and R. Guglielmetti), Plenum Publishers, New York, pp. 141–206.
101. Gendy, A.M.E., Mallouli, A., and Lepage, Y. (1980) Synthesis of 1,4-diaryl-2,3-diformylbutadienes. *Synthesis*, vol. 1, 898–899.
102. Elbe, H.-L. and Käobrich, G. (1974) *Chem. Ber.*, **107**, 1654–1666.
103. Uchida, S., Yokoyama, Y., Kiji, J., Okano, T., and Kitamura, H. (1995) Electronic effects of substituents on indole nitrogen on the photochromic properties of indolylfulgides. *Bull. Chem. Soc. Jpn.*, **68**, 2961–2967.
104. Wolak, M.A., Gillespie, N.B., Thomas, C.J., Birge, R.R., and Lees, W.J. (2001) Optical properties of photochromic fluorinated indolylfulgides. *J. Photochem. Photobiol., A*, **144**, 83–91.
105. Al-Atar, U., Fernandes, R., Johnsen, B., Baillie, D., and Branda, N.R. (2009) A photocontrolled molecular switch regulates paralysis in a living organism. *J. Am. Chem. Soc.*, **131**(44), 15966–15967.
106. Fang, L., Dhanya, R., and Urban, M.W. Colloidal films that mimic cilia. (2010) *Adv. Funct. Mater.*, **20**, 3163–3167.
107. Yamada, M., Kondo, M., Mamiya, Ji., Yu, Y., Kinoshita, M., Barrett, C.J., and Ikeda, T. (2008) Photomobile polymer materials: towards light-driven plastic motors. *Angew. Chem. Int. Ed.*, **47**(27), 4986–4988.

10
Covalent Bonding of Functional Coatings on Conductive Materials: the Electrochemical Approach
Michaël Cécius and Christine Jérôme

10.1
Introduction

This chapter describes electrochemical approaches to achieve functional organic coatings of metals, carbon, and semiconductors, and discusses the advantages and limitations in the development of the emerging strategies, particularly focusing on examples of smart or stimuli-responsive polymer coatings. The various strategies are classified on the basis of the strength of the coating/metal interactions, starting with simply deposited coatings and then describing chemisorbed films.

In the recent past, the coating of (semi)conductive inorganic materials such as metals (Au, Ag, Pt, Fe, Cu, etc.), carbon (glassy carbon, felt, microfibers, nanotubes, etc.), or semiconductors (doped Si, metal oxides, etc.) by an organic layer has attracted considerable attention because of the emergence of rapidly growing, very demanding fields such as nanotechnologies and biomedical sciences. Indeed, the wide versatility of organic and polymer chemistries allows to produce innovative coatings, such as stimuli-responsive coatings, that are of interest in the modification of surface properties of conductive materials. In particular, functional and/or responsive coatings are needed for applications such as solar cells, (bio)sensors, and biomedical implants. In addition, increasing concern about the environment tends to restrict the technologies and products to the greenest ones. Under these novel driving forces, efficient methods for the surface modification of conducting inorganic materials by polymer coatings are being developed.

For the coating of conductive or semiconductive materials, one can take advantage of their electrochemical properties to control, master, and localize the coating and its adhesion to the substrate by applying an electrochemical treatment to the substrate under well-defined conditions.

Electrochemical methods are easily applied and rapid, and are already widely used in industry. They can be often performed in aqueous media from readily available organic precursors and require relatively low energy. This technology thus appears really relevant in many cases in the creation of thin polymer films on (semi)conductors in possible greener conditions.

Handbook of Stimuli-Responsive Materials. Edited by Marek W. Urban
Copyright © 2011 WILEY-VCH Verlag GmbH & Co. KGaA, Weinheim
ISBN: 978-3-527-32700-3

10.2
Electrodeposited Coatings

The methods involving simple electrodeposition of an organic layer onto metals are long known in the scientific literature and have been used in industries such as the steel industry. The processes that are used are those of electrocoagulation and electroprecipitation that are set off by the application of a current, with the precipitation of a polymer coating from a liquid medium on the surface of a metallic substrate being used as electrode. Cataphoresis painting is probably the oldest and best known among these simple processes. It consists in the immersion of the native or pretreated conductive surface into a bath consisting of water-thinned paint, with the surface used as the cathode so that it attracts the oppositely charged paint particles on passage of current. Similarly, the immersed and polarized surface is completely covered by a polymer layer of a few tens of microns that is then thermally cured for adhesion. As indicated by the term, *cataphoresis* is a cathodic process usually performed on common metals for protection against corrosion and as an adhesion promoter for further painting.

More recently, the discovery, development, and widespread use of conjugated polymers have made the process of electrodeposition by anodic polarization more popular because of their unique electronic properties. Indeed, conductive polymers based on aromatic compounds such as polyaniline (PANI), polypyrrole (PPy), polythiophene (PThi), poly(*p*-phenylene vinylene) (PPV), and their derivatives can be very easily prepared by electro-oxidation of the parent aromatic monomer [1] as illustrated by the mechanism shown in Figure 10.1. These polymers, characterized by a conjugated π-electron system, are very rigid and exhibit strong intermolecular interactions that generally lead to insolubility either in organic solvents or in

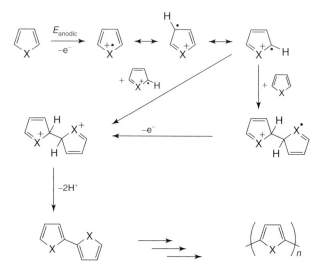

Figure 10.1 Electrochemical mechanism for the polymerization of an aromatic monomer.

water; they thus naturally precipitate at the anode as soon as the molecular weight increases. During this anodic polymerization process, doping is simultaneous to the polymer growth, leading to the formation of the polymer in its conducting state. Therefore, the electrode remains remarkably highly conductive during polymer deposition and thus the film thickness can be easily tuned by coulometry (at least up to several tens of microns), since the oxidative radical-coupling polymerization mechanism needs the withdrawal of two electrons for a new bond to be created (Figure 10.1). An advantage is that quite a few heteroaromatic rings (pyrrole, aniline, and ethylene dioxythiophene (EDOT)) are oxidized at a lower potential than water, which allows the electropolymerization to be performed in aqueous media. Acidic conditions favor the rate of polymerization of these materials. However, anodic processes generally limit the choice of substrates to so-called inert metals, that is, metals that are not oxidatively dissolved in anodic regime. These coatings are thus commonly obtained on substrates such as Pt, Au, stainless steel, carbon, and conductive oxides (e.g., indium tin oxide, ITO). These substrates are often encountered in devices such as actuators, sensors, solar cells, and electrochromic windows. In addition, it is possible, by tailoring the bath composition, to achieve deposition on substrates like iron or copper. Conjugated polymers are then helpful for protection against corrosion.

Conjugated polymers are very significant, smart, or responsive materials. Indeed, many properties (color, conductivity, volume, hydrophilicity, permeability, etc.) depend on their oxidation state and can thus be reversibly controlled by the potential applied to the material. As an example, smart windows [2] are based on the electrochromism of these coatings, that is, the ability to switch these polymers reversibly between dark and colorless state by simple control of the polymer redox state. The intensity of light transmission through a window incorporating such a layer can thus be tuned at will by external electrical stimuli.

A large number of books and reviews are already devoted to the wide topic of conductive polymers. For more details about their electrochemical synthesis and applications, the readers may consult some of them, in particular, Ref. [3].

10.3
Electrografted Coatings

If some of the above-mentioned applications of conducting polymers are efficiently achieved with simply deposited polymer coatings, the covalent bonding of the organic layer onto the conductive substrate is by far the most desirable situation. The lack of coating adhesion, which is generally the case for such interfaces, can be overcome by the strong interactions between the substrate and the organic coating. Once again, electrochemistry can provide an excellent solution to this challenging task.

The development of electrochemical methods to covalently modify inorganic surfaces by an organic layer relies on the ability of some specific organic precursors to chemisorb on metals when they are polarized at a well-defined potential. Today, the diversity of the identified precursors (amines, alcohols, (meth)acrylates, etc.) exhibiting electrochemically induced anchoring to surfaces provides them with a wide range of functionalities and properties, including responsiveness. In addition, it also allows applications to a wide diversity of substrates (metals, carbon, oxides, semiconductors, etc.). When prepared under appropriate electrochemical conditions, the stability of the resulting coatings is very high. They thus resist chemical environment aggressions (e.g., corrosion), high temperatures, and electrical and mechanical stresses. This might not be the case in the grafting of organic coatings by other types of molecules specially designed to chemisorb directly on surfaces without the need for polarization, such as the well-known self-assembled monomers of thiols on gold. In some cases, the electrochemical treatment also allows to tune the grafting density of a polymer coating, from pancakes to dense brushes, retaining the chemical composition of the coating. The main limitation of the electrochemical approach to graft organic coatings onto substrates used as electrode is that the thickness of the film cannot be easily controlled by the electrochemical conditions, since the grown film is usually insulating and this leads substrate passivation.

In the following, the most promising organic molecules are reported along with a discussion on the electrochemical conditions allowing their grafting onto specific substrates. They are divided into two categories. In the first category, the organic molecules have to be electrochemically oxidized and thus an anodic polarization is required. In the second category, the molecules are electrografted in the cathodic regime. As already mentioned, such a cathodic process is no more restricted to noble metals and can thus be easily performed on common metals.

10.4
Compounds Requiring an Anodic Process

10.4.1
Aliphatic Primary Amines

The electrochemically assisted oxidation of aliphatic primary amines with the concomitant direct bonding of an organic layer to the working electrode of the system was first reported in the early 1990s by Barbier *et al.* [4] who were interested in improving the toughness of carbon-fiber-based composites by increasing the adhesion between the carbon fibers and the epoxy-matrix by covalent bonds. Their pioneering paper describes electrochemical oxidation of (di)amine derivatives in oxygen-free and dry acetonitrile (ACN), which is an organic aprotic solvent commonly used for electrochemistry, on carbon fibers that are used as working electrode. In the case of ethylene diamine, cyclic voltammetry shows a single broad irreversible peak at $+1.3$ V versus the saturated calomel electrode (SCE), corresponding to oxidation of the amine. A second scan of the same electrode shows a clear decrease in the current intensity as compared to the first scan because of the presence of the grafted organic coating on the fibers created during the previous scan. The functionalization of the fibers was ascertained by X-ray photoelectron spectroscopy (XPS), which revealed the presence of a large amount of nitrogen only on the electrochemically treated fibers. Similar electrochemical treatment performed with 4-nitrobenzylamine rather than ethylene diamine allows immobilizing the electroactive nitro group at the carbon surface. Indeed, after electrochemical modification, cyclic voltammetry carried out in a typical electrolyte solution (ACN + NBu$_4$BF$_4$) highlighted the electroactivity of the coated carbon electrode due to the presence of nitro groups. The cyclic voltammetry curves show a reversible peak in the cathodic window around -1.2 V versus SCE, confirming the presence of nitro groups. Further studies [5] evidenced that in the case of symmetric diamines, it is not easy to reach a monolayer, rather, the formation of polymeric film is observed with only a small quantity of unreacted primary amine in the coating.

The proposed mechanism [4, 6] for the grafting of aliphatic primary amines is illustrated in Figure 10.2 and consists in the formation of a radical cation $RNH_2^{\bullet+}$ (oxidation), followed by the removal of a proton to obtain a radical $RNH^{\bullet}$ located on the nitrogen or on the α-carbon, in the tautomer. Then, the radical grafts onto the carbon surface mainly through a C–N covalent bond. This mechanistic pathway was thoroughly investigated by XPS and infrared spectroscopy by Adenier et al. in 2004 [6], who additionally showed that it is also relevant when noble metals (such as Au and Pt) are used in place of carbon. XPS spectra of Au and Pt electrochemically modified by aliphatic primary amines clearly evidence the nitrogen–metal bond and infrared spectra confirm the presence of secondary amine groups in the coating, both in agreement with the mechanism of Figure 10.2. Good resistance to sonication together with a thickness below 10 nm have also been demonstrated for these coatings.

It is worth noting that the process is inefficient with secondary amines because steric hindrance limits the access of the nitrogen radical to the solid carbon surface and severely decreases the rate of grafting. As a consequence, the coverage for these compounds is very low as compared to primary amines. The absence of hydrogen in tertiary amines makes them unsuitable for electrografting. Furthermore, the increased stability of secondary and tertiary amine radical cations is another reason for the lack of grafting which is also the case for aromatic primary amines (aniline and its derivatives).

It must be mentioned that this process strictly requires the use of organic solvents (ACN or dimethylformamide (DMF)). However, water is strongly oxidized at the potentials required to graft the amines. This decreases the industrial impact

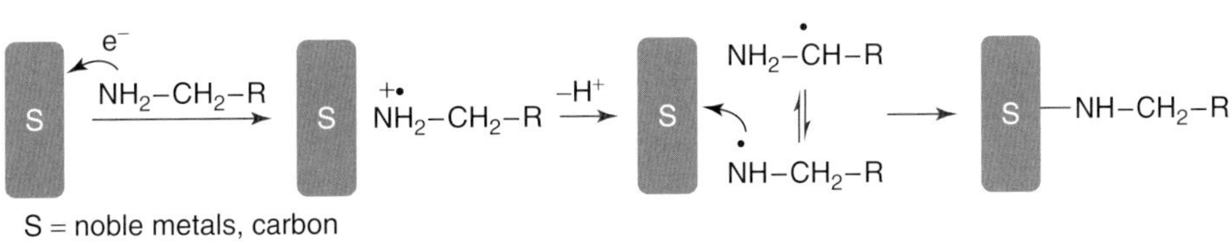

Figure 10.2 Mechanism for the electrochemical oxidation and grafting of primary amines.

of this technique. Nevertheless, the wide range of readily available functional aliphatic primary amines makes this process already quite versatile to anchor desired functionalities to the surfaces of carbon and noble metals.

As exemplified above with nitrobenzylamine, redox activity can be afforded to carbon and Au surfaces, which makes them useful as sensors in the field of modified electrodes.

10.4.2
Arylacetates

One of the oldest organic reactions in electrochemistry (nineteenth century) is the Kolbe reaction consisting of the anodic decarboxylative dimerization of carboxylates ($2RCOO^- \rightarrow R-R + 2CO_2$) via a radical reaction mechanism. As first reported by Andrieux et al. [7], when this reaction is applied to the electro-oxidation of arylacetates ($R-CH_2-COO^-$ with R = an aromatic ring) under appropriate conditions, it leads to the covalent modification by a dense organic layer of the $R-CH_2$ fragment exclusively onto carbon surfaces (glassy carbon or highly oriented pyrolytic graphite (HOPG)) with the release of carbon dioxide. A large number of functionalities can be easily imparted to the aryl ring, with the help of organic chemistry. Consequently, this method is versatile to produce carbon surfaces with selected functions.

The irreversible electrochemical oxidation of arylacetates in ACN occurs around +1.1 V versus SCE as evidenced by cyclic voltammetry [7]. In agreement with the grafting process, progressive passivation of the substrate has been observed by successive scans. Moreover, by using an arylcarboxylate bearing a nitro group on the aromatic ring, XPS clearly detected on the electrochemically modified substrate, a characteristic peak for the nitrogen atom involved in a nitro group that confirmed the grafting of the functional aryl group. The electroactivity of this group in the cathodic potential window also confirms the grafting.

The mechanism [7, 8] of this process (Figure 10.3) starts with the oxidation of the carboxylate to give $RCH_2CO_2^{\bullet}$ which is directly transformed into $RCH_2^{\bullet}$ with the concomitant release of CO_2. The potential is then sufficiently anodic to promote a second oxidation of the $RCH_2^{\bullet}$ radical to the corresponding carbocation $R-CH_2^+$ which reacts with a carbon atom from the surface. The resulting positive charge on the surface is counterbalanced by an electron from the electrode bulk; consequently, the overall reaction could be related to the reaction between one

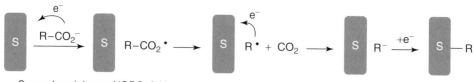

S = carbon (glassy, HOPG, felt)

Figure 10.3 Mechanism for the electrochemical oxidation of arylacetates with concomitant grafting of the aryl group onto carbon.

arylmethyl radical and the substrate. The lack of reactivity for radicals from R$_2$CHCOOH and R$_3$CCOOH explains the absence of grafting.

The possibility to electrochemically remove the film from the surface has to be mentioned. Indeed, if the coated carbon plates are polarized again in electrolyte-containing ACN solution at a potential slightly more anodic than the one for grafting, surfaces are recovered free of organic materials due to the overoxidation of the anchored layer [7, 8].

Nevertheless, with these compounds, the oxidation can be operated in aqueous media [9]. The advantage of dealing with water as a solvent is found in the increase of organic groups anchored onto the felt.

10.4.3
Primary Alcohols

Another anodic electrografting process studied by Maeda *et al.* [10] requires the anodic polarization of carbon electrode in aqueous acidic solution (1 M H$_2$SO$_4$) of an aliphatic primary alcohol. This electrografting occurs at high potential (+2.0 V vs Ag wire) and here again, successive scans between 0 and +2.0 V evidence a decrease in the current intensity due to the grafted layer. The passivation efficiency and thus coating homogeneity are demonstrated by the absence of the electroactivity signal for a solution containing Fe(CN)$_6^{3-}$ anions when the electrografted carbon is used as electrode. Indeed, after only three electrografting scans, no redox activity can be detected with such electrodes, that is, they are efficiently passivated. These observations also hold after sonication of the coated substrate in water, confirming the strong adhesion of the organic layer.

The electrografting mechanism involves a nucleophilic attack of the alcohol on a radical cation formed at the carbon surface leading to the formation of an ether linkage (C–O–R) (Figure 10.4). Contrary to the previous examples, this process requires the electroactivation of the substrate rather than the solute. This technique is very effective in terms of coating adhesion and range of functionalities available on the surface but is restricted to glassy carbon substrates.

Many applications of this process have been reported in the literature [11], one among them being the electrochemical detection of alkaline ions [12]. As mentioned above, the electrografting of 1-octanol on a carbon substrate prevents

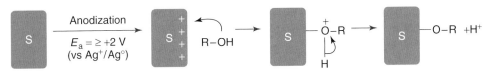

S = carbon

Figure 10.4 Mechanism for the nucleophilic attack of primary alcohols on anodized carbon surface resulting in the grafting of the R-group to the surface.

the observation of the voltammetric response to a redox probe such as $Fe(CN)_6^{3-}$. Surprisingly, however, electroactivity is restored in the presence of alkaline cations in the solution. The measured current intensity clearly depends on the type and concentration of cations, making the modified electrode a powerful sensor.

10.4.4
Arylhydrazines

Quite recently, in 2009, Malmos et al. [13] reported on the electrochemical oxidation of arylhydrazines onto glassy carbon and gold surfaces to promote the covalent grafting of a monolayer of aryl groups. Various functional arylhydrazines (bearing –OCH$_3$, –Cl, –COOH, or –NO$_2$ on the aryl ring) dissolved in aqueous medium show an irreversible peak around +0.3 V versus SCE, and demonstrate efficient passivation of the substrate as evidenced by means of the $Fe(CN)_6^{3-}$ redox probe. Electroactive coatings can also be achieved by using aryl hydrazine bearing a nitro group. Reversible voltammograms are obtained for carbon or gold grafted by the –NO$_2$-based compound.

IR shows the absence of NH$_2$ peaks evidencing the loss or alteration of the NH–NH$_2$ group during the electrografting. However, further studies are needed for a mechanism to be suggested and confirmed.

Should this last strategy require the synthesis of functional aryl hydrazines, it would have the advantage of producing strongly adhering monolayers at low anodic potential and in a wide range of arylhydrazine concentrations and pH values that would make it quite promising for further applications.

10.5
Compounds Requiring a Cathodic Process

10.5.1
Aryldiazonium Salts

The covalent bonding of functional aryl groups to a wide variety of (semi)conductive surfaces can be achieved cathodically via the electrochemical reduction of diazonium salts. This method is currently one of the most described and used since a large number of functionalities can be efficiently imparted to a broad range of inorganic substrates.

The pioneering work of Delamar et al. [14] demonstrated the coating of glassy carbon by an organic layer of 4-nitrophenyl by the electrochemical reduction of (4-nitrophenyl)diazonium tetrafluoroborate (typical concentration: 1–10 mM) from an ACN solution. Cyclic voltammetry shows two reduction phenomena in the first scan. The first peak centered at −0.04 V versus SCE is irreversible and corresponds to the diazonium reduction, whereas the second one is reversible and appears at a more cathodic potential (−1.20 V vs SCE) that is related to the electroactivity of the nitro group. A second voltammogram recorded on the same electrode shows the disappearance of the irreversible peak, leaving the reversible one intact. In spite of sonication treatment, during which the electrode is dipped into ACN with NBu$_4$BF$_4$, the reversible peak is persistent with time. These are first evidences of the strong covalent bonding between the coating and carbon. XPS analysis of the surface highlights the absence of the nitrogen atom from a diazonium group, whereas the peak is observed for the same atom derived from the nitro group.

The mechanism allowed for the electrografting is the reduction of the diazonium group resulting in the formation of an aryl radical with N$_2$ release (Figure 10.5). These radicals are then able to graft onto the surface to form a monolayer. It is worth noting that a weak polarization (−0.04 V vs SCE) is required to avoid reduction of the aryl radical into the corresponding anion that would prevent the chemisorption of the aryl radical onto the carbon surface.

Owing to the easy electroreduction of such species, the electrografting of aryl diazonium salts has been quite easily achieved in aqueous media [15]. For this purpose, the pH of the solution has to be kept below 2 (e.g., by the addition of H$_2$SO$_4$) in order to avoid degradation of the diazonium salt by its reaction with water. This advantage is of importance for both ecologic and economic impacts of the method.

In addition, if the cathodic electrografting of diazonium compounds onto carbon substrates (glassy carbon, HOPG, carbon fibers, carbon nanotubes [16], carbon felts

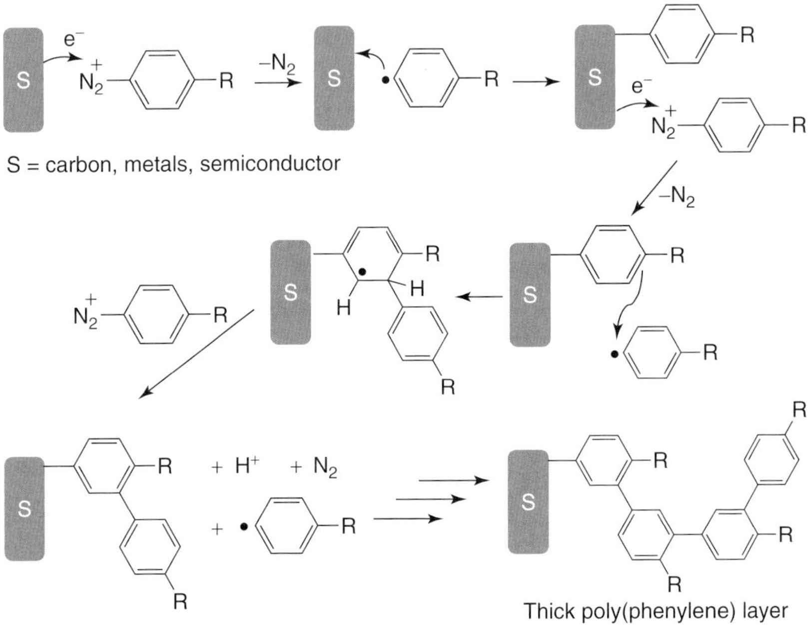

Figure 10.5 Mechanism for the electrochemical reduction of aryl diazonium followed by the thickening of the first grafted layer.

(application in combinatorial chemistry) [17], and dispersion of nanodiamonds [18] etc.) is successful, the extension to other substrates of interest such as doped silicon and common metals is also successfully achieved. Iron was the first common metal to be covalently modified by the reduction of aryl diazonium salts. The strong anchoring of functional organic layers onto this surface is ascertained by electrochemistry, XPS, IR, and other techniques [15]. For example, the formation of a covalent bond between the iron surface and aryl group (Fe–C) was evidenced by Boukerma *et al.*, with the help of XPS [19]. A crucial requirement for these substrates is the use of oxide-free metallic surfaces. Appropriate iron pretreatments such as polishing and reduction are thus applied just before electrografting. For this common metal, the major interest of the electrografting is the protection against corrosion via highly adhesive hydrophobic coatings (long alkyl chains, fluorinated chains, and polymers) [20]. However, noble metals are also successfully modified [21], illustrating the high potential of this technique. In addition, the possibility to obtain efficient coating of hydrogenated silicon surfaces (Si–H) by a film of 4-nitro or 4-bromoaryl groups was highlighted in the 1990s [22]. XPS evidenced the presence of Si–C bonds and showed typical profiles for a dense and organized monolayer. The presence of Si–C bonds rather than Si–O bonds considerably increases the chemical stability. Other studies on silicon surfaces confirmed the formation of an ordered monolayer for various functionalized aryl diazonium salts, which is the exclusive to this kind of substrate [23]. Indeed, for other substrates (carbon and metals), reaching a close-packed monolayer is very difficult because of secondary reactions which take place between the first grafted organic layer and radicals present in the vicinity of the electrode.

While the electrografting mechanism is the same for carbon and metals, a different one has been proposed for semiconductors such as hydrogenated Si. In this case, once the aryl radical is formed close to the surface, it reacts with the substrate by homolytic cleavage of Si–H bond to give a stable aromatic compound (aryl-H) and leave a silyl radical (Si•) on the surface. This silyl radical then combines with another aryl radical in the vicinity of the electrode to form the expected Si–C chemical bond.

As mentioned previously, except for hydrogenated silicon surfaces, it is not easy to produce an organized monolayer onto inorganic substrates. Coating efficiency is then explained by the progressive thickening of the organic layer [24]. As demonstrated in the case of a 4-nitrophenyldiazonium salt, the homolytic attack of another stable aryl radical is envisaged, preferentially in ortho position of the nitro group, to give a new radical with a cyclohexadienyl structure. To recover aromaticity, this radical reduces another molecule of diazonium salt present in the solution with concomitant proton and nitrogen release. Repeating this reaction results in the thickening of a poly(phenylene) film (Figure 10.5). By adjusting the diazonium concentration and the number of scans (or time of electrolysis if chronoamperometry is used), it is possible to reach coating thicknesses of several hundred of nanometers. It is worth noting that when two bulky groups, typically *tert*-butyl groups, are located in 3 and 5 positions of the aryl ring, this growth process is inhibited and only monolayers are obtained [25]. Finally, very recent results highlighted the presence of a few azo bonds (aryl–N = N–aryl) within the polyphenylene layer, which can be explained by the possible direct reaction between the aryl radical and the diazonium salt, that is, without loss of nitrogen [26].

A variety of functional aryl groups bearing a diazonium salt can be prepared and electrografted onto various substrates. These groups could be reacted to further advantage as demonstrated by the anchoring of D-glucosaminic acid, an antifouling molecule, on surfaces pretreated by amino-containing diazonium salts leading to protein-repellent surfaces [27]. Similarly, enzymes such as glucose oxidase have been successfully immobilized covalently on carbon electrodes [28]. By using 4-phenylacetic layer from the reduction of the corresponding diazonium salt, the enzyme is covalently attached via a peptidic bond as demonstrated by XPS. This modified electrode is then able to quantify glucose present in samples by electrocatalytic oxidation. The advantage of the cathodic process, as compared to the oxidative treatments when carbon substrate is considered, relies on the preservation of the smoothness of the surface. Consequently, the noise is reduced which allows the measurement of low current intensities with good precision. Since this first example, many developments of the technology in the field of electrochemical (bio)sensors have been reported and even collated in a recent review [29]. Indeed, the possibility to anchor many molecules of interest (e.g., antibodies [30]) allows the easy broadening in the choice of analytes. The patterning of carbon, gold, or silicon surfaces at the micro- or nanoscale is also possible via the local grafting of functional aryl groups [31].

Covalent grafting of conjugated polymers has also been considered by using the electroreduction of aryl diazonium salts. By using 4-aminodiphenylamine converted into the corresponding diazonium salt [32], PANI coatings have been anchored to carbon. In a first step, the aryl diazonium salt is reduced to form the covalent bond and bearing a precursor that can be converted into PANI by the polymerization of pure aniline in acidic aqueous solution. Owing to this two-step process, very adherent coatings of PANI were obtained on carbon while preserving their unique electronic properties. Thiophene oligomers could similarly be anchored to a metal surface via the reduction of the corresponding diazonium salt to produce conductive switches [33]. Molecular diodes [34] have been obtained by grafting ultrathin films of small oligomers of bithiophene.

Finally, some strategies have also been followed in order to anchor more conventional polymers (nonconjugated polymers). Indeed, some of these are also materials that have been identified today as smart or stimuli responsive and thus able to provide surfaces with, for example, pH– or thermoresponsiveness. In a recently reported strategy [35], for the first time, the electrografting onto gold surfaces of an aryl diazonium salt containing a bromide group that can be further converted into an azide function (by reaction with NaN_3) has been used to anchor preformed macromolecules (as perfluorinated ethylene glycol or oligoethylene glycol) bearing alkyne by the so-called "click" reaction. Others [36] have used such bromo groups to perform surface-initiated atom transfer radical polymerization (ATRP), growing vinylic polymer chains from the surface in a second step.

For interested readers, several reviews specifically dedicated to the modification of surfaces by using the electrochemical reduction of aryl diazonium derivatives are available for further information [37, 38].

10.5.2
(Meth)acrylic Monomers

The cathodic electrografting of (meth)acrylates is a useful method to covalently tether polymer chains, making them very adherent to a broad range of (semi)conductive inorganic materials. The broad diversity of readily available functional acrylic and methacrylic monomers, already largely applied at the industrial level, offers considerable scope of achievable functionalities. In addition, this technique not only leads to the covalent grafting of the organic coating but also simultaneously induces polymerization. In one step, polymer chains are thus tethered to the surface rather than to small molecules, imparting to the coating additional properties such as temperature-responsive properties that are specific to polymers.

In the early 1980s, the electroreduction of acrylonitrile (AN) in ACN to give grafted poly(acrylonitrile) (PAN) chains onto nickel was highlighted for the first time [39]. Further studies evidenced that during the sweeping of the cathodic potentials, two distinct reduction peaks for AN are observed [40]. Coupling cyclic voltammetry with the frequency variation of a quartz crystal microbalance (QCM) [41], studies concluded that at the potential of the less cathodic peak (the first reduction peak, about -2 V for Pt), a PAN film is chemisorbed onto the cathode, whereas at the second reduction wave, the chains desorb and grow in solution. Owing to the passivation of the electrode progressively coated by an insulating film, the achievable film thickness was quite limited. For example, when PAN films are obtained in ACN (a nonsolvent for the polymer), the film thickness remains below 25 nm. This could be improved by using DMF [42] as solvent, for which a thickness of up to 150 nm was obtained for a monomer concentration of 2 M. Indeed, DMF being a good solvent of PAN, the growing chains remain solvated during polymerization leading to improved chain propagation and consequently increased film thickness. Peeling tests, in addition to sonication treatments, clearly evidenced the high adhesion of the polymer coating formed and kept immersed in a suitable solvent. PAN films and particularly the PAN/metal interface were analyzed in depth by various surface analysis techniques in order to identify the nature and strength of interactions responsible for this adhesion. By electrografting vinylic molecules (polymerizable or not) on oxide-free surfaces, the formation of a C–metal bond was decisively highlighted by XPS [43] and atomic force microscopy (AFM) [44], in accordance with theoretical data based on density functional theory (DFT) [45]. It was shown that, by an appropriate choice of the experimental conditions, the AN monomer is either adsorbed flat on the surface and chemically bounded by a $(2p\pi)$-$(3d/4s)$ overlap where both C=C double and C≡N triple bonds of AN are involved, or grafted perpendicular to the surface by a covalent interaction between the nitrogen lone pair of AN and the $3d/4s$ of the metal.

The key parameter for the polymer electrografting to be achieved, is thus (i) to identify experimental conditions (solvent, conducting salt, potential) where the electroreduction of the vinylic monomer occurs in two steps and (ii) to keep the potential in the range of the first reduction peak. Baute *et al.* [46] showed that if electrografting of ethylacrylate (EA) fails in ACN (only one reduction peak is observed in voltammetry), it can be achieved efficiently in DMF. This important discovery widened the functionality and properties of the electrografted coatings that were restricted to PAN at that time [47]. A study of the effect of the solvent on the electrografting process showed that both the monomer and the solvent compete for adsorption at the cathode surface, preferential adsorption of the monomer being required for electrografting to occur. To choose an appropriate solvent for the electrografting of a given monomer, the solvent donor number (DN) is a helpful data, knowing that solvents with high DN have a lower ability to adsorb on the cathode. This explains why electrografting of poly(ethylacrylate) (PEA) chains fails in ACN (DN = 14.1) and occurs efficiently in DMF (DN = 26.6). In DMF, EA preferentially adsorbs to the surface in accordance with the binding energies obtained by DFT calculations [48].

By using DMF, it is thus now possible to covalently graft (meth)acrylic derivatives. Polymers containing hydrophilic or reactive groups such as protected carboxylic and hydroxyl groups, or epoxy groups have been electrografted [47]. Interestingly, an acrylic derivative with the ester activated by a *N*-succinimidyl group has been also successfully electrografted leading to functional surfaces that are quite versatile [49]. Combination of two monomers, electrografted simultaneously or successively [50], is also possible to impart the surface with multifunctionality. By further manipulation of the functionality incorporated in the ester group of the acrylic monomer, the electrografting of inimers, that is, compounds combining a polymerizable function and an initiator function within the same structure, have been also achieved. It was demonstrated that electrografting is easily combined with common controlled or living polymerization processes [51]. Acrylates bearing an initiator of (i) controlled radical polymerization (ATRP [52] and nitroxyl-mediated polymerization (NMP) [53]), (ii) ring-opening polymerization (ROP) [54], or (iii) ring-opening metathesis polymerization (ROMP) [55] have been synthesized and successfully electrografted. After the grafting of the monomer, a second family of chains can be initiated and grown from the strongly adherent first layer. Consequently, the chemical properties of the final organic coating can be extensively changed and the film thickness can be tuned while retaining the high adhesion of the coating.

In addition to these monomers of small size, the application of electrografting was recently reported for macromonomers [47], that is, preformed macromolecules bearing one or several acrylic groups. The advantage is the possibility to control the molecular characteristics of the chains (molecular weight, architecture, etc.) before their chemisorption and to apply electrografting to nonacrylic prepolymers. By this strategy, efficient protein-repellent surfaces were obtained by electrografting poly(ethylene oxide) chains end capped by an acrylic group at one or both ends [56]. Poly(ε-caprolactone) (PCL) of a well-defined molecular weight containing one acrylate per monomer unit can also be prepared by ROP of the parent dual

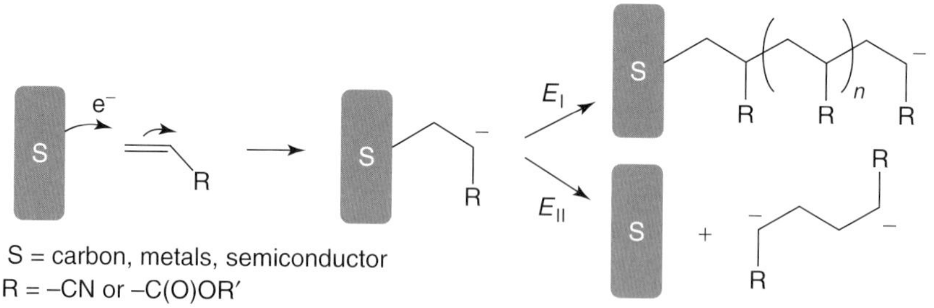

Figure 10.6 Mechanism for the electrochemical reduction of an acrylic derivative at both reduction peaks (potential E_I or E_{II}).

S = carbon, metals, semiconductor
R = –CN or –C(O)OR′

monomer, γ-acryloxy-ε-caprolactone. Under cathodic polarization, these chains are chemisorbed onto the electrode [57]. Random copolymers of ε-caprolactone and the γ-acryloxy-ε-caprolactone were also efficiently tested and found to be good primers for poly(vinyl chloride) (PVC) top coats.

The mechanism for (meth)acrylic electrografting consists in the transfer of one electron from the cathode to the monomer, leading to a radical anion that chemisorbs on the metal as long as the potential is kept within the potential range of the first reduction peak (E_I). The electrochemically chemisorbed anion is then able to propagate the polymerization and the chain growth from the surface leading to the tethered polymer chains (Figure 10.6). Chain initiation is thus an electrochemical event, whereas the propagation does not require electrical current. When a more cathodic potential is achieved (E_{II}, Figure 10.6), that is, the second wave is reached on the voltammogram, the chemisorption is no longer effective and the radical anion is free in solution. Fast coupling of two radicals then occur leaving bis-anionic dimers that are able to propagate the chains from both ends in solution.

The nucleophilic attack of the ester group of the monomer and/or polymer by the anionic propagating species of the chains growing from the surface could provide them with a complex structure. However, these chains are completely desorbed when the surface is immersed in an electrolytic solution (without monomer) and cathodically polarized in the range of the second voltammetric peak. The cross-linking of the electrografted chains can thus be precluded on the basis of their solubility upon degrafting and the release of a neat metal surface. As far as PAN is concerned, Fourier transform infrared (FTIR) analysis [58] also confirmed the absence of cross-linking, which had, however, been suggested earlier [59–61]. Nevertheless, depending on the experimental conditions, some azine intermolecular bonds were detected by Raman spectroscopy at the very first stage of electrografting [58]. Formation of these bonds that result from the nucleophilic addition of the nitrile groups requires, however, that these groups are aligned parallel to the electrode surface, which is liable to be the case for isotactic PAN. The isotacticity of these chains was experimentally confirmed by dynamic mechanical analysis and infrared spectroscopy [62]. This might be the consequence of the orientation of the nitrile dipoles in the electrical double layer of the cathode. These preliminary observations provide hints of a brushlike conformation of these electrografted chains. Recent characterization performed by AFM of approach/retraction curves has also given pertinent information about the length and conformation of the grafted chains [63]. The compression profile of the chain bridging an AFM tip to the surface actually depends on the monomer concentration originally used for the film formation by electrografting. If this solution is diluted (0.05 M), the recorded profile is typical of the isolated chain regime, where the behavior of a tethered chain is independent of the neighboring ones as result of a low grafting density. In the case of more concentrated monomer solutions (1 M), monotonically increasing repulsive forces are observed during the approach of the tip to the surface, which is characteristic of a higher grafting density responsible for a brush regime. This situation was confirmed by topographic AFM imaging. The chain length could be estimated from the tip/surface distance when the compression of the brush starts to be observed. Chain lengths of approximately 80 and 250 nm were accordingly reported for chains grafted onto silicon nitride in a monomer solution in DMF of 0.5 and 1 M, respectively. These data provide direct confirmation that the length of the grafted chains increases with the monomer concentration. In this example, the average degree of polymerization of the chains was thus 320 and 1000, respectively. The surface-grafting density onto doped silicon was also analyzed by AFM. A grafting density of $\sim$1 chain per 100 nm^2 was determined from the compression profile recorded for a surface prepared in DMF ((monomer) $=$ 1 M) and exhibiting a brush regime. The tip/surface distance at which chain rupture was observed in the retraction mode is the length of the stretched chain segment located between the grafting points on the substrate and the tip, respectively. Accordingly, an average degree of polymerization was estimated at 1016 [64]. Last but not the least, the good fit of the shape of the compression profiles and the function based on the Alexander-de Gennes scaling concept that describes the forces resulting from the compression of polymer brushes, together with the good agreement between the bridging profiles and models for a wormlike chain and freely jointed chain under tension, are strong indications that the grafted chains are essentially linear with few branching points.

Remarkably, electrografting can be applied to a broad range of substrates. Common metals such as Fe, Ni, and Cu are extremely suitable substrates for electrografting, provided the superficial oxide layer has been removed. However, this requires working under inert atmosphere, which is not the case for noble metals (Au and Pt). Zn and Al having quite resistant oxides are thus less prone to electrografting. Doped Si [65], of interest in microelectronics, is also a substrate easily modified by electrografting if pretreated by diluted HF solution. Finally, using carbon substrates leads to very stable electrografted coatings, a strong C–C bond resulting from the grafting. By adapting the electrochemical setup, electrografting can be achieved onto various allotropic forms. For example, the modification of multiwalled carbon nanotubes has been recently demonstrated [66]. Electrografting thus appears as an efficient technique for surface modification of various fillers improving the composite properties in which they are involved.

The main limitation of the cathodic electrografting of acrylates relies on the low cathodic potential required for the electrografting to occur (about -2 V for Pt). This makes the reduction of protic compounds compete with the grafting process and thus protection of such groups is needed before electrografting. In addition, DMF is one of the best solvents for electrografting to be highly efficient, but this solvent is quite toxic and costly, which prevents broad applications of the technology at the industrial level. The implementation of electrografting in greener media was thus a pending problem that has been successfully addressed recently.

Water, a green liquid medium for electrografting, could be used provided its electroreduction is shifted toward more cathodic potential than the monomer itself. Proper design of the (meth)acrylate monomers was required to reach this target. A specific polymerizable amphiphilic compound has been accordingly tailored by including (i) a long hydrophobic alkyl chain able to expel water from the electrical double layer of the cathode and to increase the electrochemical window of this solvent; (ii) the capping of this chain by a cationic hydrophilic head at one end, in order to trigger micellization and adsorption to the cathode surface; (iii) the capping of the second end of the chain by a polymerizable acrylic unsaturation. On the basis of this general concept, the amphiphilic monomer [(10-acryloyloxy)decyl] trimethyl ammonium bromide was designed, synthesized, and successfully electrografted to the cathode [67]. For this amphiphilic monomer, electrografting is thus possible in water, with the formation of an adherent polycationic coating. These positively charged films can impart specific properties to the surface, for example, bactericidal properties when a quaternary ammonium is the cationic species. Moreover, they are highly desirable anchoring layers for building up multilayered films by the well-known layer-by-layer deposition of polyelectrolytes of opposite charges [68].

The range of commercially available monomers and the ease of the electrografting process to create a strongly adhering polymer coating have accelerated the development of this process in various application fields; this has reached industrial level in some fields, at least for high value-added devices, such as cardiovascular stents [69]. Indeed, strongly adhering biocompatible coatings can be homogeneously grafted on these metallic medical devices. Other direct applications include the elaboration of antibacterial coatings [70–73] by the use of acrylic derivatives bearing quaternary ammonium groups.

In a very simple approach, the electrografting of conjugated polymers (PPy or PThi) has been achieved by synthesizing an acrylate bonded to a pyrrole or a thiophene ring, that is, N-(2-acryloyloxyethyl) pyrrole (PyA) and 3-(2-acryloyloxyethyl) thiophene (ThiA). Under cathodic polarization, the poly(acrylate) chains were chemisorbed. Upon anodic polarization, the pyrrole or thiophene rings incorporated into the grafted layer are polymerized to form adherent electroactive coatings [74]. In addition, the combination of electrografting and electrosynthesis of conducting polymers gives rise to the possible synthesis of PPy nanowires [75] or sensitive solvent sensors [76].

Taking advantage of the electrogeneration of polymer chains, smart surfaces have been targeted by electrografting of tailored monomers. For example, thermosensitive poly(N-isopropylacrylamide) (PNIPAM) surfaces [77] were prepared by immersing previously electrografted poly(*N*-succinimidyl acrylate) (PNSA) chains in isopropyl amine. The contact angle of water of the resulting coating changed dramatically when the temperature was scanned from 25 to 35 °C, that is, throughout the LCST (lower critical solution temperature) range of the polymer. Similarly, electroactive [78] and pH-sensitive [79] coatings have also been elaborated.

Polymer electrografting was applied to advantage in commercially available silicon and silicon nitride [63] AFM cantilevers with a simple polarization setup [80]. In this respect, *N*-succinimidyl acrylate (NSA) is a very versatile monomer, because of the high reactivity of the activated ester it contains and, thus, the easy and extensive derivatization of the PNSA chains electrografted onto the AFM tips. Ferrocene and proteins were chemically bonded onto a tip and used for molecular recognition merely by recording approach–retraction profiles of the tip facing the surface to be analyzed. For example, cyclodextrine available on a surface was analyzed with a ferrocene-modified tip. The recorded forces were typical of the ferrocene–cyclodextrine host–guest complex. Stronger interactions were expectedly reported in the case of avidin–biotin complexation endowed with a higher complexation constant [81]. Tips modified by polymer electrografting and further polymer derivatization are thus unique tools in the so-called (bio)molecular recognition force spectroscopy. The activated esters along electrografted PNSA chains were also used to anchor antigens of a pharmacological interest to an AFM tip, and the interactions with specific antibody receptors were monitored in water [81].

The preparation of smart probes whose surface is stimuli responsive also illustrates a remarkable application of the versatility of electrografted PNSA [79]. Thermoresponsive and pH-sensitive probes are achieved when PNSA is converted into PNIPAM and polyacrylic acid (PAA) respectively. A transition from a collapsed to a swollen brush and vice versa is then triggered by temperature or pH, the response being fast and reversible over a long period. PNSA-electrografted tips are thus quite a generic platform in the development of specific (bio)sensors. These electrografted tips also find unique application in the manipulation of single molecules and delivering them in a precisely controlled manner to a specific target [44]. Considering the rapidly increasing demand for nanoengineering operations in "bottom-up" nanotechnology, this method that operates at the molecular scale is expected to be a key player in the fabrication of organic surfaces.

For more information, a detailed review was very recently dedicated to this topic [82].

10.5.3
Iodide Derivatives

In the 1980s, Andrieux *et al.* [83] studied the electrochemical behavior of alkyl halides in DMF using cyclic voltammetry. For *t*-butyl iodide, two separate waves were recorded, each one related to the exchange of one electron. The first one was attributed to the reduction of the halide with cleavage of the C–I bond to give a *t*-butyl radical and an iodine anion. The second one is the reduction of this radical into the *t*-butyl anion. For *sec*-butyl iodide, both peaks are very close

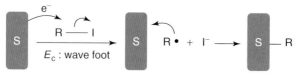

Figure 10.7 Mechanism for the electrochemical reduction of primary iodide derivatives with grafting of the alkyl group to the surface.

to each other, whereas for *n*-butyl iodide both reactions are superimposed. For bromine compounds, whatever the derivative, only a single two-electron process is observed. Radicals from secondary and tertiary species would react with the surface as diazonium salts or (meth)acrylates. Unfortunately, as in the case of secondary and tertiary amines, steric reasons hinder the approach of radicals from the cathode preventing the grafting.

More recently, Chehimi et al. [84] observed the formation of an organic layer when the potential is held at the foot of the unique wave for the reduction of primary iodide derivatives. This study was performed onto glassy carbon and metal by using ACN as solvent for two kinds of alkyliodide (1-iodohexane, IC_6H_{13}, and a semiperfluorinated iodide $I(CH_2)_2C_8F_{17}$). After sweeping the potentials until the foot of the wave, holding the potential at this value for 30 min, followed by sonication of the substrate in pure solvent, the contact angle highlighted the presence of the hydrophobic coating.

A proposed mechanism is given in Figure 10.7, even though thickness measurements by ellipsometry are in favor of a multilayer rather than a monolayer. Such multilayers might be explained by a recombination between radicals and the grafted organic layer (presence of C=C bonds). IR and XPS confirmed the film formation, owing to the presence of characteristic peaks. Nevertheless, in the present case, the passivation is less efficient than the one observed in the case of the previously discussed diazonium or (meth)acrylate derivatives. This novel electrografting process thus appears quite slow and of poor density, even if the inhibition of $Fe(CN)_6^{3-}$ electroactivity after modification confirmed the good adhesion of the layer onto the substrate. In spite of these drawbacks, the technique is worth mentioning since functional iodide derivatives are readily available compounds.

10.6
Conclusions

As far as the coating of conductive or semiconductive materials is concerned, electrochemistry is a relevant technology to easily tailor coating adhesion and functionality. Smart surfaces can thus be expensive, achieved mostly with respect to the environment. Electroactive surfaces can be easily prepared by simple electrodeposition of the aromatic rings. In addition, many strategies have been developed to impart them with stronger adhesion to the substrate by electrochemical modification of the surface in the presence of various compounds.

Among these strategies, electrografting of acrylates appears as a versatile technique to directly impart responsiveness to a surface. Indeed, this process allows the formation of a grafted polymer coating that exhibits the typical behavior of the designed macromolecules. Thermo- and pH-sensitive coatings can be grafted to metals or semiconductors by this straightforward process leading to advanced electrodes, and smart nanoprobes for sensing. The recent transfer of this process to water opens the door to further developments, particularly in the field of medical devices.

Other techniques might also be combined with polymer chemistry in the future in order to reach desired multifunctionality and responsiveness while retaining the remarkable adhesion of these organic layers.

References

1. Sabouraud, G., Sadki, S., and Brodie, N. (2000) *Chem. Soc. Rev.*, **29**, 283.
2. Mortimer, R.J. (1997) *Chem. Soc. Rev.*, **26**, 147.
3. (a) Kumar, D. and Sharma, R.C. (1998) *Eur. Polym. J.*, **34**, 1053; (b) Li, C., Bai, H., and Shi, G. (2009) *Chem. Soc. Rev.*, **38**, 2397.
4. Barbier, B., Pinson, J., Desarmot, G., and Sanchez, M. (1990) *J. Electrochem. Soc.*, **137**, 1757.
5. Antoniadou, S., Jannakoudakis, A.D., Jannakoudakis, P.D., and Theodoridou, E. (1992) *J. Appl. Electrochem.*, **22**, 1060.
6. Adenier, A., Chehimi, M.M., Gallardo, I., Pinson, J., and Vila, N. (2004) *Langmuir*, **20**, 8243.
7. Andrieux, C.P., Gonzalez, F., and Saveant, J.M. (1997) *J. Am. Chem. Soc.*, **119**, 4292.
8. Brooksby, P.A., Downard, A.J., and Yu, S.S.C. (2005) *Langmuir*, **21**, 11304.
9. Geneste, F., Cadoret, M., Moinet, C., and Jezequel, G. (2002) *New J. Chem.*, **26**, 1261.
10. Maeda, H., Yamauchi, Y., Hosoe, M., Li, T.-X., Yamaguchi, E., Kasamatsu, M., and Ohmori, H. (1994) *Chem. Pharm. Bull.*, **42**, 1870.
11. (a) Maeda, H., Kitano, T., Huang, C.Z., Katayama, K., Yamauchi, Y., and Ohmori, H. (1999) *Anal. Sci.*, **15**, 531; (b) Maeda, H., Katayama, K., Matsui, R., Yamauchi, Y., and Ohmori, H. (2000) *Anal. Sci.*, **16**, 293; (c) Downard, A.J. (2000) *Electroanalysis* **12**, 1085.
12. Maeda, H., Hosoe, M., Li, T.-X., Itami, M., Yamauchi, Y., and Ohmori, H. (1996) *Chem. Pharm. Bull.*, **44**, 559.
13. Malmos, K., Iruthayaraj, J., Pedersen, S.U., and Daasbjerg, K. (2009) *J. Am. Chem. Soc.*, **131**, 13926.
14. Delamar, M., Hitmi, R., Pinson, J., and Saveant, J.M. (1992) *J. Am. Chem. Soc.*, **114**, 5883.
15. Adenier, A., Bernard, M.C., Chehimi, M.M., Cabet-Deliry, E., Desbat, B., Fagebaume, O., Pinson, J., and Podvorica, F. (2001) *J. Am. Chem. Soc.*, **123**, 4541.
16. (a) Bahr, J.L., Yang, J.P., Kosynkin, D.V., Bronikowski, M.J., Smalley, R.E., and Tour, J.M. (2001) *J. Am. Chem. Soc.*, **123**, 6536; (b) Marcoux, P.R., Hapiot, P., Batail, P., and Pinson, J. (2004) *New J. Chem.*, **28**, 302.
17. (a) Coulon, E., Pinson, J., Bourzat, J.D., Commercon, A., and Pulicani, J.P. (2001) *Langmuir*, **17**, 7102; (b) Coulon, E., Pinson, J., Bourzat, J.D., Commercon, A., and Pulicani, J.P. (2002) *J. Org. Chem.*, **67**, 8513.
18. (a) Lud Simon, Q., Steenackers, M., Jordan, R., Bruno, P., Gruen Dieter, M., Feulner, P., Garrido Jose, A., and Stutzmann, M. (2006) *J. Am. Chem.*

Soc., **128**, 16884; (b) Dahoumane, S.A., Nguyen, M.N., Thorel, A., Boudou, J.-P., Chehimi, M.M., and Mangeney, C. (2009) *Langmuir*, **25**, 9633.
19. Boukerma, K., Chehimi, M.M., Pinson, J., and Blomfield, C. (2003) *Langmuir*, **19**, 6333.
20. (a) Adenier, A., Cabet-Delity, E., Lalot, T., Pinson, J., and Podvorica, F. (2002) *Chem. Mater.*, **14**, 4576; (b) Chausse, A., Chehimi, M.M., Karsi, N., Pinson, J., Podvorica, F., and Vautrin-Ul, C. (2002) *Chem. Mater.*, **14**, 392.
21. Bernard, M.-C., Chausse, A., Cabet-Delity, E., Chehimi, M.M., Pinson, J., Podvorica, F., and Vautrin-Ul, C. (2003) *Chem. Mater.*, **15**, 3450.
22. de Villeneuve, C.H., Pinson, J., Bernard, M.C., and Allongue, P. (1997) *J. Phys. Chem. B*, **101**, 2415.
23. (a) Allongue, P., de Villeneuve, C.H., Pinson, J., Ozanam, F., Chazalviel, J.N., and Wallart, X. (1998) *Electrochim. Acta*, **43**, 2791; (b) Allongue, P., de Villeneuve, C.H., and Pinson, J. (2000) *Electrochim. Acta*, **45**, 3241.
24. Combellas, C., Kanoufi, F., Pinson, J., and Podvorica, F.I. (2005) *Langmuir*, **21**, 280.
25. (a) Combellas, C., Kanoufi, F., Pinson, J., and Podvorica, F.I. (2008) *J. Am. Chem. Soc.*, **130**, 8576; (b) Combellas, C., Jiang, D., Kanoufi, F., Pinson, J., and Podvorica, F.I. (2009) *Langmuir*, **25**, 286.
26. Doppelt, P., Hallais, G., Pinson, J., Podvorica, F., and Verneyre, S. (2007) *Chem. Mater.*, **19**, 4570.
27. Gautier, C., Ghodbane, O., Wayner, D.D.M., and Belanger, D. (2009) *Electrochim. Acta*, **54**, 6327.
28. Bourdillon, C., Delamar, M., Demaille, C., Hitmi, R., Moiroux, J., and Pinson, J. (1992) *J. Electroanal. Chem.*, **336**, 113.
29. Gooding, J.J. (2008) *Electroanalysis*, **20**, 573.
30. Dauphas, S., Corlu, A., Guguen-Guillouzo, C., Ababou-Girard, S., Lavastre, O., and Geneste, F. (2008) *New J. Chem.*, **32**, 1228.
31. (a) Downard, A.J., Garrett, D.J., and Tan, E.S.Q. (2006) *Langmuir*, **22**, 10739; (b) Charlier, J., Palacin, S., Leroy, J., Del Frari, D., Zagonel, L., Barrett, N., Renault, O., Bailly, A., and Mariolle, D. (2008) *J. Mater. Chem.*, **18**, 3136; (c) Hauquier, F., Matrab, T., Kanoufi, F., and Combellas, C. (2009) *Electrochim. Acta*, **54**, 5127.
32. Santos, L.M., Ghilane, J., Fave, C., Lacaze, P.-C., Randriamahazaka, H., Abrantes, L.M., and Lacroix, J.-C. (2008) *J. Phys. Chem. C*, **112**, 16103.
33. Stockhausen, V., Ghilane, J., Martin, P., Trippe-Allard, G., Randriamahazaka, H., and Lacroix, J.-C. (2009) *J. Am. Chem. Soc.*, **131**, 14920.
34. Fave, C., Leroux, Y., Trippe, G., Randriamahazaka, H., Noel, V., and Lacroix, J.-C. (2007) *J. Am. Chem. Soc.*, **129**, 1890.
35. Mahouche, S., Mekni, N., Abbassi, L., Lang, P., Perruchot, C., Jouini, M., Mammeri, F., Turmine, M., Ben Romdhane, H., and Chehimi, M.M. (2009) *Surf. Sci.*, **603**, 3205.
36. (a) Guillez, A., Save, M., Charleux, B., Matrab, T., Chehimi, M.M., Perruchot, C., Adenier, A., Cabet-Delity, E., and Pinson, J. (2005) *Polym. Prepr.*, **46**, 318; (b) Matrab, T., Chehimi, M.M., Perruchot, C., Adenier, A., Guillez, A., Save, M., Charleux, B., Cabet-Delity, E., and Pinson, J. (2005) *Langmuir*, **21**, 4686; (c) Shaulov, Y., Okner, R., Levi, Y., Tal, N., Gutkin, V., Mandler, D., and Domb, A.J. (2009) *ACS Appl. Mater. Interfaces*, **1**, 2519.
37. Pinson, J. and Podvorica, F. (2005) *Chem. Soc. Rev.*, **34**, 429.
38. (a) Barriere, F. and Downard, A.J. (2008) *J. Solid State Electrochem.*, **12**, 1231; (b) McCreery, R.L. (2008) *Chem. Rev.*, **108**, 2646.
39. Lecayon, G., Bouizem, Y., Le Gressus, C., Reynaud, C., Boiziau, C., and Juret, C. (1982) *Chem. Phys. Lett.*, **91**, 506.
40. Baute, N., Calberg, C., Dubois, P., Jérôme, C., Jérôme, R., Martinot, L., Mertens, M., and Teyssie, P. (1998) *Macromol. Symp.*, **134**, 157.
41. Baute, N., Martinot, L., and Jérôme, R. (1999) *J. Electroanal. Chem.*, **472**, 83.
42. Calberg, C., Mertens, M., Jérôme, R., Arys, X., Jonas, A.M., and Legras, R. (1997) *Thin Solid Films*, **310**, 148.

43. Deniau, G., Azoulay, L., Jegou, P., Le Chevallier, G., and Palacin, S. (2006) *Surf. Sci.*, **600**, 675.
44. Duwez, A.S., Jérôme, C., and Gabriel, S. (2006) *Nature. Nanotechnol.*, **1**, 122.
45. Crispin, X., Lazzaroni, R., Crispin, A., Geskin, V.M., Bredas, J.L., and Salaneck, W.R. (2001) *J. Electron Spectrosc. Relat. Phenom.*, **121**, 57.
46. Baute, N., Teyssie, P., Martinot, L., Mertens, M., Dubois, P., and Jerome, R. (1998) *Eur. J. Inorg. Chem.*, **11**, 1711.
47. Baute, N., Jérôme, C., Martinot, L., Mertens, M., Geskin, V.M., Lazzaroni, R., Bredas, J.L., and Jérôme, R. (2001) *Eur. J. Inorg. Chem.*, **5**, 1097.
48. Crispin, X., Lazzaroni, R., Geskin, V., Baute, N., Dubois, P., Jérôme, R., and Bredas, J.L. (1999) *J. Am. Chem. Soc.*, **121**, 176.
49. Jérôme, C., Gabriel, S., Voccia, S., Detrembleur, C., Ignatova, M., Gouttebaron, R., and Jérôme, R. (2003) *Chem. Commun.*, 2500.
50. Baute, N., Geskin, V.M., Lazzaroni, R., Bredas, J.L., Arys, X., Jonas, A.M., Legras, R., Poleunis, C., Bertrand, P., Jérôme, R., and Jérôme, C. (2004) *E-Polymers.* **63**, 1–20.
51. Voccia, S., Gabriel, S., Serwas, H., Jérôme, R., and Jérôme, C. (2006) *Prog. Org. Coat.*, **55**, 175.
52. Claes, M., Voccia, S., Detrembleur, C., Jérôme, C., Gilbert, B., Leclere, P., Geskin, V.M., Gouttebaron, R., Hecq, M., Lazzaroni, R., and Jérôme, R. (2003) *Macromolecules*, **36**, 5926.
53. Voccia, S., Jérôme, C., Detrembleur, C., Leclere, P., Gouttebaron, R., Hecq, M., Gilbert, B., Lazzaroni, R., and Jérôme, R. (2003) *Chem. Mater.*, **15**, 923.
54. Voccia, S., Bech, L., Gilbert, B., Jerome, R., and Jérôme, C. (2004) *Langmuir*, **20**, 10670.
55. (a) Detrembleur, C., Jérôme, C., Claes, M., Louette, P., and Jérôme, R. (2001) *Angew. Chem. Int. Ed.*, **40**, 1268; (b) Voccia, S., Claes, M., Jérôme, R., and Jérôme, C. (2005) *Macromol. Rapid Commun.*, **26**, 779.
56. Gabriel, S., Dubruel, P., Schacht, E., Jonas, A.M., Gilbert, B., Jérôme, R., and Jérôme, C. (2005) *Angew. Chem. Int. Ed.*, **44**, 5505.
57. Lou, X.D., Jérôme, C., Detrembleur, C., and Jérôme, R. (2002) *Langmuir*, **18**, 2785.
58. Easter, P.A. and Taylor, D.M. (2009) *J. Polym. Sci., Part A Polym. Chem.*, **47**, 1685.
59. Grassie, N. and Hay, J.N. (1962) *J. Polym. Sci.*, **56**, 189.
60. Badawy, S.M. and Dessouki, A.M. (2003) *J. Phys. Chem. B*, **107**, 11273.
61. Funt, B.L. and Williams, F.D. (1964) *J. Polym. Sci., Part A Polym. Chem.*, **2**, 865.
62. Calberg, C., Mertens, M., Baute, N., Jérôme, R., Carlier, V., Sclavons, M., and Jérôme, R. (1998) *J. Polym. Sci., Part B Polym. Phys.*, **36**, 543.
63. Gabriel, S., Jérôme, C., Jérôme, R., Fustin, C.-A., Pallandre, A., Plain, J., Jonas, A.M., and Duwez, A.-S. (2007) *J. Am. Chem. Soc.*, **129**, 8410.
64. Cuenot, S., Gabriel, S., Jérôme, R., Jérôme, C., Fustin, C.-A., Jonas, A.M., and Duwez, A.S. (2006) *Macromolecules*, **39**, 8428.
65. Charlier, J., Ameur, S., Bourgoin, J.-P., Bureau, C., and Palacin, S. (2004) *Adv. Funct. Mater.*, **14**, 125.
66. Petrov, P., Lou, X.D., Pagnoulle, C., Jérôme, C., Calberg, C., and Jérôme, R. (2004) *Macromol. Rapid Commun.*, **25**, 987.
67. Cécius, M., Jérôme, R., and Jérôme, C. (2007) *Macromol. Rapid Commun.*, **28**, 948.
68. Charlot, A., Gabriel, S., Detrembleur, C., Jérôme, R., and Jérôme, C. (2007) *Chem. Commun.*, 4656.
69. Jérôme, C., Aqil, A., Voccia, S., Labaye, D.E., Maquet, V., Gautier, S., Bertrand, O.F., and Jérôme, R. (2006) *J. Biomed. Mater. Res. Part A*, **76A**, 521.
70. Ignatova, M., Voccia, S., Gilbert, B., Markova, N., Cossement, D., Gouttebaron, R., Jérôme, R., and Jérôme, C. (2006) *Langmuir*, **22**, 255.
71. Voccia, S., Ignatova, M., Jérôme, R., and Jérôme, C. (2006) *Langmuir*, **22**, 8607.
72. Ignatova, M., Voccia, S., Gilbert, B., Markova, N., Mercuri, P.S., Galleni, M., Sciannamea, V., Lenoir, S., Cossement, D., Gouttebaron, R., Jérôme, R., and Jérôme, C. (2004) *Langmuir*, **20**, 10718.

73. Ignatova, M., Voccia, S., Gabriel, S., Gilbert, B., Cossement, D., Jérôme, R., and Jérôme, C. (2009) *Langmuir*, **25**, 891.
74. Labaye, D.E., Jérôme, C., Geskin, V.M., Louette, P., Lazzaroni, R., Martinot, L., and Jérôme, R. (2002) *Langmuir*, **18**, 5222.
75. Jérôme, C. and Jérôme, R. (1998) *Angew. Chem. Int. Ed. Engl.*, **37**, 2488.
76. Jérôme, C., Geskin, V., Lazzaroni, R., Bredas, J.L., Thibaut, A., Calberg, C., Bodart, I., Mertens, M., Martinot, L., Rodrigue, D., Riga, J., and Jérôme, R. (2001) *Chem. Mater.*, **13**, 1656.
77. Gabriel, S., Duwez, A.S., Jérôme, R., and Jérôme, C. (2007) *Langmuir*, **23**, 159.
78. Gabriel, S., Cecius, M., Fleury-Frenette, K., Cossement, D., Hecq, M., Ruth, N., Jérôme, R., and Jérôme, C. (2007) *Chem. Mater.*, **19**, 2364.
79. Cecchet, F., Lussis, P., Jérôme, C., Gabriel, S., Silva-Goncalves, E., Jérôme, R., and Duwez, A.-S. (2008) *Small*, **4**, 1101.
80. Jérôme, C., Willet, N., Jérôme, R., and Duwez, A.S. (2004) *Chem. Phys. Chem.*, **5**, 147.
81. Cecchet, F., Duwez, A.-S., Gabriel, S., Jérôme, C., Jérôme, R., Glinel, K., Demoustier-Champagne, S., Jonas, A.M., and Nysten, B. (2007) *Anal. Chem.*, **79**, 6488.
82. Gabriel, S., Jérôme, R., and Jérôme, C. (2010) *Prog. Polym. Sci.*, **35**, 113.
83. Andrieux, C.P., Gallardo, I., Saveant, J.M., and Su, K.B. (1986) *J. Am. Chem. Soc.*, **108**, 638.
84. Chehimi, M.M., Hallais, G., Matrab, T., Pinson, J., and Podvorica, F.I. (2008) *J. Phys. Chem. C*, **112**, 18559.

Index

a
ab initio emulsion polymerization   13
acetonitrile (ACN)   250, 251, 252, 258, 264
acetylcholinesterase (AChE)   184
acrylamides   8
3-acrylamidophenylboronic acid (APBA)   31
acrylates   231
acrylonitrile (AN)   258
10-acryloyloxy decyl trimethyl ammonium bromide (acry–C10)   262
activators generated by electron transfer (AGET)   13
adenosine triphosphate (ATP) synthase   151
admantanes   230
aggregachromic dyes   117, 120
Alexander-de Gennes scaling   261
aliphatic primary amines   250–251
alkanethiol   141
alkyl halogenides   4
alkylimmonium   232
4-di(2-methoxyethyl) aminobenzylidene malononitrile (AODCST)   199, 207
4-aminodiphenylamine   257
3-aminopropyl triethoxysilane (APTES)   146
aminosilane-grafted 3-(ethoxydimethylsilyl) propylamine (APDMES)   146
Angstrom-level repairs
– covalent bonding   99–104
– hydrogen bonding   104–107
anodic polymerization   248–249
antigen-responsive polymers   33–35
apo-glucose oxidase (apo-GOx)   151
aptamers, switches based on   154
Arginine–Glycine–Aspartic acid (RGD) peptides   141, 142
Arrhenius equation   227
aryl diazonium salt   257
arylacetates   252–253

arylhydrazines   254–255
atom transfer radical polymerization (ATRP)   2, 4, 5, 13, 19, 31, 36, 37, 42, 149, 257
atomic force microscopy (AFM)   179, 258, 261, 263
avidin   183
azine intermolecular bonds   260
azobenzenes   44, 45, 142–143, 178–179, 226, 229–231, 232

b
backbone configuration   2
bacteriorodophsin   230
Bässler model   196
battery-free motions   240
Belousov–Zhabotinsky (BZ) gels   59, 60, 63, 79, 82, 84–86
– confinement effect on dynamics of   69–77
– regular oscillations in   68
– response, to nonuniform illumination   77–87
bis(*p*-methylbenzoate) (2,9,16,23-tetra-*tert*-butyl-phthalocyaninato) silicon (SiPc)   194, 208
– chemical structure of   209
benzodithiole   232
benzopyrans   232
benzoquinone   141
benzothiazoline   232
1,2-bis(2-methylbenzothiophene-3-yl)-1,2-dicyanoethene (BTCN)   225
1,2-bis(2-methylbenzothiophene-3-yl) maleic anhydride (BTMA)   225
1,2-bis(2-methylbenzothiophene-3-yl) perfluorocyclopropene (BTF6)   225
benzylidene   237
4-(azepan-1-yl) benzylidenemalononitrile (7-DCST)   207, 208

Handbook of Stimuli-Responsive Materials. Edited by Marek W. Urban
Copyright © 2011 WILEY-VCH Verlag GmbH & Co. KGaA, Weinheim
ISBN: 978-3-527-32700-3

bidentate metal ion chelation, intramolecular   231
bioelectronics   151
biologically responsive polymer systems   28
– antigen-responsive polymers   33–35
– enzyme-responsive polymers   32–33
– glucose-responsive polymers   28–32
– redox-/thiol-responsive polymers   35–39
biosensing systems and fabrication   183–184
biotins   183
birefringence   214
– orientational birefringence (OB)   202, 203
bisphenol-A-polycarbonate   196
bithiophene   257
block copolymers   1, 5, 7, 31
boronic acid-diol complexation   30–32
Bragg reflector   172, 173
butanediol (BDO)   128
butyl benzyl phthalate (BBP)   199
butyrylcholinesterase (BChE)   184

c

C1-RG dyes   124, 125, 126, 127, 128
C1-YB dyes   124
C12OH-RG dyes   128–129
C12-YB dyes   121
C18-YB dyes   121
C18-RG dyes   120, 126, 127, 128
C60   194
Caeonorhabditis elegans   239
cantilever-based sensors   155
carbaldehyde aniline   192
carbon nanotubes   151
carrier mobility determination   214
carrier photogeneration, in polymers   196
cataphoresis painting   248
catenanes   152–153
Cauchy–Green (Finger) strain tensor   62
cavitation   43
chain transfer agent (CTAs)   7–8, 13
chain-end recombination   101
charge generating (CG) and transporting molecules   193–200
– chemical structures   194, 197–199
charge-shifting polymers, hydrolysis of   170
chemical sensing with excimer-forming dyes   133–135
chemisorption   259–260
chitin   102
chitosan (CHI)   102, 104, 105
chitosan   183
chromogenic dye molecules   117
– chemical sensing with excimer-forming dyes   133–135

– excimer-forming sensor molecules   118–120
– fluorescent mechanochromic sensors   121–129
– thermochromic sensors   129–132
chromophores   44, 45
cilia-like materials   239
cinnamoyl groups and healing   103
cinnamylidene acetic acid (CAA) molecules   145
click reactions   3, 6, 257
C–O bond cleavage   231
colloidal dispersions   11–20
concanavalin A (ConA)   29–30, 183
conjugated polydiacetylenes   231
conjugated polyelectrolytes (CPEs)   178, 180, 181
continuum equations   61–63
controlled free radical polymerization   3–11
controlled radical polymerizations (CRPs)   1, 12, 31
covalent bonding   99–104, 102, 247
– aliphatic primary amines   250–251
– arylacetates   252–253
– aryldiazonium salts   254–258
– arylhydrazines   254–255
– electrodeposited coatings   248–249
– electrografted coatings   249–250
– iodide derivatives   263–264
– (meth)acrylic monomers   258–263
– primary alcohols   253–254
crack healing   102, 110
craze healing   110
critical micelle concentration (CMC)   12
crown ethers   230, 231
cucurbituril   151
CX-RG dyes   121
1,2-bis(2-cyano-1,5-dimethyl-4-pyrrolyl) perfluorocyclopentene   234
cyano ester styrene (CEST)   208
1,4-bis(α-cyano-4(12-hydroxydodecyloxy) styryl)-2,5-dimethoxybenzene (C12OH-RG)   133
cyano-OPVs   119, 120, 121, 124, 127–129, 134
cyclic voltammetry   250, 251, 252, 254, 258, 263
cyclodextrin   141, 230
cyclodextrine   263
cyclohexane   226

d

dendrimers   144, 231
density functional theory (DFT)   258

deprotonation 151
DGEBA 134, 135
diarylethenes 225, 234–236
diazonium salts 254–257, 264
(N,N-di-n-butylaniline-4-yl)
    l-dicyanomethylidene-2-cyclohexene
    (DBDC) 210
2-dicyanomethylen-3-cyano-5,5-dimethyl-4-
    (4′-dihexylaminophenyl)-2,5-dihydrofuran
    (DCDHF-6) 208
dicyanostyrene, fluorinated 192
dicyclopenta diene (DCPD) 112, 113
Diels–Alder (DA) reaction 102, 104
N,N-diethyl-substituted pNA 200
4-(N,N′-diethylamino) benzaldehydediphenyl
    hydrazone 196
3-fluoro-4-N,N-diethylamino-β-nitro-styrene
    201
4-N,N-diethylaminocinnamonitrile 201
diethylenetriamine 107
N,N-diethyl-substituted pNA 200
diffraction efficiency and response time
    216–218
diffusion 110–111
1,2-dilauroyl-sn-glycero-3-phosphocholine
    (DLPC) 15
N,N-dimethylacrylamide (DMA) 29, 31
2-(N,N′-dimethylamino) ethyl methacrylate)
    (DMAEMA) 15, 17, 18, 19
4-(dimethylamino)-4′-nitrostilbene (DANS)
    200
N,N-dimethylamino propylacrylamide
    (DMAPA) 31
2-(1-(2,5-dimethyl-3-furyl) ethylidene)-
    3-(2-adamantylidene) succinic anhydride
    228
2,5-dimethyl-3-furyl moiety 237
2,5-dimethyl-4-(p-nitrophenylazo) anisole
    (DMNPAA) 201, 202, 207, 208
dimethyl sulfoxide (DMSO) 111
dimethylformamide (DMF) 251, 258, 259,
    262, 263
dip pen lithography 157
N,N′-diphenyl-N,N′-bis(3-methylphenyl)-
    [1,1′-biphenyl]-4,4′-diamine 196
diphenyl phthalate (DPP) 211
Disperse Red 1 200
disulfides 35
– reduction 36–37
dithenylethenes 235
dithia-(dithienylethena) phane 235
dithiothreitol (DTT) 36, 37
divinylbenzene (DVB) 13

N-(DL)-(1-hydroxymethyl)
    propylmethacrylamide (DL-HMPMA)
    15, 17
DNA 170
– damages 96, 183
DNA-based surface switches 153–154
donor number (DN) 259
donor/acceptor interface 180
dye aggregation 124, 126–127

e
elastin-like polypeptides (ELPs), switchable
    surfaces of 155–157
electroactive materials
– high-density and low-density self-assembled
    monolayers 139–141
– self-assembled monolayers, with
    hydroquinone incorporation 141–142
electrochemical impedance 155
electrochemical methods 247
electrochemical stimuli 181–183
electrocoagulation 248
electrodeposited coatings 248–249
electrografting 251, 253, 254, 255–256, 257,
    258–263
– coatings 249–250
electro-optic (EO) effect 191, 200–204, 207,
    213–214
electropolymerization 249
electroprecipitation 248
electroresponsive polymers 39–40
ellipsometry 264
emulsion polymerization 11, 12–13
energetic disorder 196
enzyme-responsive polymers 32–33
epoxy polymer particles 110
– coatings 134
ester exchange reaction, in polycarbonate
    101
esters 231
ethers 231
9-ethyl carbazole 192
N-ethylcarbazole (ECZ) 196, 202, 207, 208
ethylacrylate (EA) 259
ethylene diamine 250
ethylene-co-methacrylic acid (EMMA) 109
ethylene-glycol-1-acrylate-2-cinnamic acid
    (HEA-CA) molecules 145
1-(2′-ethylhexyloxy)-2,5-dimethyl-4-(4′-
    nitrophenylazo) benzene (EHDNPB) 200
5-ethylidene-2-norboroene (ENB) 112
N-ethylmethylacrylamide (EMA) 8
E-to-Z isomerization 237

excimer-forming sensor molecules   118–120
external diffraction efficiency   216

f
fatigue problem   228
fibrinogen adsorption   143
fibroblasts   148
Fickians diffusion   98
field-responsive polymers
– electroresponsive polymers   39–40
– magnetoresponsive polymers   40–42
– photoresponsive polymers   43–46
– ultrasound-responsive polymers   42–43
figure of merit (FOM)   203
flavin adenine dinucleotide (FAD)   151
Flory model   62
Flory–Huggins form   62
fluorenyltriamine   151
fluorescein isothiocyanate (FITC)   35
fluorescent ligands   117
fluorescent mechanochromic sensors   121–129
fluorophores   121, 153
fluoropolymers   124, 127
four-wave mixing (FWM) geometry   216
– beam paths in   217
Friedel–Crafts acylation   194
fulgicides   237–238
fulgides   225
fullerenes   180
functional groups   32
functionalized polymer systems   211–213
furans   234

g
gating system schemata   144
gel lattice spring model (gLSM)   63–67
glucose oxidase   184
glucose-responsive polymers   28
– based on boronic acid-diol complexation   30–32
– based on concanavalin A (ConA)   29–30
– based on glucose-GO_X 28–29
D-glucosaminic acid   257
glucosyl-terminal poly(ethylene glycol) (G-PEG) insulin   29–30
Glutathione (GSH)   35, 37
glycogen   183
glycopolymers   29
gold nanoparticles   178
guest-diffusion technique   124, 132
guest–host polymer systems   206–211

h
helical peptides, switches based on   155
hemoglobin   149
heptadecafluorodecyl methacrylate (FMA)   14, 15, 19
heterogeneity, of colloidal reactions   13–14
hexamethylene di-isocyanate (HDI)   102
hexane   228
high-density self-assembled monolayers (HDSAMs)   140
highest occupied molecular orbital (HOMO)   193, 204
high-performance photorefractive polymers
– functionalized polymer systems   211–213
– guest–host polymer systems   206–211
holographic time of flight (HTOF)   214
humidity-sensing polymer   133
hyaluronan   183
hydrogels   29, 30, 31, 32–33, 40
hydrogen bonding   104–107, 168, 174
– molecular healing mechanism of urea isopyrimidone (Upy) network in   107
hydrogenated soybean phosphatidylcholine (HSPC)   15
2-hydroxyarenealdehyde   232
4-hydroxybutyl-2-bromoisobutyrate (HBBIB)   7
hydroxyethyl methacrylate (HEMA)   5
hydroxyl ethylene diamine triacetic acid (HEDTA)   109

i
imides   231
2-iminobiotin   183
i-motif   154
inclusion complexes (ICs)   141
indium tin oxide (ITO)   146, 147, 213
indoles   234
infrared irradiation   46
iniferter   3
initiator   4
interconversion, of thiols and disulfides   35
interfacial diffusion   110
internal diffraction efficiency   216
internal quantum efficiency   213
interpenetrating polymer network   35
inverse miniemulsion photoinitiated polymerization   13
ionic interactions   109–110
irradiation   43, 44
isocyanates   231
isomerization   44, 45
isopropylidene   237

k

kinesin 151
Kolbe reaction 252

l

layer-by-layer (LBL) self-assembly technique 165–166, 175, 177, 181, 184
lectins 29
ligands 4
light-activated shape memory polymers (LASMPs) 112
light-driven plastic motors 240
light-guided motion and complex paths 81–87
linear low-density polyethylene (LLDPE) 124–126
linear poly(ethylenimine) (LPEI) 170, 182
linear stability analysis, in limiting cases 69–72
liquid crystallines
– dendritic compounds 122–123
– polymers 225–226
low-density polyethylene (LDPE) 109
low-density self-assembled monolayers (LDSAMs) 140–141
lower critical solution temperature (LCST) 9, 37, 147–148, 156, 168, 175, 177, 263

m

macroemulsions 13
macroligand effectiveness 10
macromonomers 259
magnetoresponsive polymers 40–42
main-chain liquid crystalline elastomers (MC-LCEs) 150
malonic acid (MA) 69
merocyanine (MC) isomers 226
mesitylmethylene 237
mesophases 234
metal–ligand coordination 107–109
metal-to-ligand charge transfer (MLCT) 212
methacrylates 231
methacrylic acid 29
4,4′-methyl-1,2-di-(2-dimethyl-5-phenylthiophen-3-yl) perfluorocyclopentene 229
N,N-methylenebis-acrylamide (MBA) 31, 35
methyl methacrylate (MMA) 13, 15, 19, 111
methyl methacrylate, poly(NSP-co-MMA) 143
4,4′-methylene-bis(phenyl isocyanate) (MDI) 128
micellar nucleation 12
microbubble 43
microemulsions 13
microgels 175
micrometer-level repairs
– diffusion 110–111
– relaxation and shape memories 111–112
millimeter-level repairs
– multiphase systems 112–113
miniemulsions 13
moisture sensors 133
molecular doping 196
molecularly doped polymers (MDPs) 196
multicomponent networks 17
multilayer stimuli-responsive polymeric films and layer-by-layer (LBL) self-assembly technique 165
– electrochemical stimuli 181–183
– fabrication of multilayer polymer coatings 167–169
– light radiation 178–181
– pH alternation 170–174
– salt concentration change 169–170
– specific molecular interaction 183–184
– temperature variation 175–177
myosin 151

n

nanometer-level repairs
– ionic interactions 109–110
– metal–ligand coordination 107–109
nanoparticles 151
nanopatterning 143
nanowires 262
natural rubber 124
n-butyl acrylate (nBA) 14, 15, 18, 19, 111
near-infrared (NIR) 178
neo-Hookean elasticity 65
NEST 184
neuropathy target esterase (NTE) 184
N-heterocyclic carbenes (NHCs) 111
N-isopropylacrylamide (NIPA) 67
N-isopropylacrylamide (NIPAM) 29, 31, 35
N-isopropylacrylamides (NIPAAms) 8
4-nitroazobenzene 226
nitrobenzospiropyran 143
4-nitrobenzylamine 251
nitrocellulose 238
4-nitrophenyl diazonium tetrafluoroborate 254
4-(4-nitrophenylazo) aniline (DO3) 211
nitroxide radicals 3
nitroxide-mediated radical polymerization (NMRP) 1, 41

noncontact operation mechanical devices 240
nonlinear optical chromophores 200, 201
nonlinear optics (NLO) 192, 202, 203–204, 207, 208, 210, 211, 212
norbornenes 2
n-π^* transitions 229, 230
N-succinimidyl acrylate (NSA) 263
N-vinylcarbazole (NVK) 196, 197

o

oligo(p-dioxanone diol) (ODX) 149
oligo(ε-caprolactone) diol (OCL) 149
ω-hydroxy-functionalized dye 128
optomechanical studies, in situ 126, 127, 129
Oregonator model 61
organic bulk heterojunction photovoltaics (OBHPVs) 180
organic photovoltaic (OPV) devices 180
organoboron polymers 31
organometallic ferrocene polymers 231
orientational birefringence (OB) 202, 203
orientational enhancement 202
oscillating magnetic field (OMF) 110
oxetane (OXE) 102, 104, 105
oxonium 232

p

para-nitroaniline (pNA) 200
partially confined sample behavior 74–77
peeling tests 258
4-4'-n-pentylcyanobiphenyl (5CB) 202
peptide-pendant copolymers 2
PFO-PBD-NMe^{3+} 181
[6,6]-phenyl C61-butyric acid methyl ester (PCBM) 194
phenyl rings 224
phenyls 234
1,2-bis(5-methyl-2-phenylthiazol-4-yl) perfluorocyclopentene 235
1,2-di(2-dimethyl-5-phenylthiophen-3-yl) perfluorocyclopentene 228, 229
phospholipids (PLs) 14
photo irradiation 165, 168
photochromic polymeric systems 223
– azobenzenes 229–231
– diarylethenes 234–236
– environmental effects 226–227
– fulgicides 237–238
– molecular structure influence 224–226
– photoreaction kinetics 227–229
– spiropyrans 231–234
photoconductive sensitivity 214

photoconductivity 191, 195, 208, 209, 213–214
– in carbazole-containing polymers 197
photo-cross-linking 112
photocyclization reactions 235
photoexcitation 223
photogeneration of carriers, in polymers 196
photoisomerization 46, 223, 224, 226, 228
– trans–cis, in azobenzenes 230
photorefractive (PR) polymers 191
– charge generation and transport 193–200
– diffraction efficiency and response time 216–218
– electro-optic response 200–204
– high-performance 206–213
– photoconductivity and electro-optic responses 213–214
– photorefractive gratings formation 204–205
– two-beam coupling effect 205–206, 214–216
photorefractive gratings formation 204–205
photoresponsive materials
– azobenzenes 142–143
– shape-memory polymers 144–145
– spiropyrans 143–144
photoresponsive polymers 43–46
photosolvolysis 46
photovoltaics (PVs) 180
pH-responsive materials 145–147
– self-assembled monolayers (SAMs), switchable surfaces based on 146
– switchable surfaces based on polymer brushes 146–147
pH-sensitive dyes 117
phthalocyanines (Pcs) 194
piezochromic optical sensors 121
4-piperidinobenzylidene-malononitrile (PDCST) 215
4-piperidin-4-ylbenzylidenemalononitrile 201
π-π^* transitions 229, 230
plasticization of polymers 133, 134
plasticizers 198, 202–203
– flexible oligomer 225
platelet adhesion 143
p-methyl methacrylate/n-butylacrylate/ heptadecafluorodecyl methacrylate (p-MMA/nBA/HDFMA) 110
PMSF 184
poly(2-acrylamido-methane-2-propanesulfonic acid) (PAMPS) 182
poly(acrylic acid) (PAA) 29, 44, 45, 147, 168, 171–172, 174, 176, 177, 180, 263

poly (acrylic tetraphenyldiaminobiphenyl) (PATPD) 209
poly(acrylonitrile) (PAN) 258, 260
poly(allylamine hydrochloride) (PAH) 167, 168, 170, 171–172, 173, 174, 175, 176, 178, 182, 184
polyaniline (PANI) 182, 248, 257
polycaprolactam (Nylon-6) 228
polycyclics 230
poly(cyclooctene) (PCO) 132
polydimethylsiloxane 234
poly(2-(dimethylamino) ethyl methacrylate) (PDMAEMA) 149
polydispersity (PDI) 2
poly(E,E-[6,2]-paracyclophane-1,5-diene) 199
polyelectrolyte complex (PEC) 169, 170
polyelectrolyte hydrogels 40
polyelectrolyte multilayer (PEM) films see multilayer stimuli-responsive polymeric films
poly(ε-caprolactone) (PCL) 259
polyethylene glycol grafted polysiloxane (PEG-g-PS) 225
polyethylenes (PEs) 124, 225
(poly(3,4-ethylenedioxythiophene) (PEDOT) 181, 182
poly(2-ethylhexyl acrylate/arylic acid) (p(2EHA/AAc)) elastomers 17
poly(ethyl methacrylate)-co-poly(methyl acrylate), P(EMA)-co-P(MA) 143
poly(ethylacrylate) (PEA) 259
poly(ethylene oxide) (PEO) 13, 109, 168
poly(ethylene terephthalate) (PET) 124, 228
poly(ethylene terephthalate glycol) (PETG) 124, 130, 131
poly(ethylene) glycol (PEG) chain 9, 37, 39, 46, 174
poly(ethylene-co-butylene) (PEB) 109
poly(L-glutamic acid) (PSLG) 226
poly(hexylviologen) 182
polymer brushes
– pH-switchable surfaces based on 146–147
– temperature-dependent switching in 149
polymer solids 20
polymeric micelles 31, 35, 37, 43, 45
polymeric tris(bipyridyl) ruthenium(II) 181
polymerization 231, 236, 239
– of aromatic monomer 248
polymer–solvent interaction 62
poly(methacrylic acid) (PMAA) 174, 175
– -graft-ethylene glycol) gels 29
poly(methyl-bis-(3-methoxyphenyl)-(4-propylphenyl) amine) siloxane (MM-PSX-TAA) 198

poly(methyl methacrylate) (PMMA) 110, 130, 196, 225, 226, 227, 228, 238
poly(methyl methacrylate/n-butyl acrylate/heptadeca-fluorodecylmethacrylate) (p(MMA/nBA/FMA) colloidal dispersions 19
poly(methyl phenylsilane) (PMPS) 226
poly(n-butylmethacrylate) (PnBMA) 226, 227
poly-(N-(2-carbazolyl) ethyl acrylate) 197
poly(N-[(2,2-dimethyl-1,3-dioxolane) methyl]acrylamide) (PDMDOMAAm) 5
poly(2-(N-ethyl-N-3-tolylamino) ethyl methacrylate) 199
poly(N-hydroxypropyl methacrylamide) (PHPMA) 37, 44, 45
poly(N-isopropylacrylamide) (PNIPAM) 3, 5–6, 8, 9, 10, 13, 31, 37, 39, 44, 45, 147–148, 168, 175, 176, 177, 262, 263
– thiol (PNIPAAm-PEG-thiol) 148
poly(N-isopropylacrylamide coacrylic acid) (PNIPAM-co-AA) 175
poly[(4-nitronaphthyl)[4[-2-(methacryloyloxy) ethyl]ethylamino]phenyl]diazene] (pNDR1M) 224
poly(N,N-dimethylaminoethyl methacrylate) (PDMAEMA) 39, 44
poly(2-(N,N-dimethylamino) ethyl methacrylate-co-n-butylacrylate-co-N,N-(dimethylamino) azobenzene acrylamide) 239
poly(bis-phenol A carbonate) (PC) 130
poly(N-succinimidyl acrylate) (PNSA) 263
poly(N-vinylcaprolactam) (PVCL) 175
poly(N-vinylcarbazole) (PVK) 194, 195, 196, 197, 202, 204, 207, 208, 209
– CT complex, with TNF 207
poly(N-vinylcarbazole-co-styrene) 197
poly(N-vinyldiphenylamine) 199
polyolefins 127
poly[o(p)-phenylenevinylene-alt-2-methoxy-5-(2-ethylhexyloxy)-p-phenylenevinylene] (p-PMEH-PPV) 211
polyoxometalates (POMs) 182
poly(phenylene-ether) s (PPEs) 101
poly(phenylene ethynylene)-type (PPE) 178
poly(phenylene vinylene) (PPV) 180, 181, 248
polyphenylene vinylene poly(2,5-bis[3-N,N-diethylamino)-1-oxapropyl]-1,4-phenylenevinylene) (DAO-PPV) 131, 132
poly(propylene sulfide) (PPS) 39
polypyrrole (PPy) 248, 262
polystyrene 196, 226, 238

poly(styrene sulfonate) (PSS)   167, 173, 174, 178, 181
poly(tetramethylene glycol) (PTMG)   128
polythiophene (PThi)   248, 262
poly(*trans*-1-(3-vinyl carbazolyl)-2-(9-carbazolyl) cyclobutane)   197
polyurea (PUA)   104
polyurethane (PUR)   102, 104
– synthetic steps in creation of self-healing   105–106
polyurethane elastomers   225
poly(vinyl alcohol) (PVA)   31, 196
poly(vinyl methyl ether) (PVME)   175
poly(vinylarylamines)   199
polyvinylchloride   196
poly(2-vinylpyridine) (P2VP)   147
poly(vinylpyrrolidone) (PVP)   168
positional disorder   196
primary alcohols   253–254
protonation, of MC-phenoxide moiety   231
Prussian Blue (PB) nanoparticles   181
pseudorotaxane   151
pseudostillbenes   230
pullulan   234
puncture reversal   110
push–pull molecules, in PR systems   203
bis(pyridinium) DTE   239
N-(2-acryloyloxyethyl) pyrrole (PyA)   262
pyrroles   234

q

quantum yield, and photoisomerization   224, 225
quartz crystal microbalance (QCM)   258

r

$RCH_2^\bullet$ 252
redox enzymes   151
redox-/thiol-responsive polymers   35–39
refractive index   200, 202, 203
reptation model, in polymeric network solutions   98
resins   134
responsive coatings   165, 167
responsive polymers   27
– biological   28–39
– field   39–46
reversible addition fragmentation chain transfer (RAFT)   2, 7–8, 9, 31, 36, 37
– inverse microemulsion polymerization (RAFT-IMEP)   13
reversible chain termination   1
reversible network schema   108
ring cyclization   224, 225

ring-opening metathesis polymerization (ROMP)   2, 112–113
RNA   142
$RNH^\bullet$ 251
rotaxane shuttles   151–152
Ru-complex   212

s

saturated calomel electrode (SCE)   250, 252
sec-butyl iodide   263
seed   12
self-assembled monolayers (SAMs)   139–140
– high-density (HDSAMs)   140
– with hydroquinone incorporation   141–142
– low-density (LDSAMs)   140–141
– and smart coatings   165
– switchable surfaces based on   146
self-assembled monolayers (SAMs)   151
– pH-switchable surfaces based on   146
self-healing   105–106, 109
self-oscillating gels   59
– autonomous motion towards dark region   78–81
– BZ gels photosensitivity modeling   77–78
– continuum equations   61–63
– gel lattice spring model (gLSM)   63–67
– light-guided motion and complex paths   81–87
– linear stability analysis, in limiting cases   69–72
– model parameters and correspondence between simulations and experiments   67–69
– oscillations, induced by confinement release   72–73
– partially confined sample behavior   74–77
self-repairing polymeric materials   93, 98, 114
– Angstrom-level repairs   99–107
– damage and repair mechanisms in mammals   94–95
– micrometer-level repairs   110–112
– millimeter-level repairs   112–113
– nanometer-level repairs   107–110
shape memories   111–112, 117
shape-memory polymer (SMP)   111, 131, 132
– light-activated (LASMPS)   112
– photoresponsive   144–145
– thermoresponsive   149–150
shearing   121
Si–C bonds   256
Si–H bond   256
siloxane polymer   198
smart coatings   165
smart polymers   28

smart windows   249
sodium dioctyl sulfosuccinate (SDOSS)   15, 19
sonication treatments   258
sonoporation   43
spiropyran-grafted dextran   234
spiropyrans   143–144, 231–234
– derivatives   44, 45
– polymer matrix effect on   227
– ring-opening and ring-closing in   233
squarylium dye   194
Stille coupling reaction   212
stimuli-responsive polymers   1
– colloidal dispersions   11–20
– controlled free radical polymerization of   3–11
– fluorescent   117, 134
stress tensor   62
styrene–butadiene–styrene (SBS)   226, 228
sulfadimethoxine   29
sulfonated poly(thienothiophene)   182
supramolecular polymers   107
– schema of self-healing   109
supramolecular shuttles, switchable surfaces based on   150
– catenanes   152–153
– rotaxane shuttles   151–152
surface conformation   143
surface plasmon resonance (SPR)   155
surface relief grating (SRG)   179
surface-initiated polymerization (SIP)   146
surface-localized ionic clusters (SLICs)   19
swelling–deswelling, of microgel   175
switchable surface approaches   139
– DNA and peptide monolayers   153–157
– electroactive materials   139–142
– photoresponsive materials   142–145
– pH-responsive materials   145–147
– supramolecular shuttles   150–153
– thermoresponsive materials   147–150
synthetic materials, schema of damages and repairs in   100
synthetic polymers   27–28, 46

t

tacticity   2
t-butyl iodide   263
terpolymers   31
7,7,8,8-tetracyanoquinodimethane   194
tetraphenyldiaminobiphenyl (TPD)   192
tetrathiafulvalene (TTF)   152–153
Texin 985   128
thermal relaxations   226, 227
thermal reversibility   102, 227, 234
thermal stability   235
thermochromic sensors   129–132
thermoplastic materials   15, 16
thermoplastic polyurethanes (TPUs)   124, 129
thermoresponsive materials
– shape-memory polymers   149–150
– temperature-dependent switching based on poly(N-isopropylacrylamides) (PNIPAAms)   147–148
– temperature-dependent switching in polymer brushes   149
thermosetting materials   15, 16
thiazole   234
thiols   35
thionium   232
1,3-(2-acryloyloxyethyl) thiophene (ThiA)   262
thiophene   234, 257
3D element, schematic of   64
thrombin   155
time-gated holographic imaging (TGHI)   192
time-of-flight (TOF) technique   214
time–temperature indicators (TTIs)   129, 130, 131
trans isomers   224
trans-azobenzenes   179
transglutaminase   32
transition temperature (T_g)   98–99, 110, 111–112, 130, 133, 201, 202, 203, 207, 211, 214, 215, 217, 225, 226
traps   199, 204, 214
l-β-(4-trifluoromethyl benzoyloxy) ethyl]-3,3-dimethyl-spiro[indoline-2,3′-[3H]-naphtho[2,1-b]-1,4-oxazine) (SO)   228
7-trifluoromethylquinoline group   231
1,3,3-trimethylindoline-6′-nitrobenzopyrylospiran (SP)   226
2,4,7-trinitrofluorenone (TNF)   194, 207, 208
triphenyldiamine   199
1,1,1-tris-(cinnamoyloxymethyl) ethane (TCE)   102
two-beam coupling (TBC)   207
– effect   205–206, 214–216
– gain coefficients   191, 209

u

ultrasound-responsive polymers   42–43
ultraviolet (UV) light   142, 143, 145, 178, 235, 237, 240
upper critical solution temperature (UCST)   8, 9
urea isopyrimidone (Upy)   107
urethanes   231

v

valine–proline–glycine–valine–glycine (VPGVG) 44
4-vinylbenzyl chloride (VBC) 13
visible light 45, 46
voids, localized 15

w

wettability
– control 143, 151
– hydrophilic/hydrophobic 165

Woodward–Hoffmann pericyclic photoreaction 237

x

X-ray photoelectron spectroscopy (XPS) 251, 252, 255, 256, 257, 258
(2,9,16,23-tetra-*tert*-butyl-phthalocyaninato) zinc (ZnPc) 194, 208
– chemical structure of 209